SCIENCE POLICY, ETHICS, AND ECONOMIC METHODOLOGY

K. S. SHRADER-FRECHETTE

Department of Philosophy,
University of Florida

SCIENCE POLICY, ETHICS, AND ECONOMIC METHODOLOGY

Some Problems of Technology Assessment and Environmental-Impact Analysis

D. REIDEL PUBLISHING COMPANY

A MEMBER OF THE KLUWER  ACADEMIC PUBLISHERS GROUP

DORDRECHT / BOSTON / LANCASTER

Library of Congress Cataloging in Publication Data

Shrader-Frechette, K. S., 1944–
Science policy, ethics, and economic methodology.

Includes bibliographies and indexes.
1. Technology assessment. 2. Environmental-impact analysis.
3. Decision-making. I. Title.
T174.5.S48 1984 333.7'1 84–15916
ISBN 90–277–1806–7
ISBN 90–277–1845–8 (pbk.)

Published by D. Reidel Company,
P.O. Box 17, 3300 AA Dordrecht, Holland.

Sold and distributed in the U.S.A. and Canada
by Kluwer Academic Publishers,
190 Old Derby Street, Hingham, MA 02043, U.S.A.

In all other countries, sold and distributed
by Kluwer Academic Publishers Group,
P.O. Box 322, 3300 AH Dordrecht, Holland.

All Rights Reserved
© 1985 by D. Reidel Publishing Company, Dordrecht, Holland
No part of the material protected by this copyright notice may be reproduced or
utilized in any form or by any means, electronic or mechanical
including photocopying, recording or by any information storage and
retrieval system, without written permission from the copyright owner.

Printed in The Netherlands

333.71
S561 S JAF
228640

For my parents

TABLE OF CONTENTS

IV. STEPS TOWARDS SOLUTIONS

PREFACE

If indeed scientists and technologists, especially economists, set much of the agenda by which the future is played out, and I think they do, then the student of scientific methodology and public ethics has at least three options. He can embrace certain scientific methods and the value they hold for social decisionmaking, much as Milton Friedman has accepted neoclassical economics. Or, he can condemn them, regardless of their value, much as Stuart Hampshire has rejected risk-cost-benefit analysis (RCBA). Finally, he can critically assess these scientific methods and attempt to provide solutions to the problems he has uncovered.

As a philosopher of science seeking the middle path between uncritical acceptance and extremist rejection of the economic methods used in policy analysis, I have tried to avoid the charge of being "anti science". Fred Hapgood, in response to my presentation at a recent Boston Colloquium for the Philosophy of Science, said that my arguments "felt like" a call for rejection of the methods of risk-cost-benefit analysis. Not so, as Chapter Two of this volume should make eminently clear. All my criticisms are constructive ones, and the flaws in economic methodology which I address are uncovered for the purpose of suggesting means of making good techniques better. Likewise, although I criticize the economic methodology by which many technology assessments (TA's) and environmental-impact analyses (EIA's) have been used to justify public projects, it is wrong to conclude that I am anti-technology. Technological advances surely have been a necessary condition for many benefits, including economic growth, and this growth, in turn, has been a necessary condition for other advances, notably greater social and political equality. My real target in these essays is neither science nor technology, but uncritical use of deficient scientific methodologies, especially if these methodologies can be employed to serve the technological demands of a bureaucratic elite, rather than the technological needs of humanity as whole.

An outgrowth of my work on TA and EIA teams, these essays are philosophical analyses of the methodological, epistemological, and ethical problems facing assessors, especially those brought about by use of the economic method of RCBA. To the extent that neoclassical economics rests on a positivistic philosophy of science, then my remarks are, in part, criticisms of this type of philosophy of science. If I am correct in my rationalistic belief that certain scientific methods and ethical presuppositions help to determine assessment conclusions, and that assessment conclusions help to determine public policy, then much is to be gained by using applied philosophy of science and applied ethics as tools of policy analysis. I realize, however, that many scientists claim that little is to be learned from methodological and conceptual investigations of TA and EIA. Lynton Caldwell, for example, maintains that the problems besetting these forms of policy analysis are best resolved not by methodological modifications but by proper administration of TA and EIA.

While it is true that politics and the quality of governmental administration help determine both assessment conclusions and the use that is made of them, natural and social scientists who use assessment techniques sometimes fail to understand and to employ them in any critical and systematic way. Because of this failure, they often view methodological discussions as tiresome constraints upon the urge to produce figures. My thesis is that methodological discussions often produce better figures and greater awareness of the limits of their validity and use.

In the first chapter, I outline TA, EIA, and one of their preeminent methods, RCBA. Since no one has ever published an analysis of the role of the philosopher in TA and EIA, I point out how a philosopher of science and an applied ethicist might contribute to assessment work. In Chapter Two, I defend analytical assessment techniques in general, and RCBA in particular. (This defense sets the stage for the constructive criticisms of TA, EIA, and RCBA in Chapters Three–Seven and for the solutions offered in Chapters Eight and Nine.)

With this introduction to TA, EIA, and RCBA accomplished, Chapters Three and Four, respectively, analyze two general assumptions central to most assessments. The first, the subject of Chapter Three, is the principle of complete neutrality, the epistemological presupposition that pure science and wholly objective TA and EIA are possible. The second assumption, the subject of Chapter Four, is the fallacy of unfinished business, the metaphysical presupposition that sound public policy demands continuing the scientific and technological status quo, rather than rethinking the ethical and social commitments underlying technology- and environment-related policies. The main thrust of Chapters Three and Four is that, in general, assessors have ignored the ethical and social dimensions of TA and EIA; I use a number of case studies both to argue for this claim and to illustrate it.

Perhaps the best example, of many assessors' tendency to ignore the ethical and social dimensions of TA and EIA (Chapters One–Three), has been their inclination to avoid considering the ethical and policy consequences of various economic methods used in assessment. Chapter Five reveals how uncritical acceptance of the concept of compensating variation means that assessment conclusions frequently fail to take account of key ethical notions, such as the claims that market price ≠ value and that preferences are not infallible indicators of authentic welfare. Chapter Six shows how use of the assumption of partial quantification is likely to result in bias against adequate consideration of nonmarket factors, such as social costs, and in aggregations of benefits and costs which often bear little relation to authentic societal well-being or its absence. Finally, Chapter Seven illustrates how routine employment of the classical method of RCBA typically causes TA's and EIA's not to take account of policy-induced distributive inequities.

In order to address some of the faulty presuppositions of general assessment methodology and the undesirable consequences of the use and misuse of particular economic methodologies, Chapters Eight and Nine offer two proposals. Both are necessary (but neither sufficient, even together) for improving TA, EIA, and the scientific methods underlying them. Chapter Eight proposes a methodological change in welfare economics: use of alternative, ethically weighted RCBA parameters. Chapter Nine advocates a modification in the legal and political procedure for determining science- and technology-related policy: adoption of a judicial procedure of adversary assessment, which I describe as the "technology tribunal".

With these two changes, in the scientific methods employed in assessment and in the legal procedure for making use of assessment, TA and EIA can be modified and employed so as to avoid many of the methodological, epistemological, and ethical difficulties uncovered in the earlier chapters. I hope that my criticisms and proposals for change can lead us to more objective scientific methodology and to more democratic public policy. At the least, I hope that they provide grounds neither for pessimism nor for optimism but for cautious hope about technological and environmental progress.

The University of California, Santa Barbara K. S. S-F.
October 1983

ACKNOWLEDGMENTS

This book owes much to many. Thanks to grants from the National Science Foundation (Program in History and Philosophy of Science and the Division of Policy Research and Analysis), I was able to enjoy a long period of uninterrupted work on several of the chapters in this volume. (Any opinions, findings, conclusions or recommendations expressed in this publication, however, are those of the author and do not necessarily reflect the views of the National Science Foundation.)

Other chapters in this volume were written because of the impetus afforded by speaking invitations. I am especially grateful to the 1982–1983 American Philosophical Association (APA) Western Division Program Committee for their request to speak at the Berkeley meeting, where I presented an early version of Chapter Two, and to the 1983–1984 APA Eastern Division Program Committee for their request to speak at the Boston meeting, where I delivered an early version of Chapter Eight. An excerpt from Chapter Three was read in Atlanta in October 1981, at the annual conference of the Humanities and Technology Association, and an early version of Chapter Four was published in *Environmental Ethics* in Spring 1982 (vol. 4, no. 1). Chapter Five is an outgrowth of a short paper, commissioned for presentation in Toronto in October 1980, at the joint meeting of the Philosophy of Science Association and the Society for the Social Studies of Science, and published in *Science, Technology, and Human Values* in Fall 1980, vol. 6, no. 33 (Copyright 1980 by MIT and the President and Fellows of Harvard College who have granted me nonexclusive permission to publish an expanded version of the article). Chapter Six is previously unpublished, except for a small section which was read in Bad Homburg, West Germany, in May 1983 at the Biennial Meeting of the Society for Philosophy and Technology; this section was published (under another title) by Vieweg Verlag in 1982 (in German) in *Technikphilosophie*, edited by F. Rapp and P. Durbin, and published by Reidel in 1983 in *Philosophy and Technology*, edited by R. Cohen and P. Durbin. Chapter Seven grew out of a short paper presented in December 1980 in connection with the Boston APA meetings and published in 1982 in *Research in Philosophy and Technology* (vol. 5). Chapter Nine, in shortened form, was delivered in September 1983, in New

York, at the Biennial Meeting of the Society for Philosophy and Technology.

A number of people have helped make this book better than it might have been. Betty Myers Shrader provided flawless typing, proof reading, editorial assistance, and humor, and many friends at the University of California, Santa Barbara and the University of Louisville gave me clerical assistance and collegial support. I am especially grateful to Stan Carpenter, Paul Durbin, Ken Goodpaster, Ed Lawless, Isaac Levi, Alex Michalos, Carl Mitcham, Holmes Rolston, Maurice Shrader-Frechette, and several anonymous referees for Reidel Publishers, each of whom has provided constructive criticisms of earlier versions of one or more of the chapters, and to Robert Deacon, Steven Strasnick, and Maurice Shrader-Frechette, for helpful discussions of the problem of numerical representation of a lexicographic ordering.

My greatest debt is to my husband, Maurice, my brightest and most endearing critic, and our children, Eric and Danielle. They make everything worthwhile.

PART I

INTRODUCTION

CHAPTER ONE

AN OVERVIEW OF TECHNOLOGY ASSESSMENT AND ENVIRONMENTAL-IMPACT ANALYSIS

1. INTRODUCTION

Historians tell us that some forms of technology assessment and environmental-impact analysis may date back more than 4000 years. In ancient Mesopotamia, for example, the priests routinely evaluated the social merits of proposed technological projects.[1] Despite repeated attempts by societies since that time to control the technologies which they have implemented, there has been a relentless flow of technological failures: pesticides in Christmas cranberries; nerve gas stored on the flight path of the Denver airport; regional electricity blackouts; faulty automobiles; and baby clothing treated with carcinogens. Even when technologies do not cause widespread public alarm, they are still under suspicion whenever someone claims that they have been inadequately assessed. Respected government researchers have revealed that many of our allegedly most beneficial technologies have been insufficiently evaluated as to their societal and environmental risks and benefits. Computed tomography, mammography, and electronic fetal monitoring, for example, are widely used in developed countries, despite real doubts as to both their efficacy and their safety.[2]

One reason why evaluations of technological and environment-related projects are often inadequate is that they require one to ascertain, ahead of time, what *present* actions are desirable, given possible *future* impacts. Determining what policy actions are reasonable is useful only *ex ante*, however, but possible mostly only *ex post*. Necessary conditions for rational decisionmaking are almost never present in contemporary environmental and technological dilemmas, frequently because of information gaps. Even experts simply do not know what the future holds. In the absence of this knowledge, their forecasts are often little more than 'guestimates'. Policy-makers 125 years ago, for instance, were seriously troubled by the fact that their projections showed that levels of horse manure would reach unmanageable proportions in the cities of the next century. Little did they realize that the invention of the automobile would help to solve the problem of citizens' being inundated with the feces of animals used for transportation.

Evaluation of proposed environmental and technological projects is difficult, in general, not only because of the inadequacy of information needed to make good policy, but also because the analysis requires the skills of many different types of natural scientists, social scientists, and humanists. As one person put it: "Quality assessment requires the artful blending of the quantitative and the qualitative, of the objective and the subjective, and of science and politics . . . [it is] a form of socio-moral (therefore philosophical) reflection on the large-scale, unintended consequences of technology at large".[3]

Given the difficulty of adequately evaluating many societal projects and a number of widely publicized failures of applied science, many persons have ignored the enormous successes and benefits of technology. Especially in the last two decades, the number of critics of technology has swelled. They have created a neo-Luddite revolt of unparallelled dimensions and have charged their opponents with scientism.

2. THE CONCEPTS OF TECHNOLOGY ASSESSMENT AND ENVIRONMENTAL-IMPACT ANALYSIS

If there is a reasonable alternative either to Luddite panic or to mindless enslavement to technology, perhaps it can be found in an invigorated approach to technology assessment (TA) and environmental-impact analysis (EIA). The first stage of this rejuvenation of TA and EIA began in the sixties. In 1964, IBM endowed a program of research at Harvard University to study the impact of technological change on people and the environment. In 1966, US Congressman Emilio Daddario, the first Director of the US Office of Technology Assessment, introduced the phrase 'technology assessment'. In the last twenty years during which this term has been used, it has come to mean the systematic study of the impacts on society and the environment that occur when a technology is introduced, extended, or modified,[4] Environmental-Impact Analysis, a subset of TA, is the systematic study of the environment-related consequences of implementing a particular project or technology.

The purpose of TA is to provide competent, unbiased information concerning the physical, biological, economic, social, political, and environmental effects of applications of technology.[5] More specifically, the goal of TA is to anticipate these effects, to scrutinize a diverse range of impacts, to identify the various stakeholders who may be affected by implementation of the technology, and to analyze the effects of alternative policies. Realizing these goals is especially difficult since the intended effect of a project is rarely the only impact it has on persons and on the environment. The real challenge of TA is to anticipate unintended fifth- and sixth-order consequences arising from use of a particular technology. In a classic article, one of the leaders of the TA movement, Joseph Coates, gives an instructive example illustrating how implementation of automobile technology can lead to unforeseen higher-order consequences.

First-order consequences:	People are able to travel rapidly, easily, cheaply, privately, door-to-door.
Second-order consequences:	People often patronize stores farther away from their homes, and these stores are generally larger ones with bigger clienteles.
Third-order consequences:	Residents of a community do not meet so often and do not know each other so well.
Fourth-order consequences:	Because they are often strangers to each other, community members find it difficult to unite to deal with common problems. People become increasingly isolated and alienated from their neighbors.
Fifth-order consequences:	Since they are often isolated from their neighbors, family members depend more on each other for satisfaction of most of their psychological needs.
Sixth-order consequences:	When spouses are unable to meet the heavy psychological demands that each makes on the other, frustration and divorce often occur.[6]

Although the core effort of both TA and EIA is analysis of impacts, especially higher-order or unforeseen ones, the scope of these assessments varies. Some studies are very broad, while others are quite specific. Generally TA/EIA may take any one of the following forms: (1) *project* assessment, which focuses on the impacts of a particular, localized undertaking, such as a sewage-treatment plant; (2) *problem-oriented* assessment, which addresses means of solving some specific societal problem, e.g., the energy shortage; and (3) *technology-oriented* assessment, which examines some new technology, e.g., ultrasound, and analyzes its impacts on society and the environment. (3) is the form of TA/EIA which is broadest in scope, while (1) is the most limited.

2.1. *TA and EIA: Similarities and Differences*

As was already mentioned, TA's generally tend to be somewhat more comprehensive than EIA's since environmental impacts are only a subset of the effects assessed in TA. TA and EIA nevertheless remain close in purpose and in major features. Both are intended to anticipate *future* consequences as well as immediate impacts, and both address technology-related issues. Although TA's and EIA's must be bounded in some way, in principle they are systematic and comprehensive. Also, in principle (thought often not in practice) they consider the pros and cons of alternative policies and projects. Likewise both TA and EIA are highly variable in scale; they can range (in terms of work required to complete the assessment) from several person-months to ventures requiring the effort of many people over a number of years.[7]

In addition to the fact that TA is generally broader in scope than EIA, there are a number of other differences between the two. TA tends to have an international or national focus, while EIA often deals with purely local or regional considerations. Perhaps because of its more global emphasis, TA frequently is more policy-oriented than EIA. As the latter becomes more sophisticated, however, this difference may disappear. The format for the two types of assessment also varies; regulations of the US Council on Environmental Quality (CEQ) require uniform EIA's, but there is no specific format for a TA.

Perhaps the greatest difference between the two types of assessment has to do with time horizon. TA is in principle intended to address both immediate and long-term policy options, but EIA usually deals with technologies or environmental projects close to the point of completion. EIA's also usually rely more on primary data gathering than do TA's and they are somewhat more likely to be performed in-house, as the responsibility of some US agency. TA's, on the other hand, are slightly more likely to be contracted out to universities and research firms.[8]

2.2. *The Components of TA and EIA*

Having discussed the basic characteristics of TA and EIA, let us turn now to the question of how they are done. Numerous authors and agencies (for example, Martin Jones, Joseph Coates, Alan Porter, the National Science Foundation, the US Council on Environmental Quality, the National Academy of Sciences, and the US Office of Technology Assessment) have provided lists of steps to be undertaken by an interdisciplinary assessment team. The overall similarity of their strategies is striking, and they typically define TA/EIA in terms of ten components:

1. problem definition
2. technology description
3. technology forecast
4. social description
5. social forecast
6. impact identification
7. impact analysis
8. impact evaluation
9. policy analysis
10. communication of results.[9]

The first of these ten components involves defining the scope and depth of the assessment and identifying the 'parties at interest' to the project – those persons likely to gain or to lose as a result of particular impacts. The second component of TA/EIA is a thorough description of the technology being evaluated. At the next stage, the assessors attempt to anticipate the character and timing of changes in the technology and others related to it. The purpose of this study component is to reveal factors such as likely future cost savings, new applications of the technology, and possible future scientific breakthroughs. In the fourth component of the work, the assessment team attempts to describe those aspects of society (economic conditions, political climate, government regulations) likely to interact with the technology under consideration. Various social indicators and surveys are useful at this stage. On the basis of the social-description component, assessors next seek to represent the most plausible future configurations of society and to project possible changes in it. With these projections in mind, the team is able to accomplish the sixth aspect of TA/EIA: identifying both direct and higher-order impacts of the proposed project or technology.

At the seventh stage of TA/EIA, the interdisciplinary team studies the likelihood and magnitude of the various impacts identified during work on the previous assessment component. Here expertise essential to particular disciplines plays a great role. Economists may use cost-benefit analysis to investigate certain financial impacts, while psychologists may employ psychometric surveys to analyze possible social consequences. The next step in the assessment is to use these analyses both to evaluate the impacts and to determine their significance relative to the technology and to societal goals. After all the impacts are suitably evaluated, assessors at the ninth stage compare options for implementing technological developments and for dealing with their consequences. On the basis of this analysis, explicit policy recommendations may or may not be made. Finally, at the last stage, the team determines ways in which the results of its study can be communicated to persons or groups most likely to benefit from it.[10]

2.3. *The Uses and Users of TA's and EIA's*

A US Senate study noted that federal executive departments, state legislatures, universities, and high-technology industries were those most likely to use completed TA's and EIA's. Most importantly, the agency or group which prepared or commissioned the study is ordinarily the primary user of it, although a proprietary assessment by industry would not be handled in the same way as would a public study done under government auspices or contract.

If a TA or EIA is done well, it conceivably could lead to a number of possible outcomes. These include modification of the project; specification of a necessary program of environmental or social monitoring; stimulation of research and development, especially as regards risks of the project or alternatives to it; identification of regulatory channels to promote or control the technology; definition of techniques for reducing negative impacts or emphasizing positive ones; delaying the project or technology; stopping it; and providing a data base for persons interested in the issues which were assessed.[11]

On a more general level, TA or EIA can lead to a number of benefits. It is a useful tool for dealing with the growing complexity of society and our increasing power over nature, as well as an important vehicle for helping to create a better future. It provides us with a means of protecting ourselves against extreme environmental degradation, dealing with value changes influenced by technological development, and formulating sound public policy. As Congressman Emilio Daddario, first director or the US Office of Technology Assessment, put it: "if assessment is delayed, or incomplete, or incompetent, then public policy is inadequate".[12]

3. THE INSTITUTIONALIZATION OF TA AND EIA

Despite the importance of TA and EIA and the value of their outcomes, numerous organizations in society are committed to developing technologies, and to initiating environment-related projects, while very few are devoted solely to evaluating them.[13] Through agencies like the Office of Technology Assessment (OTA), the National Science Foundation (NSF), and numerous regulatory bodies such as the Food and Drug Administration (FDA), the Nuclear Regulatory Commission (NRC), and the Environmental Protection Agency (EPA), however, the US government is attempting to "buy this capability" by various "one-shot contracts and grants".[14]

To the extent that such government-financed attempts are inadequate in scope, purpose, funding, or methodology, to that same degree is there danger that unguided and self-directed technology will provide a new analogue for the *laissez-faire* economics of the nineteenth century. We have learned that economic developments need to be guided so as neither to falter nor to injure people.[15] We have been slower to realize that scientific and technological innovations need to be assessed and managed for the same reason. If they are not, they could undermine democratic institutions, disjoin societal power and responsibility, and cause what Langdon Winner terms 'reverse adaptation', a situation in which technology is treated as the *end* of progress rather than as the means to it.[16] In fact, if Winner is correct, there may be reason to believe that technology already is out of control, as evidenced by "the decline of our ability to know, to judge, or to control our technical means".[17]

3.1. *The National Environmental Policy Act (NEPA)*

Public concern over whether we could in fact understand and control the technology we were creating culminated in 1969 with the passage in the US of the National Environmental Policy Act (NEPA). The first attempt at federal legislation to provide comprehensive environmental policy was introduced into the US Senate in 1959, but it was later abandoned. As public pressure for environmental legislation mounted, NEPA passed both houses of Congress and was signed into law on January 1, 1970. It includes at least five important provisions:

1. a declaration that the federal government should use all practicable means to preserve environmental quality;
2. a mandate to establish a Council on Environmental Quality;
3. a definition of the duties of the members of the Council on Environmental Quality;
4. a stipulation that the President submit annual reports to Congress on environmental quality in the nation; and
5. a requirement that all federal agencies prepare an environmental impact statement (EIS) every time they consider proposals for federal actions significantly affecting the quality of the environment.[18]

Because it initiated some significant new procedures, e.g., preparation of EIS's, NEPA has been called "the Magna Carta of environmental protection".[19] It provided numerous opportunities for citizen participation in environmental decisionmaking, owing to the fact that draft EIS's are circulated (for comments) to public-interest groups as well as to government agencies. Philosophically, the emphases of NEPA contrast sharply with previous US commitments to untrammelled economic growth and development. The act mandates development in the context of minimal environmental degradation and states that the US should attempt to attain the *maximum* recycling of depletable resources. In a classic and often-quoted prescription, NEPA affirms that each generation has responsibilities as trustee of the environment for succeeding generations.[20]

3.2. *The US Office of Technology Assessment (OTA)*

During the same year (1969) that NEPA became law, Congressional hearings culminated in a bill to create the OTA. In 1972, after passage by both houses of the US Congress, the President signed the Technology Assessment Act. Its purpose was "to provide early indications of the probable beneficial and adverse impacts of the applications of technology and to develop other coordinate information which may assist the Congress".[21] To accomplish this purpose, the act created the OTA. Governed by a Board of six senators and six representatives, evenly divided between the two political parties, the OTA and its technical staff perform technology assessments and authorize their performance by various consulting firms and research groups. Requests for assessments can originate with a congressional committee, the OTA Board, or its Director. In overseeing the preparation and circulation of various technology assessments, the OTA was directed, by the Technology Assessment Act of 1972, to attempt to accomplish eight objectives. It has been charged to:

1. identify existing or probable impacts of technology or technological problems;
2. where possible, ascertain cause-and-effect relationships;
3. identify alternative technological methods of implementing specific programs;
4. identify alternative programs for achieving requisite goals;
5. make estimates and comparisons of the impacts of alternative methods and programs;
6. present findings of completed analyses to the appropriate legislative authorities;
7. identify areas where additional research or data collection is required to provide adequate support for the assessments and estimates . . . ;
8. undertake such additional associated activities as the appropriate authorities . . . may direct.[22]

The OTA, as an analytical support agency of the US Congress, helps federal legislators deal with many kinds of issues confronting our scientifically sophisticated society. Since its creation in 1972, the OTA has sponsored a wide range of technology assessments. Some of its completed studies include those on

Residential Energy Conservation
The Direct Use of Coal
Application of Solar Technology
Gas Potential From Devonian Shales
Environmental Contaminants in Food
Drugs in Livestock Feed
Nutrition Research Alternatives
Cancer Testing Technology and Saccharin
Effects of Nuclear War
Nuclear Proliferation and Safeguards
Renewable Ocean Energy Sources
Coastal Effects of Offshore Energy Systems
Oil Transportation by Tankers
Community Planning for Mass Transit
Solar Power Satellite Systems
Oil Shale Technology
Taggants in Explosives
Impacts of Applied Genetics
Technology and World Population and
Telecommunication Policy.[23]

3.3. *TA/EIA Outside the OTA*

In addition to the OTA, several other US groups have been engaging in assessment activities. Among federal executive agencies, the National Science Foundation has played a leadership role in TA/EIA. It has supported more than 30 comprehensive assessments, in order to provide essential information for policymaking, and it has initiated several studies of TA methodology. A handful of state governments (those in California and New York, for example) have undertaken assessments, and approximately 20 private companies have accomplished TA/EIA in one form or another.[24]

Outside the US, Sweden and Japan probably have the widest range of TA/EIA activities. In Sweden, assessment is conducted by the Secretariat for Future Studies under the Office of the Prime Minister. Many Japanese TA's/EIA's are accomplished by the (government) Science and Technology Agency, but public participation in assessment activities and policymaking is very minimal. France has shown more interest than any other country outside the US in establishing a national office of assessment and has used the US OTA as a model. Although most TA activity throughout the world takes the form of EIA of specific project proposals, assessment is rapidly becoming more sophisticated and more widespread.TA/EIA activities are currently taking place even in areas such as Indonesia, Ghana, Egypt, Mexico, and many third-world countries.[25]

4. TA/EIA METHODOLOGY

Despite the fact that TA/EIA is widely practiced throughout the globe, there is no *generally* accepted approach to which the terms 'technology assessment' and 'environmental impact analysis' refer. (Of course, as was noted earlier, in section 2.1, the US CEQ has its own format for EIA.) In the US Office of Technology Assessment, the method employed, personnel involved, and the skills tapped depend on the issues under study, the requesting client, the time for and setting of the project . . . and the resources available for a study. [26] Consequently, there are "no agreed upon techniques for carrying out a technology assessment".[27]

One of the main reasons for the absence of a uniform methodology is that, despite the fact that "science . . . underlies technology and provides the fundamental knowledge for technological application",[28] a technology or environmental impact ordinarily is evaluated as successful, acceptable, or desirable for quite different reasons than is a scientific theory. Scientific theories are usually assessed on the basis of fairly well-established criteria for their *truth*, e.g., simplicity, explanatory or predictive power. Technological inventions, on the other hand, are judged in the context of the *purpose* for which they are used, e.g., employing nuclear power to generate electricity needed by an aluminum plant.[29] Their success must be evaluated in a way analogous to judging the correctness of a particular route for a journey. As the Cheshire Cat responded, when Alice asked which way she ought to go: "that depends a good deal on where you want to get to".

Because TA/EIA is bound up with evaluating the *purposes* of particular technologies and projects, and with assessing the wisdom of certain *ends*, as well as the *means* to accomplish them, it is not purely a scientific, engineering, or disciplinary enterprise. As Joseph Coates, and many other leaders of the TA movement in the US, put it: "It is essentially an art form".[30] TA and EIA are in a class of wholistic studies which attempt in some sense to embrace everything which is essential to a technology or to a particular environmental impact. To accomplish such a wholistic enterprise, members of the assessment team are drawn from a variety of disciplines relevant to the project under consideration. An interdisciplinary team evaluating a particular CT scanner, for example, might include a radiologist, a health physicist, a biophysicist, an attorney, an ethicist, a psychologist, a political scientist, an economist, and a systems analyst. Since a good assessment must search out not only the impacts of a particular project but the interactions among them, the TA/EIA team must also work together, in an interdisciplinary mode, as well as individually, on particular sub-studies related to their own individual fields of expertise.

Because TA and EIA are wholistic, multidisciplinary efforts to gain a very wide and comprehensive analysis of a particular proposal or project, they have no unique methodology. Rather assessment practitioners use a battery of analytic tools and techniques, e.g., cost-benefit analyses, sensitivity analyses, dynamic modelling, opinion surveys, relevance trees, and so on.

In general, TA/EIA methods are of three types: analytical, empirical, and synthetic. *Analytical* methods are those which employ formal models from which one deduces insights about the problem at hand; they utilize little raw data. *Empirical* methods begin with information gathering and inductively build models to explain or to predict impacts. *Synthetic* methods combine the analytical and the empirical approaches so that models are based on empirical information, and data gathering is structured by the preexisting models.

Each of these three methods is used in particular situations, rather than in others, and each of them is associated with a number of techniques. Analytic methods are generally used in studies whose problems are well defined conceptually. One such study, for example, might be simulation of an electronic system. Empirical methods tend to be employed when the assessment problem is well defined and when there is available data on it. For example, empirical approaches might be used to forecast a specific development, such as use of solar technology for hot-water heating. Synthetic methods, finally, are often used for more complex problems which are frequently ill-defined and for which complete data are not available, e.g., deciding how best to meet energy needs in the future.

Some of the techniques associated with analytic methods include simulation modelling, substitution analyses, and relevance trees. Techniques such as regression analyses, trend extrapolation, monitoring, opinion measurement, and calculation of probabilities are used in connection with empirical methods. Still other techniques, such as use of checklists, cost-benefit analyses, cross-effect matrices, decision analyses, development of scenarios, and sensitivity analyses, are associated with synthetic methods. Even though each of these three approaches tends to be used in connection with certain techniques and in specific problem situations, it should be emphasized that many different techniques and methods are often needed to study each of the many problems addressed in an assessment. Just as the ten components of a TA/EIA (see section 2.2 earlier in this chapter) are not necessarily linear steps to be followed in chronological order, so also there is no simple methodological prescription for which techniques to use in a given problem situation in TA/EIA. Often however, assessors use opinion measurement and trend extrapolation, for example, for social forecasting (component 5 of TA/EIA, as given earlier in section 2.2 of this chapter), while they use relevance trees and checklists for impact identification (component 6 of TA/EIA). For impact analysis (component 7 of TA/EIA), assessors are likely to use techniques such as simulation models, cross-effect matrices, and sensitivity analyses.[31]

5. THE PREEMINENCE OF RISK-COST-BENEFIT ANALYSIS IN TA/EIA

Of all the techniques repeatedly employed in TA and EIA, the one to which the most decisionmaking weight is given is almost certainly risk-cost-benefit analysis (RCBA). A variant of cost-benefit analysis, RCBA incorporates notions of probability and uncertainty as a basis for estimating technology- and environment-related risks and for determining their values as costs. RCBA is preeminent among assessment methods for a number of reasons. *First*, as section 3.1 of this chapter revealed, the 1969 NEPA requires an EIS for all legislative and federal-project proposals, and "the courts have specifically interpreted section 102(2) (C) [of NEPA] to require a cost-benefit assessment".[32] The National Environmental Policy Act demands that assessors employ "a systematic interdisciplinary approach" in order to achieve a "balanced weighting of costs and benefits" of proposed projects (see note 20). Although the courts unanimously agree that NEPA does not require the risks, costs, and benefits of each environmental impact of a project to be converted to actual monetary values, there is disagreement among them over whether assessors should attempt to quantify (in some non-monetary fashion) many of the values involved.[33] Hence, although the risks, costs, and benefits of a project are not represented in terms of money, assessors routinely attempt to estimate all these parameters and to quantify as many of them as possible.

Besides its being mandated by NEPA, another reason why RCBA dominates technology-related decisionmaking is that the problems addressed by TA and EIA are often neither clearly defined nor those for which adequate data are available. This means that use of analytical and empirical assessment methods (see section 4 earlier in this chapter) is inappropriate, and practitioners of TA and EIA must rely on synthetic approaches, such as RCBA, to deal with problems which are more complex and ill-structured. Of all the synthetic methods available to assessors, RCBA is one of the most comprehensive (in terms of the number of parameters it can take into account), and the most theoretically developed. Although it is beset with various problems, there are a number of strategies for handling them (see Chapters Two, Eight, and Nine in this volume).

A final reason for the supremacy of RCBA among assessment methods is that the impetus for technological development and the initiation of environment-related projects is often in private industry, e.g., as in the case of construction of a liquefied natural gas facility. Because the central investment determinant, for private enterprise, is whether a proposed project will turn a profit, the key component of *industry*-related TA's and EIA's is almost always RCBA. Moreover, to the extent that government adopts the decision criteria (whether inappropriate or not) of private industry, then to that degree are assessments for government-related projects also likely to emphasize RCBA over all other methods. This is why Galbraith, Mishan, Schumacher, and many other experts have affirmed that, apart from whether it should be, RCBA is "the final test of public policy".[34]

5.1. *The Origins of RCBA*

The goal of risk-cost-benefit analysis (RCBA) is to determine whether the benefits of a proposed project outweigh the risks and costs involved. An offshoot of decision analysis which is widely touted as a 'developing science', RCBA is used to help administrators make decisions about how to spend money and how to make policy.[35] For example, experts might wish to determine, given the risk of nuclear core melt and the possible resultant release of radioactive materials, whether additional containment (for reactor vessels) would be cost-effective.

The use of RCBA to analyze the economic impacts of a specific project has its origins in the investment decisions of private firms. Companies weigh the proposed financial commitment against the expected return before making a decision on whether or not to invest.

In the sixties, RCBA was integrated into the planning and budgeting efforts of many federal agencies in the defense, aerospace, and energy fields. In the seventies, a number of US regulatory agencies controlling environmental and occupational hazards likewise began to use the technique. Currently nearly all US regulatory agencies (with the exception only of the Occupational Safety and Health Administration, OSHA), consistently use RCBA to help determine their policies.[36]

5.2. *The Techniques of RCBA*

Risk-cost-benefit analysis incorporates a number of techniques of economic evaluation in a common framework. Most good RCBA's are structured in terms of some variant of the following four steps:

1. definition of the subject of the analysis, including the parties at interest;
2. identification and description of all the anticipated *risks* (e.g., increased cancer rates), *costs* (e.g., construction outlays for a project), and *benefits* (e.g., in the case of a power plant, generation of needed electricity);
3. measurement of as many as possible of the risks, costs and benefits, including direct costs and benefits, externalities, and higher-order impacts; and
4. evaluation of the various economic impacts.[37]

The *direct costs* measured at stage 3 are actual outlays, or charges incurred, by the investor, e.g., for purchasing the land on which a particular facility might be built. *Direct benefits* are received by the investor as receipts, e.g., in the case of a power plant, as payment for electricity obtained by consumers. *Externalities* are social risks, costs, and benefits borne by the public, for which they are neither compensated nor charged. For example, a negative externality for which the public is not compensated by the investor is the air pollution caused by coal-generating plants. Likewise, for instance, a positive externality of the power plant is the job opportunities it provides. *Higher-order impacts* are delayed, often unintended, risks, costs, and benefits, arising from the effects of the initial or foreseen consequences. (For an example of a higher-order impact, see section 2 earlier in this chapter.) In other words, in step 3 of RCBA, analysts attempt to measure *indirect* risks, costs and benefits (externalities) and *delayed* risks, costs, and benefits, as well as direct and immediate ones.

Assessors usually evaluate the various economic impacts (step 4) by selecting decision criteria and a discount rate, by actually performing the economic calculations, and then by doing a sensitivity analysis. (Discounting is the means by which one compares current dollar investments with future returns; the discount rate measures how much more resources used now are valued over resources used in the future. For example, if an individual felt that $106 a year from now was as acceptable as $100 today, the discount rate would be 6 percent.)[38] Several of the more important decision criteria include: (1) net present value, (2) annual equivalent; (3) internal rate of return, (4) payback period, (5) benefit: cost/risk ratio, and (6) composite economic and factor profiles. Decision criteria (1)–(5) differ in the way they present the results of discounting the time stream of risks, costs, and benefits. All five of these criteria, however, are comparable in that they provide ways of evaluating the (quantitative) monetary risks, costs, and benefits. The sixth decision criterion, composite economic and factor profiles, helps assessors to consider both quantitative and qualitative factors and to weigh the merits of each. This criterion emphasizes the fact that economic decisions should be based on the differences among policy alternatives.[39]

Another important tool for evaluating economic impacts (stage 4 of RCBA) is sensitivity analysis. In using it, assessors alter the risk, cost, or benefit variables, their values, or their interrelationships, in order to observe how these modifications affect the values and range of the impacts. By using sensitivity analysis, assessors hope to determine how dependent their conclusions are on their assumptions and on the accuracy of their risk-cost-benefit data. They also hope to identify the leverage points on which to base strategies for modifying their conclusions.[40]

6. THE ROLE OF THE PHILOSOPHER IN TA AND EIA

Although evaluation of economic impacts, as just described, is largely the task of economists, a significant role must also be played by assessors from other disciplines. This is because economic impacts are also likely to have, for example, psychological, medical, and ethical impacts on the parties at interest, and these impacts are perhaps best evaluated, respectively, by psychologists, physicians, and philosophers. Numerous other examples could be given to illustrate that, even in an allegedly purely disciplinary enterprise, such as analysis of economic impacts, broad interdisciplinary expertise is required in order to evaluate adequately those impacts.

This observation about the necessity of *evaluative*, interdisciplinary work, even in an allegedly disciplinary task within TA and EIA, raises a broader question. What ought the evaluation of impacts to be like? Surely it is not reducible merely to use of a particular quantitative decision criterion of RCBA (see section 5.2 earlier in this chapter), for example, because every such criterion has its own evaluative presuppositions and consequences which themselves need to be evaluated. This suggests that, if assessment is to be appropriately critical and comprehensive, two of the components which every good TA and EIA ought to include are *methodological* evaluation of the various assessment techniques employed, and *ethical* analysis of the presuppositions in, and consequences following from, the conclusions reached at each step of the TA or EIA. For example, component 1 of TA/EIA is definition of the assessment problem (see section 2.2 earlier in this chapter). Unless a philosopher provides an analysis of the normative presuppositions inherent in a particular definition of the TA problem, then it is conceivable that its ethical dimensions might be ignored. As a consequence, project team members might explore neither these normative aspects nor possible means of avoiding ethically questionable impacts, such as violations of rights. Yet, if the recent history of technology is to be believed, then the ethical dimensions of many projects are often among the most important. Rachel Carson, for example, observed that those who implemented pest-control technology, several decades ago, often ignored the ethical dimensions of their activities.[41] Policymakers, likewise, were apparently unaware that one of the difficulties they faced (with allowing use of certain chemical pesticides) was that their application might violate persons' rights to equal protection from carcinogenic and mutagenic injury.

Admittedly, besides arguing that one ought not to ignore *ethical* presuppositions, one could just as well maintain that assessors ought not to neglect the chemical, the sociological, or some other disciplinary dimensions of a particular TA/EIA. Why ought one be so concerned with emphasizing the philosophical aspects of an assessment? There are at least two main reasons why it is important to spell out the role of philosophers in TA and EIA. The most obvious one, as will be argued in Chapters Three and Four later in this volume, is that the philosophical (ethical, methodological, and logical) dimensions of all the assessment components are usually ignored. Interdisciplinary TA/EIA teams rarely include philosophers, and when they do, their contributions are often discounted. A second reason for emphasizing philosophical analysis, even though all disciplines ought to play a role in TA and EIA, is that there is no literature on the role of the philosopher as assessor. This may be in part because the field of applied philosophy is relatively new, and because TA and EIA often have been viewed as purely technical studies. There is, however, a highly developed literature on the assessment roles of, for example, sociologists, chemists, political scientists, and economists. Good books on TA and EIA often will devote an entire chapter to each of the disciplinary classes of impacts, e.g., to psychological impacts, social impacts, economic impacts, legal impacts, and so on.[42] Yet nowhere, in this rapidly growing body of assessment literature, does anyone deal at length with the philosophical aspects of TA/EIA. Since this volume is one first and modest step to remedy that deficiency, it is important to understand precisely what a philosopher might contribute to assessment.

6.1. *The Role of the Ethicist in TA and EIA*

Although there are various tasks for which a person trained in philosophy might be suited, two appear to be of primary importance. These are ethical analysis and methodological analysis. Let's discuss the role of the ethicist first.

Persons trained to do ethical analysis have a key role in TA and EIA precisely because so many of the major problems posed by medical, energy, and chemical technologies are ethical problems. Does a newborn have more right to life than a fetus? Does a defective human often have no more right to life than an animal? Does generation and storage of radioactive waste violate the rights of future generations? Do the current legal impacts of implementation of liquefied-natural-gas technologies violate persons' due-process rights? Do persons have an absolute right to protection against carcinogens to which they might be exposed without their consent? Do some newer medical technologies, e.g., in-vitro fertilization, threaten the sacredness of life? These and other moral issues are at the heart of much technological controversy and hence ought not be ignored in TA and EIA. And since they ought not be ignored, there are several reasons why moral philosophers or ethicists might have some advantage over others in discussing them.

First, since philosophers receive general training in understanding logical arguments and detecting fallacies, it follows that they might be adept at understanding the logic and the fallacies imbedded in alternative arguments about the ethics of proposed assessment policies and projects. *Second*, philosophers are taught to analyze specific moral concepts by virtue of their education in what is known as 'metaethics'. Clarity about moral concepts is of course not sufficient to resolve difficult ethical matters, but it is often a necessary condition for a good decision. It will help one avoid dealing with those issues in a muddled way. *Third*, philosophers have spent a significant amount of time studying moral theories, such as utilitarianism, egalitarianism, libertarianism, and natural-law views. They also study theories of rights and of justice. Apart from which moral theory, or which theory of rights, one adopts, it is important to know the implications and presuppositions of each and the various objections that can be brought against each of them, since this knowledge is often helpful in discussing practical or technological issues which have ethical dimensions. Without expert knowledge of ethical theories, one is likely unable to recognize the normative presuppositions inherent in one's stand on particular technological issues or the ethical consequences following from his adoption of a specific TA/EIA methodology. Cost-benefit analysis, for example, a widely used assessment technique, has been criticized for its allegedly utilitarian presuppositions. Whether or not one agrees with those presuppositions, or with the ethical theory from which they emerge, assessors need to recognize them, to point out their significance, and to describe alternatives to them or to the methods containing them. Only by so doing will they produce a TA/EIA which evaluates the ethical, as well as the technical, alternatives to proposed projects.[43]

Moral philosophers ought to be able to take an active role in TA and EIA by providing ethical analyses of the concepts, methods, assumptions, models, and theories used in each of the 10 components of assessment. They ought to be able to investigate both a particular technology or environmental impact and to evaluate specific TA's or EIA's with respect to their ethical presuppositions and likely consequences. Admittedly the skills of the moral philosopher may not be sufficient to resolve the ethical quandries generated by technology- and environment-related controversy. Nevertheless, greater clarity, the avoidance of logical fallacies, a better understanding of moral theories and concepts, and a critical examination of alternative ethical arguments on policy matters is sure to take assessors further toward rational analysis than ignoring these elements of ethical evaluation.

6.2. *The Role of the Applied Philosopher of Science*

Because assessors utilize the data, models, and theories of the natural and social sciences in order to provide information concerning the effects of these applications, applied philosophers of science also have a clear role to play in TA and EIA. According to Emilio Daddario, "assessment work will be directed by the performers and managers of applied science. The value framework against which technological consequences are judged will be erected by the social sciences, the arts, and humanities. . . "[44]

There are several reasons why the evaluations and methodological analyses characteristic of good scholarship in the philosophy of science are particularly needed in technology assessment and environmental impact analysis. For one thing, scientific theory becomes conceptually impoverished when used as a means to achieve practical ends.[45] Bunge notes, for example, that the scientific concepts, theories, and models employed in technological studies are much less carefully worked out than those characterizing pure science. He says that this is because the former are intended for practical persons, interested in making things 'work' for them, rather than in how "things of any kind really are".[46] Since technological theories and studies are much more guilty (than scientific theories) of "oversimplification" and "superficiality",[47] methodological analyses of them are needed all the more. In fact, both governmental and industrial policymakers admit that the biggest problem with TA and EIA is "the lack of conceptual thinking on the technology assessment function, including alternative notions of adequacy of assessment".[48] The difficulty, they say, is *not* what "research must be done, but how to do it" in terms of "methodological requirements".[49]

Another reason why the methodological analyses characteristic of good work in philosophy of science are so needed in contemporary TA and EIA is that there is a great tendency to perform the assessments "as though they are engineering studies".[50] This is a serious difficulty because teams composed only of engineers and natural scientists often attempt either to deal with issues outside their areas of professional expertise, or to ignore them; as a consequence, technological problems are frequently defined in terms "too narrow to do justice to the complexity of real situations".[51] Such limited definitions often exclude consideration of relevant normative, epistemological, logical, or methodological assumptions affecting the validity of the conclusions in a given technology assessment or impact analysis. Hence it is not surprising that one government spokesperson commented: "the experimental skills of scientists and engineers do not necessarily coincide with those [conceptual, methodological] skills necessary for identifying and analyzing the consequences of new or existing technologies".[52] This is in part because "the higher the policy level [of the TA or EIA], the less strictly technical are the factors that have to be taken into account".[53]

If assessment work is left merely to engineers and natural scientists, this might mean more than the fact that certain critical logical, normative, epistemological, or social issues will go unanalyzed or unrecognized. It might also be a source of liability to practical decisionmakers and an embarrassment to philosophers who could have had a role to play in developing and altering the conceptual and technological framework in which TA and EIA are done.[54] Admittedly, philosophic thought without application is blind. If some of the insights developed in this discussion are correct, however, then TA and EIA without philosophy may place society in a worse situation.

Philosophers of science ought to be able to take an active role in applied work by analyzing the concepts, methods, assumptions, models, and theories of particular sciences, both with respect to their inherent viability and with respect to the legitimacy of their employment in TA/EIA. These types of analysis may be said to constitute either pure or applied philosophy of science, respectively, depending on what questions they address. Queries about the *per se* logical or epistemological validity of particular scientific assumptions of models typically have been in the domain of *pure* or classical philosophy of science. Other questions, however, lie in the applied realm of the discipline. Some of these include: whether, apart from their *a priori* viability, particular scientific concepts, assumptions, or models may legitimately be used, in a given situation, to assess a specific technology; whether they have been applied correctly in cases where it appears legitimate to do so; and whether employment of certain scientific concepts, assumptions, or models, in a specific TA or EIA, leads to methodologically justifiable consequences.

Understood in this sense, one may use applied philosophy of science either to evaluate a technology or environmental impact or to analyze a certain TA or EIA. Each task, however, presupposes a clear understanding, both of philosophy of science and of the particular science used to describe, analyze, or predict a given impact. This is because, despite the political values often motivating controversy over a certain technology assessment, such conflicts almost always involve technical questions.[55] As Feigl and Brodbeck put it: "if the philosophy of science is taken seriously, then it is not to be expected that philosophical discussion of the more advanced and technical branches of a science can itself be elementary and wholly nontechnical".[56]

Specifically, philosophers of science can contribute to assessment and hence ultimately to the guidance of environment- and technology-related public policy on at least three levels, that of conceptual analysis, that of scientific methodology, and that of concrete applications, where the concepts and methods are employed in specific contexts.[57] At the level of *conceptual analysis*, one scrutinizes critically the central notions of philosophy of science (e.g., 'prediction') or of some particular science (e.g., 'Pareto optimality' in economics) in order to provide clarification, to evaluate claims of logical connection, or to explore the foundations of scientific argument. At this level, the inquiries of the philosopher of science may often be used to support or question forms of practical argument (e.g., in a technology assessment) which appeal to conceptual warrants for their justification.

At the second level, that of *analysis of scientific methodology,* the philosopher of science focuses on articulation, criticism, or defense of assumptions, rules, or principles used to guide substantive deliberation about science. Here questions are not so much about the meanings of scientific terms, as about the difficulties attending their use. One specific problem, for example, is under what modelling circumstances one ought to employ the simplifying principle known as 'exclusion'. When one appeals to this principle, he attempts to justify his assumption that the influence of any factor not included in the model is unimportant in affecting the conclusions.[58]

This second level of philosophical analysis of science is particularly important because it is here that one can often uncover what went wrong with some specific predictions in a technology assessment or environmental-impact analysis. The assumptions made (and often not recognized) by the modellers in applied situations (e.g., energy-scenario assessment) determine the results of the models. In fact, this is precisely what the authors of the recent Harvard Business School assessment of energy technologies discovered. They concluded that a great amount of theoretical, methodological work needed to be done to spell out assumptions and thus 'demystify' a model *before* it was used in an assessment. Otherwise the utility and limitations of certain models would never be clear.[59]

Since methodological and conceptual issues often remain programmatic and general, the philosopher of science also has a more specific role to play, at the third level, that of *concrete applications*. Here the conceptual analysis of level one and the methodological analysis of level two are employed to guide substantive deliberation about specific science policy or particular TA's and EIA's. Here the more theoretical questions of the earlier levels make contact with questions of fact and public policy regarding the assessment. The goal at the third level is to employ critically methods of philosophy of science in contexts calling for practical responses.

For example, on the basis of a conceptual analysis of 'prediction' and a methodological analysis of relevant mathematical techniques, a philosopher of science might evaluate a well-known practical argument. This argument, that a core melt can be predicted to occur once in every 17 thousand reactor years, is made in the allegedly best assessment of nuclear technology.[60]

Likewise, for example, on the basis of a conceptual analysis of the principle of 'exclusion', and a methodological and logical analysis of the various assumptions employed in models used to predict future energy demand, a philosopher of science could be helpful in investigating another specific claim. Made in a recent assessment (of various energy technologies) on whose project team the author served, this claim is that only nuclear- and coal-generated energy would be able to help meet electricity demand by the year 2000.[61]

As these two examples of analysis indicate, there are several tasks at which the applied philosopher of science may be helpful. He may help ascertain the extent to which a given scientific (e.g., econometric) model adequately depicts the technological or environmental situation being analyzed. Or he may undertake a methodological analysis of the procedures, limitations, models, and assumptions employed by the scientists and engineers working on an assessment team.[62] In this way, a philosopher of science could help to encourage the practice (almost never followed in TA and EIA) of stating explicitly, even highlighting, the limitations of a particular assessment or model.[63]

7. CRITICISM OF TA/EIA AND AN OUTLINE FOR REFORM

Since the last several pages have outlined the potential contributions of philosophers to assessment work, it should not be surprising to discover that many of the limitations of TA/EIA arise from ignoring the ethical and methodological dimensions of the assessment task. In fact, the purpose of the remaining chapters in this volume is to uncover several serious flaws in TA/EIA and to provide some specific proposals for improving both the theory and practice of assessment.

Because the focus of these essays is on constructive criticism, and on improving, rather than abandoning, TA and EIA, it is important to provide a proper context for understanding assessment and the nature of the charges levelled against it. Chapter Two presents a sustained defense both of TA/EIA and of its preeminent but highly controversial method, risk-cost-benefit analysis (RCBA). This defense accomplished, Chapters Three and Four analyze some general methodological problems in TA and EIA. Chapter Three explains and finds fault with the positivistic doctrine that wholly neutral assessments are possible and desirable; it reveals the questionable scientific, epistemological, and ethical presuppositions underlying this view. Chapter Four criticizes the tendency of most practitioners of TA and EIA to define the scope of the assessment problem in terms of the technological *status quo*, rather than in terms of alternative scenarios demanding sophisticated philosophical, social, scientific, and political analyses. Both these chapters deal with assessors' assent to two highly doubtful methodological tenets. Chapter Three analyzes (what the author calls) 'the principle of complete neutrality', while Chapter Four discusses one variant of (what Keniston calls) 'the fallacy of unfinished business'.

The next three essays in the volume focus on specific methodological problems associated with the use of RCBA in TA and EIA. Chapter Five uncovers the implicit ethical and methodological presuppositions built into the notion of 'compensating variation' and the questionable policy consequences following from its use. Chapter Six discusses the methodologically controversial assumption that nonmarket parameters need not be quantified in RCBA, and uses case studies to illustrate the fact that following this assumption in TA and EIA distorts the evaluation of actual risks, costs, and benefits and often results in methodologically suspect policy conclusions. Chapter Seven takes up the question of distributive justice, so often a target of critics of RCBA. Its analysis of this issue, however, is from a wholly new perspective, that of regional equity. The argument of this chapter is that RCBA ought to be amended, so as to take account of the geographical distribution of technological and environmental risks, costs, and benefits.

In response to the problems uncovered and criticized in the previous five chapters, the last two essays offer some suggestions for improving TA and EIA. To address the difficulties discussed in Chapters Three, Five, and Seven, Chapter Eight proposes that assessors employ ethically weighted versions of RCBA. Speaking to the problems analyzed in Chapters Four, Five, Six, and Seven, Chapter Nine recommends that TA/EIA be accomplished by means of adversary assessment, using a 'technology tribunal'.

Although it is unlikely that the problem analyses and solutions proffered here will be sufficient, alone, to effect an improvement in the theory and practice of TA and EIA, they should provide a useful framework for continuing interdisciplinary discussion of assessment. Out of such interdisciplinary cooperation, perhaps we can create a global vision of a just, participatory, sustainable society in which just, participatory, and sustainable assessments are the norm.

NOTES

[1] L. H. Mayo, 'The Management of Technology Assessment', in *Technology Assessment* (ed. by R. G. Kasper), Praeger, New York, 1972, p. 73; hereafter cited as: Mayo, TA, in Kasper, *TA*.

[2] H. D. Banta and C. J. Behney, *Assessing the Efficacy and Safety of Medical Technologies*, US Office of Technology Assessment, Washington, D.C., 1978, pp. 4, 23–58.

[3] A. L. Porter, F. A. Rossini, S. R. Carpenter, and A. T. Roper, *A Guidebook for Technology Assessment and Impact Analysis*, North Holland, New York, 1980, p. 255 (hereafter cited as: Porter, *Guidebook*); and H. Skolimowski, 'Technology Assessment as a Critique of Civilization', in *PSA 1974* (ed. by R. S. Cohen *et al.*), Reidel, Boston, 1976, p. 459 (hereafter cited as: TA).

[4] This definition follows J. F. Coates, 'Technology Assessment', *The Futurist* 5 (6), (December 1971), p. 225; hereafter cited as: TA.

[5] Congress of the US, Office of Technology Assessment, *Annual Report to the Congress for 1978*, US Government Printing Office, Washington, D.C., 1978, p. 113; hereafter cited as OTA, *AR 78.*

[6] Coates, TA, pp. 228–229.

[7] This comparison relies heavily on Porter, *Guidebook*, p. 50.

[8] See the preceding note in this chapter.

[9] Porter, *Guidebook*, pp. 54–60. See also, for example, Coates, TA; Martin Jones, *A Technology Assessment Methodology*, 7 vols, MITRE Corporation, Washington, D.C., 1971; National Academy of Sciences, Panel on Technology Assessment, *Technology: Processes of Assessment and Choice*, Report to the Committee on Science and Astronautics, US House of Representatives, US Government Printing Office, Washington, D.C., 1969; and L. H. Tribe, 'Technology Assessment and the Fourth Discontinuity', *Southern California Law Review* 46 (3), (June 1973): 3–5; hereafter cited as: Tribe, TA.

[10] My discussion of the ten components of TA/EIA relies heavily on J. Golden, R. Ouellette, S. Saari, and P. Cheremisinoff, *Environmental Impact Data Book*, Ann Arbor Science, Ann Arbor, 1979, esp. pp. 4–10 (hereafter cited as: Golden, EIDB), on M. Soroush, K. Chen, and A. Christakis, *Technology Assessment*, North Holland, New York, 1980, pp. 46–52 (hereafter cited as: Boroush, TACF), and on Porter, *Guidebook*, pp. 54–60. See also L. Canter and Loren Hill, *Handbook of Variables for Environmental Impact Assessment*, Ann Arbor Science, Ann Arbor, 1979 (hereafter cited as: Canter: HV), and P. A. Erickson, *Environmental Impact Assessment*, Academic, New York, 1979 (hereafter cited as: Erickson, EIA). Finally, see R. K. Fain, L. V. Urban, and G. S. Stacey, *Environmental Impact Analysis*, Van Nostrand Reinhold, New York, 1981, pp. 36–71 (hereafter cited as: Jain, EIA.

[11] S. Arnstein and A. Christakis, *Perspectives on Technology Assessment*, Science and Technology Publishers, Jerusalem, 1975, p. 16.

[12] E. Q. Daddario, 'Foreword', in *Second-Order Consequences: A Methodological Essay on the Impact of Technology* (ed. by R. A. Bauer), The MIT Press, Cambridge, Massachusetts, 1969, p. v. Hereafter cited as: Bauer, *SOC*. See Coates, TA, p. 226.

[13] D. E. Kash, Director of the Science and Public Policy Program, University of Oklahoma, in Congress of the US, *Technology Assessment Activities in the Industrial, Academic, and Governmental Communities*. Hearings Before the Technology Assessment Board of the Office of Technology Assessment, 94th Congress, Second Session, June

8–10, 14, 1976, US Government Printing Office, Washington, D.C., 1976, p. 201. Hereafter cited as: Congress, *TA in IAG.*

[14] Kash in Congress, *TA in IAG*, p. 201.

[15] W. H. Ferry, 'Must We Rewrite the Constitution To Control Technology?' in *Technology and Society* (ed. by Noel de Nevers), Addison-Wesley, Reading, Massachusetts, 1972, p. 11, makes this same point. Hereafter cited as: Ferry, Constitution, in de Nevers, *Technology.*

[16] See Langdon Winner, *Autonomous Technology: Technics-out-of Control as a Theme in Political Thought*, The MIT Press, Cambridge, Massachusetts, 1977, p. 251. Hereafter cited as: *AT.* See also Joseph Haberer, 'Scientific Leadership and Social Responsibility', in *Philosophical Problems of Science and Technology* (ed. by A. C. Michalos), Allyn and Bacon, Boston, 1974, p. 511 (hereafter cited as: Haberer, Leadership, in Michalos, *Problems*); H. G. Rickover, 'A Humanistic Technology', in de Nevers, *Technology*, p. 32; and D. J. Rose, 'New Laboratories for Old', in *Science and Its Public: The Changing Relationship* (ed. by G. Holton and W. Blanpied), D. Reidel, Boston, 1976, p. 151 (hereafter cited as: Rose, New, in Holton and Blanpied, *Science*).

[17] Winner, *AT*, p. 30.

[18] See Porter, *Guidebook*, pp. 27–30.

[19] I. G. Barbour, *Technology, Environment, and Human Values*, Praeger, New York, 1980, p. 188; hereafter cited as: *Technology.*

[20] *The National Environmental Policy Act of 1969*, USCA, 42 sec. 4321 ff.

[21] Cited in Barbour, *Technology*, p. 197.

[22] Congress, OTA, *AR 1978.*

[23] OTA, *Current Assessment Activities*, Congress of the US, OTA, Washington, D.C., 1980, pp. 4–23.

[24] For a complete list of all groups who sponsor TA's/EIA's see S. Herner (ed.), *EIS: Cumulative 1980*, Information Resources Press, Arlington, Virginia, 1981; Golden, EIDB, pp. 1–9; Marc Landy (ed.), *Environmental Impact Statement Directory*, Plenum, New York, 1981, pp. 9–149; and Porter, *Guidebook*, pp. 35–36.

[25] See M. Srinivasan (ed.), *Technology Assessment and Development*, Praeger, New York, 1982; G. J. Stöber and D. Schumacher (eds.), *Technology Assessment and Quality of Life*, Elsevier, New York, 1973; T. J. Knight, *Technology's Future*, Krieger, Malabar, Florida, 1982, pp. 25–54; G. Wandesforde-Smith, "International Perspectives on Environmental Impact Assessments", *Environmental Impact Assessment Review* 1 (1), (March 1980), 53–69; M. Cetron and B. Bartocha (eds.), *Technology Assessment in a Dynamic Environment*, Gordon and Breach, London, 1973; and M. Boroush, K. Chen, and A. Christakis, *Technology Assessment: Creative Futures*, North Holland, New York, 1980; hereafter cited as: Boroush, TACF. See also Porter, *Guidebook*, pp. 37–40.

[26] Congress, OTA, *AR 1978*, p. 74.

[27] Congress of the United States, Office of Technology Assessment, *Technology Assessment in Business and Government*, US Government Printing Office, Washington, D.C., 1977, p. 13. Hereafter cited as: Congress, OTA, *TA in B and G.* This same point is made in numerous other analyses, for example, in Congress of the US, Office of Technology Assessment, *Annual Report to the Congress for 1976*, US Government Printing Office, Washington, D.C., 1976, pp. 66–67; hereafter cited as: Congress, OTA, *AR 1976.* See also William Fisher, Assistant Secretary for Energy and Minerals, US Department of the Interior, in Congress, *TA in IAG*, p. 27; R. G. Kasper, 'Introduction', in *Technology*

Assessment: Understanding the Social Consequences of Technological Applications (ed. by R. G. Kasper), Praeger, New York, 1972, p. 4 (hereafter cited as: Kasper, *TA*); and J. R. Ravetz, *Scientific Knowledge and Its Social Problems*, Clarendon Press, Oxford, 1971, pp. 369–70 (hereafter cited as: Ravetz, *SK and SP*).

[28] R. C. Dorf, *Technology, Society, and Man*, Boyd and Fraser, San Francisco, 1974, p. 20; hereafter cited as: *TSM*. See Don Ihde, *Technics and Praxis*, D. Reidel, Boston, 1979, pp. xviii–xxvi, 109 ff., for a discussion and critique of the claim that technology is merely applied science.

[29] Ravetz, *SK and SP*, pp. 361, 426–432, makes this same point about purpose being central to the evaluation of a technology. Although this is not the place to engage in a lengthy discussion of the technology–science relationship, it does seem quite clear that because of their differing focus on purpose versus truth, technology may not naively be thought of simply as 'applied science'. Even the development of technologies reveals that inventions are not merely fruits on some tree of theoretical science. A study of the origins of 84 British technological innovations revealed, for example, that 'demand pull', for some industrial or governmental purpose, occurred much more frequently than 'discovery push' from theoretical science. (J. Languish, 'The Changing Relationship Between Science and Technology', *Nature* 250 (5468), (August 23, 1974), 614.)

[30] Quoted in S. R. Arnstein, 'Technology Assessment', *IEEE Transactions on Systems, Man, and Cybernetics* 7 (8), (August 1977), 572; hereafter cited as Arnstein, TA.) For this same opinion, see also S. Enzer in Congress, *TA in IAG*, p. 222, and D. Kash in Congress, *TA in IAG*, p. 194.

[31] Much of this discussion of methods and techniques is based on Golden, EIDB, pp. 23–71; Boroush, TACF, 201–312; P. N. Cheremisinoff and A. C. Morresi, *Environmental Assessment and Impact Statement Handbook*, Ann Arbor Science, Ann Arbor, Michigan, 1977, pp. 75–92; R. E. Munn (ed.), *Environmental Impact Assessment*, SCOPE, Toronto, 1975, pp. 42–67; P. Black, *Environmental Impact Analysis*, Praeger, New York, 1981, pp. 45–52; and L. Canter, *Environmental Impact Assessment*, McGraw-Hill, New York, 1977, pp. 173–219, 294–310; hereafter cited as: Canter, EIA. Finally, see Jain, EIA, pp. 72–94; J. Rau and D. Wooten (eds.), *Environmental Impact Analysis*, McGraw-Hill, New York, 1980, pp. 8-1 through 8-29 (hereafter cited as: Rau, EIA); Canter, HV; and Porter, *Guidebook*, p. 79, and on I. Mitroff and M. Turoff, 'Technology Forecasting and Assessment', *IEEE Spectrum* 10 (3), (March 1973), 62–71; hereafter cited as: TFA. For more specific definitions and discussions of each of the techniques mentioned (e.g., cross-impact analysis, simulations modelling), see Porter, *Guidebook*, pp. 77–87.

[32] J. R. Luke, 'Environmental Impact Assessment for Water Resource Projects', *George Washington Law Review* 45 (5), (August 1977), 1107; hereafter cited as Luke, EIA.

[33] Luke, EIA, p. 1108.

[34] J. K. Galbraith, *The New Industrial State*, Houghton Mifflin, Boston, 1967, p. 408; hereafter cited as: NIS; see also Ravetz, *SK and SP*, p. 396, and E. J. Mishan, *Welfare Economics*, Random House, New York, 1969, p. 5; hereafter cited as *WE*. Finally, see E. F. Schumacher, *Small Is Beautiful*, Harper, New York, 1973, p. 38.

[35] S. Levine, 'Panel: Use of Risk Assessment', in *Symposium/Workshop . . . Risk Assessment and Governmental Decisionmaking* (ed. by Mitre Corporation), Mitre Corporation, McLean, Va., 1979, p. 634, hereafter cited as: Mitre, *Symposium*.

[36] L. Carter, 'Dispute over Cancer Risk Quantification', *Science* 203 (4387), (1979),

1324–1325; C. Starr and C. Whipple, 'Risks of Risk Decisions', *Science* 208 (4448), (1980), 1114–1119; and S. Gage, 'Risk Assessment in Governmental Decisionmaking', in Mitre, *Symposium*, p. 13.

[37] This and the ensuing discussion of RCBA rely heavily on Rau, EIA, pp. 2-1 through 2-76; L. Gsellmann, *Cost-Benefit Analysis and R and D Planning*, M77-12, Mitre Corporation, McLean, Virginia, 1977; Golden, EIDB, pp. 31–33; Jain, EIA, pp. 157–168; Canter, EIA, pp. 163–172; Erickson, EIA, pp. 246–258; and Porter, *Guidebook*, pp. 254–278.

[38] Golden, EIDB and Porter, *Guidebook*, pp. 270–272.

[39] For discussion of alternative decision criteria, see Golden, EIDB, pp. 26–57; Boroush, TACF, pp. 201–302, 359–378; Canter, EIA, pp. 220–232; L. Caldwell, *Science and the National Environmental Policy Act*, University of Alabama Press, Birmingham, 1982; J. Armstrong and W. Harman, *Strategies for Conducting Technology Assessments*, Westview, Boulder, 1980, pp. 57–76; D. O'Brien and D. Marchland (eds.), *The Politics of Technology Assessment*, Heath, Lexington, Mass., 1982, and Porter, *Guidebook*, pp. 272–278.

[40] For a good discussion of sensitivity analysis, see P. Frank, *Introduction to System Sensitivity Theory*, Academic Press, New York, 1978.

[41] Houghton Mifflin, Boston, 1962, pp. 12–13.

[42] See Porter, *Guidebook*, for example.

[43] Much of this discussion of the value of applied ethics relies on Peter Singer, 'How Do We Decide?', *The Hastings Center Report* 12 (3), (June 1982), 9–10.

[44] E. Q. Daddario, 'Foreword', *SOC* (note 12), p. vi.

[45] Mario Bunge, 'Towards a Philosophy of Technology', in Michalos, *Problems* (note 16), p. 32; hereafter cited as: Bunge, Philosophy, in Michalos, *Problems*.

[46] Bunge, Philosophy, in Michalos, *Problems*, pp. 51–52.

[47] Bunge, Philosphy, in Michalos, *Problems*, p. 32.

[48] Mayo, TA, in Kasper, *TA*, p. 99. See also Congress, *TA in IAG*, p. 229.

[49] Congress, OTA, *A Review of the US Environmental Protection Agency Environmental Research Outlook: FY 1976 through 1980*, US Government Printing Office, Washington, D.C., 1976, pp. 8, 53, 97–98; hereafter cited as: Congress, OTA, *Review EPA*.

[50] F. P. Huddle, 'The Social Function of Technology Assessment', in Kasper, *TA*, p. 170.

[51] Dorothy Nelkin, *The Politics of Housing Innovation*, Cornell University Press, Ithaca, 1971, p. 88; hereafter cited as: Nelkin, *PHI*.

[52] Cited in C. H. Danhof, 'Assessment Information Systems', in Kasper, *TA*, p. 23.

[53] M. D. Reagan, *Science and the Federal Patron*, Oxford University Press, New York, 1969, p. 97.

[54] See K. E. Goodpaster and K. M. Sayre, 'Introduction', in *Ethics and the Problems of the 21st Century* (ed. by K. Goodpaster and K. Sayre), University of Notre Dame Press, Notre Dame, 1979, p. vii; hereafter cited as: Goodpaster and Sayre, *Ethics*; K. E. Goodpaster, 'From Egoism to Environmentalism', in Goodpaster and Sayre, *Ethics*, p. 21; and Ferry, 'Constitution', in de Nevers, *Technology*, p. 17, who also make this point. J. L. Huffman, 'Individual Liberty and Environmental Regulation: Can We Protect People While Preserving the Environment?', *Environmental Law* 7 (3), (Spring 1977), 447, cites a famous statement of Albert Schweitzer, from *The Philosophy of Civilization*,

in this regard: "So little did philosophy philosophize about civilization that she did not even notice that she herself and the age along with her were losing more and more of it. In the hour of peril the watchman who ought to have kept us awake was himself asleep and the result was that we put up no fight at all on behalf of our civilization".

[55] Dorothy Nelkin, 'Science, Technology, and Political Conflict: Analyzing the Issues', in *Controversy: Politics of Technical Decisions* (ed. by D. Nelkin), Sage, Beverly Hills, 1979, p. 16; hereafter cited as: Nelkin, *Controversy*. This point is substantiated, not only in the article, but also throughout the book, *Controversy*, which treats numerous technology-related case studies.

[56] H. Feigl and M. Brodbeck, 'Preface', in *Readings in the Philosophy of Science* (ed. by Feigl and Brodbeck), Appleton-Century-Crofts, New York, 1953, p. v., hereafter cited as: *Readings*.

[57] These three levels follow closely those outlined by K. E. Goodpaster and K. M. Sayre, 'Introduction', in Goodpaster and Sayre, *Ethics*, pp. vii–viii. Here they describe the three areas in which moral philosophers may make a contribution to social action.

[58] Sergio Koreisha and Robert Stobaugh, 'Appendix: Limits to Models', in *Energy Future: Report of the Energy Project at the Harvard Business School* (ed. by Robert Stobaugh and Daniel Yergin), Random House, New York, 1979, pp. 237–238, discuss 'exclusion' and make several of the same points investigated here; hereafter cited as: Koreisha and Stobaugh, 'Appendix', in Stobaugh and Yergin, *EF*.

[59] In this regard, see Koreisha and Stobaugh, 'Appendix', in Stobaugh and Yergin, *EF*, pp. 234–265, esp. pp. 236–237. Here the authors use three studies, (1) The Kennedy–Houthakker World Oil Model, (2) The MIT Energy Self-Sufficiency Study, and (3) The Federal Energy Administration's 'Project Independence' Report to show how methodological assumptions undercut the value of various models' predictions.

[60] See K. S. Shrader-Frechette, *Nuclear Power and Public Policy: Social and Ethical Problems of Fission Technology*, Reidel, Boston, 1983, pp. 83–85; hereafter cited as: *Nuclear Power*.

[61] See J. J. Stukel and B. R. Keenan, *Ohio River Basin Energy Study*, I-A, Grant Number R804848-01, US Environmental Protection Agency, Washington, D.C., November 1977, pp. 12–15; hereafter cited as *ORBES*. As Koreisha and Stobaugh point out, use of 'exclusion' in technological studies or technology assessments, often takes the form of excluding analyses of social-science-related parameters and concentrating on purely technical ones. They cite an example of a methodologically questionable use of this principle by quoting a *Chemical Week* headline: 'Oil from Coal: It could be a gusher by the 1990's'. The entire article focused on the fact that the technology is now available, while the authors admitted (but ignored) the fact that "the only barriers [to oil from coal] are political, social, and economic". Another excellent example of use of the principle of exclusion is found in the *MIT Energy Self-Sufficiency Study*. Here, say Koreisha and Stobaugh, the MIT authors admitted that politics was a crucial factor in assessing feasible energy technologies for the future, and that problems such as nuclear safety and environmental protection were obviously affected by politics. After this admission, however, the MIT group simply stated: "Such issues, though both appropriate and important to the debate ... are beyond the scope of this report". (Koreisha and Stobaugh, 'Appendix', in Stobaugh and Yergin, *EF*, pp. 263–264.

[62] In this regard, see K. S. Shrader-Frechette, 'The Ohio River Basin Energy Study: Methodological and Ethical Porblems', in *Ohio River Basin Energy Study* (ed. by J. J.

Stukel and B. R. Keenan), IV, Grant Number R804848-01, US Environmental Protection Agency, Washington, D.C., 1978, pp. 50 ff.

[63] One way to determine the limiting assumptions of a particular model, for example, is by testing the validity of the predictions. This might be done, for instance, by discovering what happens when specific changes are made in key parameters, and then by determining whether the resultant changes are plausible or not. An implausible, substantial change could indicate faulty assumptions or incorrect logical relationships in the model. This technique, of changing parameters in order to test a model's predictions, is also mentioned and employed in the Harvard Business School assessment of energy technologies. See Koreisha and Stobaugh, 'Appendix', in Stobaugh and Yergin, *EF*, pp. 234–265, esp. p. 265. See also the discussion of sensitivity analysis in section 4 of this chapter.

CHAPTER TWO

ASSESSING RISK-COST-BENEFIT ANALYSIS, THE PREEMINENT METHOD OF TECHNOLOGY ASSESSMENT AND ENVIRONMENTAL-IMPACT ANALYSIS

1. INTRODUCTION

Several centuries ago, Edmund Burke observed that the age of chivalry had gone and that of sophisters, economists, and calculators had succeeded.[1] More recently, Stuart Hampshire lamented that the contemporary quantitative methods used in economics and social planning have brought a coarseness and grossness of moral feeling.[2] If Burke and Hampshire are to be believed, then society has a great deal to lose by employing techniques such as risk-cost-benefit analysis as a basis for public policymaking regarding environmental and technological projects.

An offshoot of decision analysis, risk-cost-benefit analysis (RCBA) has been widely touted as a "developing science".[3] Used to help administrators make decisions on how to spend money, this set of procedures is directed at calculating whether the expected benefits from a proposed activity or policy alternative outweigh its expected risks or costs. For example, experts might wish to determine, given the risk of nuclear core melt and the possible resultant release of radioactive materials, whether the cost of additional containment (for reactor vessels) would be offset by its benefits. To make such a determination, one must assume either that individuals attach a numerical quantity of desirability (cardinal utility) to alternatives or that they rank for preferability each of the alternatives open to them (ordinal utility). The agent is then assumed to choose that combination of available goods, from among those which he believes is available, which he most prefers.[4]

In the sixties, RCBA was integrated into the planning and budgeting efforts of many federal agencies in the defense, aerospace, and energy fields. In the seventies, a number of regulatory agencies controlling environmental and occupational hazards likewise began to use the technique. Currently, nearly all US regulatory agencies (with the exception only of the Occupational Health and Safety Administration, OSHA) routinely use RCBA to help determine their policies.[5]

Despite the fact that RCBA dominates current US decisionmaking regarding science and technology, the technique continues to draw much criticism.

According to a number of interpreters, this controversy is merely the latest stage of the many historical ups and downs of utilitarianism. Once accepted by many philosophers and economists, utilitarianism lost support early in this century. Later, it staged a comeback, allegedly in part because of von Neumann's and Morgenstern's 1944 classic on the *Theory of Games and Economic Behavior*, and in part because of the practical advantages provided by using risk-cost-benefit techniques as a basis for government decisionmaking.

Although the National Environmental Policy Act (NEPA) of 1969 requires that RCBA be used to evaluate all proposed environment-related federal projects,[6] opponents of this technique often view its practitioners as dehumanized numerators engaging in a kind of economic Philistinism. Amory Lovins compares RCBA to the street lamp under which the proverbial drunkard searched for his wallet "not because he lost it there but because that was the only place he could see"; Lovins claims that RCBA's, like such street lamps, "only make the surrounding darkness darker".[7] Fred Hapgood continues the same line of objection and criticizes the "meaningless specificity" of RCBA calculations. How, he asks, can one shadow price sexual experience at $40, or life after a heart attack at 0.8 of its normal "full-function" monetary value? His answer is that we can cost such activities only by practicing RCBA, a technique used by "mean-minded and heartless bureaucrats thinking like computers".[8]

Formulated less emotively, the objections to RCBA generally come down to several problems. The most basic difficulty with any use of the technique is (1) that there is no widely accepted theory of rationality, yet such a theory appears necessary if one is to take RCBA seriously. Moreover, as Milnor and Arrow have both suggested, we are not likely to find a theory of social rationality.[9] Numerous scholars have also criticized use of RCBA on a variety of specific charges: (2) that the democratic process, not some mathematical-economic theory, ought to determine public policy;[10] (3) that RCBA ignores factors such as the equity of distribution and the incommensurability of various parameters;[11] (4) that its data base is inadequate;[12] (5) that there are fundamental methodological disagreements among RCBA experts;[13] and (6) that various political, ethical, and moral attitudes cannot be taken into account by the method.[14]

Rather than focus on the pros and cons of variants of each of these objections, in this chapter, I ask a more basic question. What arguments *ought to be counted as decisive* in the case for and against RCBA? In answering this question, I discuss several criteria for such arguments and attempt to show why certain philosophical objections to RCBA fail to meet these criteria.

In the first part of the chapter, I claim that what I call 'the deficiency argument' ought not to be counted as decisive in the case against RCBA. Used (in a variety of forms) by Hubert and Stuart Dreyfus, Alan Gewirth, Alasdair MacIntyre, and Douglas MacLean, as well as by Amory Lovins and Robert Socolow, this argument is that since RCBA is seriously deficient in a number of ways, it should not be used routinely for societal decisionmaking regarding technological or environmental projects.

In the second part of the chapter, I argue that there are at least five reasons why the deficiency argument fails to provide decisive grounds for rejecting RCBA. Proponents of the argument appear: (1) to confuse necessary with sufficient conditions for showing that RCBA ought not to be used; (2) to oppose all systematic, rational forms of environment- and technology-related decisionmaking; (3) to ignore the fact that every policy alternative has some residual, unaccounted for by the theory of comparison; (4) to overestimate the potential for formulating rational policy; and (5) to ignore many of the constraints on real-world decisionmaking.

2. THE DEFICIENCY ARGUMENT

In defending a modified version of RCBA, I do not want to focus on the pros and cons of all the objections that can be brought against this technique. Obviously, at least some of these objections are substantial ones. Even the most neoclassical economist and the most utilitarian philosopher would have to admit, it seems to me, that serious difficulties beset RCBA. For this reason, I would like to ask a more basic question. What arguments *ought to be counted* as decisive in the case against RCBA? Unless proponents and opponents of this technique agree on the criteria for resolving their disputes over the adequacy of RCBA for environment- and technology-related decisionmaking, then no consensus on the methods for making public policy seems possible. As it is now, much of the debate is at cross purposes. Both opponents and proponents of this technique appear to be arguing for theses which those who disagree with them do not regard as decisive on the issue of whether or not to use RCBA.

One particular argument used by opponents of RCBA appears to be little more than a 'red herring'. I would like to expose it for its true colors and, in the process of examining the clouded waters where economics, ethics, and policy science mix, attempt to point out what alternative types of arguments might be both relevant to, and decisive for, the cases for and against use of RCBA.

This so-called red-herring argument is that since RCBA is deficient in a number of serious ways, it should not be used routinely for societal decision-making regarding technological or environmental projects. More phenomenologically oriented proponents of the deficiency argument, such as Dreyfus, claim (and rightly so) that RCBA cannot model all instances of "human situational understanding".[15] More Kantian advocates of the deficiency argument, such as MacIntyre, maintain that RCBA exhibits some of the same defects as "classical utilitarianism".[16] As Alan Gewirth puts it, RCBA should be rejected because "it makes persons' lives and health matters of bargaining or purchase" when it should view them "as basic goods and rights not subject to such cost-benefit calculation".[17]

Although none of the proponents of the deficiency argument formulates it in terms of explicit premises, they all appear to be employing a simple, four-step process. For the more Kantian philosophers, these steps are as follows:

1. RCBA has a utilitarian structure.
2. Utilitarianism exhibits serious defects.
3. Typical applications of RCBA, e.g., in technology assessment and in environmental impact analysis, exhibit the same defects.
4. Just as utilitarianism should be rejected, so should the use of RCBA techniques.[18]

The more phenomenological variants of the deficiency argument consist of four similar steps:

1. RCBA is unable to model all cases of human situational understanding and decisions.
2. The inability to model all cases of human situational understanding and decisions is a serious defect.
3. Typical applications of RCBA, e.g., in technology assessment and in environmental impact analysis, exhibit the same defect.
4. Just as faulty models of human situational understanding and decisions should be rejected, so should the use of RCBA techniques.

This four-step argument is significant, not only because persons such as Dreyfus, Gewirth, MacIntyre, and MacLean appear to be employing it, but also because it seems to reflect the thinking of many (if not most) philosophers who practice applied ethics.

After taking a closer look at both the phenomenological and Kantian variants of the deficiency argument, I will attempt to establish two claims:

I. Even were all the premises of both versions of the deficiency argument true, because the *inference* from them to the conclusion is not obviously *valid*, it is not clear that the conclusion (4) ought to be accepted.
II. Although the second and third premises of the Kantian variant of the deficiency argument are likely true, because the *first premise* is not obviously *true*, it is not clear that the conclusion (4) ought to be accepted.

2.1. *Phenomenological Variants of the Deficiency Argument*

Let's examine first the phenomenological variants of the deficiency argument. Along with many economists and philosophers of economics, proponents of this version of the deficiency argument usually assume that what is rational for individual decisionmaking is rational for societal decisionmaking.[19] Hence they conclude that because RCBA is not rational for individual decisionmaking, it is not rational for societal decisionmaking.

Although proponents of the deficiency argument would probably admit that RCBA is useful for some cases of individual decisionmaking, e.g., determining whether to repair one's old car or buy a new one, they claim that, in most instances, individuals do not use RCBA to do a "point count".[20] Dreyfus and Tribe maintain, instead, that they use intuition.[21] Socolow and MacLean claim that they employ open discourse, argument, or debate.[22] In any case, Dreyfus, Socolow, Tribe and MacLean are agreed that any formal model such as RCBA is unable to capture what goes on when, in a given situation, someone understands something or makes a decision. They maintain that such formal models fail in being too narrow and oversimplified.[23] For proponents of the deficiency argument, all formal models encourage policymakers to believe, erroneously, that decisions are a matter of transparent rationality and scientific know-how, rather than a matter of "mysterious" wisdom and intuition.[24]

As Dreyfus put it, much policymaking is "beyond the pale of scientific decisionmaking". It requires "wisdom and judgement going beyond factual knowledge", just as chess playing and automobile driving require "expertise" and "human skill acquisition" not amenable to RCBA.[25] The expert performer, so the objection goes, plays chess or drives the automobile "without any conscious awareness of the process". Except during moments of breakdown, "he understands, acts, and learns from results", without going through any routine like RCBA. Indeed he not only *does* not go through any routine like RCBA, he *could* not. And he *could* not, according to Coburn, Lovins,

MacIntyre, Self, and other proponents of the deficiency argument, because he is often unable to tell what is a cost from what is a benefit; much of the time he doesn't know either the probability or the consequences of certain events.[26] As one theorist put it:

> Men and women, in cafes and shops, do not behave like calculating machines. Even if they would, they could not do this, for psychical magnitudes are not precise.[27]

According to proponents of the deficiency argument, people do not use RCBA for a great many problem-solving situations. A chess player contemplating castling, for example, could not answer the question, "what is the probability that you will eventually wish your king were more centrally located?" Likewise, says Dreyfus, a driver could not answer the question, "what probability *p* would make you indifferent between (a) breaking your back ... and (b) getting past that other car without incident ... ?" Instead people readily "see" what must be done, by means of "intuition and know-how" and "strong feelings".[28] Hence, goes the argument, because RCBA cannot model many cases of human situational decisionmaking, it is unrealistic to believe that it can model societal decisionmaking. And because such models "cannot succeed",[29] they ought to be rejected.

2.2. *Kantian Variants of the Deficiency Argument*

More Kantian proponents of the deficiency argument likewise wish to reject use of RCBA for societal decisionmaking, but their emphasis is on the claim that RCBA *ought* not to succeed, rather than that it *cannot* succeed, as a basis for policymaking. Although Alasdair MacIntyre is perhaps the best-known proponent of this Kantian variant of the deficiency argument, Gewirth, Lovins, MacLean, and Tribe also appear to subscribe to it.

The gist of their objections to RCBA is that the technique forces one to assume that "everybody has his price". MacLean and Lovins, however, are quick to point out that some things are priceless and hence not amenable to RCBA calculation.[30] In a classic essay on one's right not to be caused to develop cancer, Gewirth argues that such rights cannot be costed and then traded off for certain benefits. Tribe, MacLean, MacIntyre, and Gewirth all argue that moral commitments, rights, and basic goods (like the sacredness of life) are inviolable[31] and incommensurable and hence cannot be bargained away for any alleged benefits revealed in a RCBA. In a nutshell, MacIntyre, Gewirth, and others allege that RCBA shares the same defects as classical utilitarianism; it cannot account for crucial items such as distributive justice.[32] Hence they conclude that, in a society where decisionmaking ought to be based on moral commitments to rights and the values such as distributive justice, public policy ought not to be based on RCBA.

3. ASSUMPTIONS UNDERLYING THE MAIN INFERENCE OF THE DEFICIENCY ARGUMENT

With these outlines of the deficiency argument in mind, let us examine what is perhaps its most problematic aspect. This is that, even were all three premises of the argument true, it is not clear that its conclusion ought to be accepted since the relevant inference is not obviously valid. This inference, recall, is that since RCBA exhibits a number of deficiencies, and since these deficiencies are quite serious, the technique should be rejected.

Persuasive as it sounds, this inference is flawed because proponents of RCBA are likely to admit, with their opponents, that RCBA has serious shortcomings. They would likely agree that societies committed to human rights, as well as expert chess players and automobile drivers, do not believe that RCBA alone is sufficient grounds for deciding their next moves. For them, the technique is one way of clarifying and facilitating democratic decisionmaking, not a substitute for it. Proponents of RCBA maintain that, although MacIntyre, Self, Dreyfus, and others are *correct* in pointing out deficiencies in the method, these flaws alone do not constitute *decisive* reasons for abandoning it. Why not? There seem to be at least five reasons why the main inference of the deficiency argument is highly questionable and hence why the argument fails to provide conclusive grounds for rejecting RCBA.

3.1. *The Assumption That RCBA Deficiencies Are a Sufficient Condition for Rejecting It*

For one thing, the problems with RCBA are not the issue; the real issue is far more complex and more pragmatic. It is whether RCBA represents the best, or the least objectionable, of all methods used for policymaking regarding environmental and technological projects. This means that it is not sufficient for a proponent of the deficiency argument to find fault with RCBA and then simply infer that he ought to reject such a faulty technique. He must, in addition, show that there are viable alternatives, and that at least one of these other options, as a tool for democratic decisionmaking, has fewer (or less significant) deficiencies than RCBA.

Admittedly all policymaking is not enhanced by RCBA, especially if no technological or environmental projects are involved. And admittedly critics such as MacLean, Lovins MacIntyre, and Gewirth have done us all a great service by pointing out the flaws in RCBA. It is useful to know that a technique is faulty even if one does not have a substitute. Such knowledge spurs the search for something better and it discourages inquiry in unprofitable directions. Surely it has been the brilliant criticisms of people like MacIntyre that have awakened us to the implicit, unacknowledged, and highly questionable philosophical underpinnings of bureaucratic discourse and methods. My quarrel with proponents of the deficiency argument is not that they *erroneously* criticize RCBA (since, in large part, the deficiencies they allege are real), but that they erroneously infer that the deficiencies they have uncovered are alone sufficient grounds for rejecting use of RCBA in all forms of decisionmaking. To the extent that proponents of the deficiency argument make this inference (without arguing for it) and to the degree that its acceptance underlies much contemporary work in applied philosophy, then philosophers might be hostage to the neo-Luddite and anti-scientific tendencies of our contemporary society.

In one recent text in environmental ethics, for example, which I just finished reviewing for a journal, all four articles in the section on cost-benefit analysis pounded away at its deficiencies, and at least two of the authors appear to have accepted (without argument) the inference that these RCBA deficiencies were adequate grounds for rejecting the technique.[33] Surely, however, the matter is not so simple. I can think of a number of decision-theorist philosophers, e.g., Ron Giere, Alex Michalos, who could have given these four authors grounds for questioning their conclusions.

One reason why recognition of RCBA deficiencies is a necessary, but not a sufficient, condition for inferring that it ought not be used in policy-making is that, problem-laden as it is, RCBA provides at least some alleged benefits to society. If one wishes to argue that it ought to be rejected, then one ought to show either that some alternative is preferable to it or that no methodological alternative is needed, even if RCBA is rejected. If one argues that some alternative is preferable, then one ought to argue either that the alleged benefits of RCBA are not in reality beneficial or that some alternative possesses at least as many benefits as RCBA but has fewer defects. In any case, one ought not to ignore the issue of the alleged benefits of RCBA, avoid the question of alternatives to it, and then assume that pointing out RCBA deficiencies is a sufficient condition for rejecting it. Rather, if RCBA is rejected, then one ought to show how society might achieve the alleged benefits which (RCBA proponents claim) were obtained through use of the technique. One important function allegedly served by RCBA is that it enables government policymakers to cope with externalities. Since firms seeking to maximize profits will likely ignore the social costs of their decisions, state intervention is required to create a system of taxes and subsidies to bring about equality between private and social costs. Without such equalization, government would have an even greater bias toward commercially viable projects.[34] Proponents of RCBA claim that the technique is essential for providing estimates for the application of social-cost pricing, so that prices can be set equal to marginal social costs,[35] so that government can set standards to regulate output, and so that it can determine the optimum size of projects. If RCBA were rejected as a basis for policymaking, they argue, then the consequences would be disastrous. We would have to abandon not only techniques used for regulating externalities and for achieving social benefits such as traffic control, but also the means of insuring competitive pricing.[36]

If proponents of the deficiency argument believe that there is some alternative which is superior to RCBA, then they ought either to analyze and defend this belief or to refrain from making the inference that RCBA ought to be rejected. If, for example, they favor 'muddling through' democratically as an alternative to use of RCBA within normal democratic processes, then they ought to analyze the assets and liabilities of using purely political means of policymaking. They ought to ask questions such as whether *ad hoc* regulation can be consistent and equitable and whether there ought to be any criteria for successful policy. Likewise, if they believe that planning is a desirable alternative to RCBA,[37] then they ought either to defend their position or to refrain from making the inference that RCBA ought to be rejected. If they choose to defend planning, then they would need to consider how one might evaluate particular plans, assess the public interest, and avoid (what critics of planning call) "obscurantist intuitionism".[38] In other words, proponents of the deficiency argument need to recognize that calling attention to RCBA deficiencies is a necessary, but not a sufficient, condition for proving that it ought not be used in policymaking. Surprisingly, many proponents of this argument, with the exception of perhaps Hare and Self, do not discuss what might constitute sufficient conditions for rejecting RCBA. Instead, they appear content merely to point out deficiencies in the method.[39]

3.2. *The Assumption That Any Systematic, Rational Form of Policymaking Ought to Be Rejected*

A second reason why this 'deficiency argument', at least in its phenomenological variants, does not provide decisive grounds for rejecting RCBA is that, in many cases, its acceptance would preclude *all* systematic forms of societal decisionmaking regarding technological and environmental projects. If the phenomenological argument proves anything at all, it proves far too much, viz., that the deficiencies rendering RCBA questionable likewise count against all systematic societal decision methods used in similar situations. But if they count against all such methods, and if at least one such method is required for consistent and equitable societal governance, then of course the deficiencies cannot be compelling.

To understand what is meant by the claim that, if the phenomenological variants of the deficiency argument are correct, then any systematic, rational form of policymaking regarding technological and environmental projects is impossible, let's look more closely at some of these arguments and at the presuppositions underlying them. Consider first Dreyfus' charge that "the analytic decomposition of a decision problem into component probabilities, utilities, and tradeoffs" is misguided because true understanding of policymaking cannot be made explicit; decisionmakers, so the argument goes, "know how to do things, but not why they should be done that way".[40] MacIntyre says much the same thing: "Moral [and policy] arguments are in our culture generally unsettleable".[41] Or as Socolow puts it, "time is not well spent pondering the available analyses"; he speaks grimly of "the failure of technical studies to assist in the resolution" of environmental and technological problems.[42] Self, MacIntyre, Coburn, Lovins, and other proponents of the phenomenological variant of the deficiency argument claim, as was already noted,[43] that people can't tell what is a cost and what is a benefit of a particular technological proposal. And since we are unable to sort out costs from benefits, Dreyfus, Tribe, and MacLean claim that good policymaking is not the result of some RCBA method, but the result, instead, of intuition, subjectivity, and open discourse.[44]

If all such claims were correct, however, if technology- and environment-related policymaking were purely intuitive and subjective, because people know only *how* and not *why* they do things, because policy disputes are generally unsettleable, and because costs cannot be distinguished from benefits, then the phenomenological variants of the deficiency argument have some grim consequences. For example, if policymakers knew only *how* and not *why* they do things, then their ignorance would undercut even any nonquantitative policy methods, as well as any rational analysis of them. Presumably one would have to know *why* he did a particular thing before he could evaluate his doing it. Hence his not knowing why would undercut any rational policy analysis, not just RCBA. Likewise if someone could not distinguish costs from benefits, could not distinguish negative from positive impacts of a policy proposal, then any way of rationally evaluating the proposal would be undercut, not merely RCBA. In appealing to intuition, subjectivity, and wisdom as the basis for policy analysis, proponents of the phenomenological variant of the deficiency argument appear to preclude systematic rational analysis of any kind, not just RCBA. And in precluding systematic, rational analysis, their appeal to intuition and subjectivity really comes down to sanctioning skepticism and relativism. Why is this so?

If even alleged experts know *how* but not *why* public policy should be made in a particular way, then criteria for policymaking must remain *implicit* and, by their very nature, cannot be made explicit. But if so, then how is any rational, societal decisionmaking possible? If no one understands policy-making, and if no one understands the criteria for its success, then *any* rational form of social decisionmaking, whether quantitative or nonquantitative, scientific, democratic, or legal, is undercut. This is because any *non-arbitrary* form of decisionmaking requires specification of policy goals and criteria for their success. Any democratic form of decisionmaking requires, *further*, rendering these goals and criteria *explicit*, so that they can be recognized and evaluated by the body politic. If variants of the deficiency argument are correct in their assertion that decision criteria are implicit and subjective, that they are not explicable and cannot be made explicit, then policymaking cannot be nonarbitrary (and democratic) and therefore rational. Instead of condemning RCBA on grounds that would indict *any* systematic, analytic, decision methodology, policymakers would do better to argue for a particular method, on the grounds that it was superior to all known alternatives, and to work at analyzing, criticizing, and improving all available policy methods. Indeed, any other approach would be illogical, and would amount to begging the question as to the inadequacy of a particular method.

3.3. *The Assumption That Not All Policy Alternatives Have Theoretical Residuals*

A third reason why the main inference in the deficiency argument does not provide decisive grounds for rejecting RCBA is that it fails to take account of the fact that *any* theory of decisionmaking providing a way of comparing policy alternatives is deficient by virtue of leaving some residual 'why' unaccounted for by the theory of comparison. For Gewirth and MacLean, for example, the main theoretical residual of RCBA is that it allegedly cannot account for the overarching significance given to values such as rights and the sacredness of life. However, even if one used some nonquantitative, purely philosophical evaluation criterion (for assessing policy alternatives), for example, 'maximizing equity', then any opponent could always charge that another criterion was able to account for some reasonable considerations not adequately addressed by the equity criterion. This is because *any* evaluative means for decisionmaking must employ certain simplifying assumptions, apart from whether that scheme is quantitative or not. Any simplifying assumptions, in turn, are bound to have a 'residual' not accounted for by the theory of which they are a part. One can think immediately of how various assumptions of exclusion, aggregation, range, and reversibility, for example, are both unavoidable in policy models and responsible for various theoretical residues.[45] This is why arguments focusing merely on the deficiencies of RCBA miss the point. The point is which residuals are more or less important and which decision alternatives have the least deficiencies.

The more Kantian variants of the deficiency argument fail, for example, because often they merely rehash the old problems with utilitarianism. Instead, they should recognize that both utilitarian and deontological policy alternatives have theoretical residues, and that the best policy is the one with the least significant such residue. One common tactic of proponents of the deficiency argument is to explicitly or implicitly adopt some Rawlsian stance toward policymaking and to argue for the deficiency of RCBA because of its allegedly utilitarian underpinnings. Coburn, for example, is one of the many who take this approach.[46] Yet, as Sen points out, Bentham and Rawls capture two different aspects of interpersonal welfare considerations. Both provide necessary conditions for ethical judgments, but neither alone is sufficient. Utilitarians are unable to account for the theoretical residue of how to evaluate different *levels* of welfare, and Rawlsians are unable to account for the theoretical residue of how to evaluate *gains and losses* of welfare.[47]

Likewise when Coburn, Gewirth, Lovins, MacIntyre, MacLean, and Self criticize RCBA for its inability to measure certain incommensurables,[48] they appear to forget that alternative Rawlsian policy structures, for example, also have problems of measurement and comparison.[49] This particular deficiency of many deficiency arguments, that they are not based on comparative policy analyses and instead focus merely on criticisms of RCBA, is all the more significant since Rawls, for example, recently admitted that the social-welfare-function approach (used in RCBA) is perhaps more convincing than the approach which he took in *A Theory of Justice.*[50] Hence, whether Rawls is correct or not, his admission reinforces a crucial point: mere discussion, even of serious deficiencies, is itself not sufficient grounds for inferring that one ought to reject RCBA. Merely citing such deficiencies misses the key point.

In particular, the Dreyfus' claim that one does not normally employ RCBA to play chess or to drive an automobile, but rather relies on unconscious expertise, is a red herring and misses the point.[51] Likewise MacLean's claim, that he does not use RCBA to value his antique Russian Samovar, misses the point. Of course one doesn't rely on RCBA to do these things, but why should the inadequacy of the technique for these purposes be decisive? It is not surprising that RCBA would not be used by those who are expert at playing chess or driving an automobile. One reason is that one's own experiences, values, and goals are unified, and hence provide a single, integrated basis from which to make decisions about valuing his possessions, choosing chess strategies, and evaluating automotive alternatives. RCBA, however, has been proposed for *societal* decisionmaking precisely because the disparate members of society have no unifying experience, values, and goals which provide an integrated basis from which to make decisions about how to spend funds or assess technologies. Collective choices, exercised through government, are more difficult than individual ones. As Mandelbaum argued so effectively, "if we are to explain certain forms of individual behavior we must use societal concepts, and these concepts are not . . . translatable without remainder into terms which only refer to the behavior of individuals".[52] Societal choices often must follow a different 'logic' because there is no single, unified, and agreed-upon experiential base which provides clear decision criteria for societal policies. Some analytic 'logic' or method is needed in order to reconcile and unify the diverse experiential bases of all the members of society. Hence the fact that RCBA is not used by *individuals* in playing expert chess, in expert automobile driving, or in pricing valued Samovars, does not count against the desirability of using RCBA in making *societal* decisions. The individual-skill cases are fundamentally disanalogous with the societal-choice cases. Hence a convincing argument, or better, a decisive argument against RCBA (or any other decision strategy) would have to be based on showing that social-choice examples were not amenable to RCBA, not that individual-choice examples were not amenable to it. Moreover, any argument that a particular social-choice situation was not explicable in terms of RCBA would have to demonstrate, not that persons do not *in fact* use RCBA in such situations, but that persons ought not and cannot use RCBA in them, perhaps because of uncertainty or because the notion of expected utility is problematic. As it is, most proponents of the deficiency argument have not made the latter sort of argument, but have attempted merely to refute RCBA on conventional, rather than on scientific and rational, grounds.

Admittedly, however, there is no way to show in advance that RCBA, when refined and developed, will or will not get us close to being able to explain, in principle, some of our obviously correct Kantian intuitions about ethics or phenomenological intuitions about choice. This being so, it seems inappropriate for proponents and opponents of RCBA to provide unqualified support, respectively, for acceptance or rejection of RCBA. Rather, arguments for or against RCBA ought to aim at showing why one ought or ought not to work toward an allegedly adequate RCBA theory and why one ought or ought not continually to test its promise and power.[53] Not to take this line of argumentation is merely to beg the question of what might or might not provide the soundest basis for public policymaking. Kenneth Boulding astutely noted that, in "the social sciences . . . almost all we can really know is what we create ourselves", and that social-scientific knowledge "can be achieved only by setting up consciously created systems".[54] If he is correct about our only avenues of social scientific knowledge, then he has provided an additional reason for formulating arguments pro or con RCBA in terms of whether or not continued use of, and repeated attempts to improve, that technique make sense. If all we know in economics is via the systems we create, then those systems must be weighed against each other. Hence it must be argued, not assumed, that RCBA does or does not have a more significant theoretical residue than alternative means of policy analysis.

3.4. *The Assumption That Wholly Rational Decisionmaking Is Possible*

Closely related to the presupposition, that RCBA has more theoretical deficiencies than other approaches to policymaking, is the assumption that wholly rational decisionmaking is possible. Indeed, if proponents of the deficiency argument reject RCBA solely on the basis of its flaws, then they must be assuming that a wholly rational (or at least a more rational) approach is possible. They appear to have unrealistic expectations for what a policy technique can accomplish. This brings us to a fourth reason why many variants of the 'deficiency argument' do not provide convincing grounds for rejecting RCBA: they underestimate the benefits of using an explicit, clearly defined decision technique, and they overestimate the possibility for rational policymaking when no procedure such as RCBA is used. A society pleading for policymaking based solely on Dreyfus' expertise, intuition, and wisdom, or on MacLean's 'open discourse', rather than also on RCBA or on any other analytic method, is like a starving man pleading that only steak will satisfy him. To the extent that society is starved for rational decisionmaking, and to the extent that much current policy is arbitrary, irrational, and based on purely political considerations, then to that same degree is at least some use of any systematic, analytic methodology for decisionmaking desirable. This is simply because use of a given 'system' is likely to render the arguable aspects of controversial decisions clearer and more explicit, and thus more susceptible to correction than they might be if decisionmaking were purely arbitrary, irrational, or political. Hence just as any food is better for a starving man than no food, so some use of any analytic system (like RCBA) for policymaking is better than none.

One reason why some use of RCBA is better than none is that failure to use a well-defined system leaves one open to the charge of both methodological and substantive arbitrariness. At least in the sense that its component techniques and underlying value judgments are clear and consistent, use of an analytic decision procedure is rarely arbitrary in a methodological sense, although it may be controversial in a substantive sense.

For example, a decision (based on RCBA) to use a particular technology for storing low-level radwastes may be *substantively* controversial in the sense, for instance, that different persons have opposed views on what levels of the waste ought to be permitted to enter the biosphere within a given time frame. If the RCBA were done properly, however, its conclusion would not be *methodologically* arbitrary in the sense that entire classes of benefits or risks were excluded from consideration. Likewise it would not be methodologically arbitrary in the sense that someone who examined all the calculated risks, costs, and benefits would be unable to tell *why* a particular policy were said to be more or less beneficial, or more or less cost-effective, than another. A proper RCBA is not methodologically arbitrary, at least in the sense that its decision criterion, however faulty, is explicit; all the bases for its conclusions, including individual weights and numbers assigned to particular values, are there for everyone to see and evaluate.

If someone were to use a nonanalytic decision procedure, such as 'intuition' or 'open discourse' (as suggested by many proponents of the deficiency argument), however, then the resulting conclusions would likely be both methodologically arbitrary and substantively controversial. In the case of the radwaste example, a decision based wholly on some nonanalytic approach would likely be substantively controversial in the sense that persons might not agree on what levels of emitted waste ought to be deemed acceptable. In addition, it would likely also be methodologically arbitrary because it would be impossible to ascertain whether the policymaker(s) had taken all relevant parameters into account and whether they were accorded a particular importance. There would be no way to analyze and evaluate a particular intuitive decision; that is the whole point with something's being intuitive: it is directly apprehended and is not capable of being known through some reasoning process. Hence any decision based on intuition must be methodologically arbitrary at least in the sense that the grounds for having made it are not in principle capable of being broken down into component parts and made explicit so that they can be analyzed. In this sense, a systematic technique (such as RCBA), even with known flaws, is less methodologically arbitrary than an intuitive approach whose assets and liabilities cannot, in principle, be the object of some reasoning process. In other words, it is likely better to use some well-defined analytic procedure rather than none because at least the methodological bases for generating conclusions are clear, and therefore accessible for criticism, even if their truth is controversial. In the absence of such a procedure, both the methodological bases, if any, and the resultant conclusions are likely to be unclear, and hence less available for constructive criticism.

Because they appear to ignore the value of admittedly flawed, systematic approaches (such as RCBA), over nonsystematic, nonanalytic ones (such as intuition and open discourse), proponents of the deficiency argument underestimate the benefits of using RCBA. Likewise they appear to overestimate the possibility of rational policymaking when no analytic procedure is part of the process. This means that any decisive argument against RCBA ought to explain why the known flaws of this technique are more significant than the flaws of other approaches, and why rational policymaking appears more likely in the absence, rather than the presence, of RCBA. Since they have not provided such arguments, proponents of the deficiency argument are like persons who own a faulty bathroom scale,[55] but who do not attempt to discover and correct the ways in which their scale typically errs, e.g., by consistently showing a weight which is 10 percent above or below the correct reading. Proponents of the deficiency argument are like those who prefer to determine their weight either by their intuitions or by their feelings about their fat or by open discourse, by discussing their weight with other people.

After repeated trials, however, it might be possible to determine exactly the respects in which the scale consistently erred, just as many have argued that RCBA consistently errs in failing to take account of considerations such as rights and distributive equity.

Of course, the advocate of the deficiency argument is likely to respond that he doesn't *know* that his bathroom scale and RCBA err in consistent and predictable ways. Maybe, he would say, the scales are completely broken and the weight shown follows no particular function, just as RCBA is not flawed only in consistent and predictable ways, but is seriously deficient in ways which are not wholly predictable. Hence the opponent of RCBA is likely to believe that use of a seriously deficient, analytic decision procedure, incorporated within the democratic process, is not necessarily preferable to using no decision procedure or to relying on the democratic process alone. He would likely argue that, in the absence of some decisionmaking system, people would be forced to confront the arbitrariness of their social choices. Thus confronted, goes the objection, they would be forced to seek 'wisdom' and intuitive 'expertise' as the only possible bases for policymaking. In fact, this is exactly what Dreyfus has claimed, and what Holmes Rolston argued, when he criticized an earlier version of this paper. "Don't you know", Rolston said, "using RCBA is like weighing hogs in Texas". Down there, they put the hog in one pan of a large set of scales, put rocks in the other pan, one by one, until they exactly balance the weight of the hog. Having done that very carefully, they guess how much the rocks weigh. Just as we guess the magnitude of risks, costs and benefits in RCBA. According to Rolston, they ought to guess the weight of the hog in the first place. Likewise he says we ought to use intuition and discourse, through the democratic process, and not bother with RCBA, which is a more sophisticated way of guessing the merits of various policy options.

Such a response does not work, however, and for several reasons. *First*, one ought to distinguish the *method* of valuation from the precision with which any factor can be calculated by means of that method. If consistent methods, such as RCBA, yield imprecise results, these results are not mere guesses, as Rolston suggests. They are not extraordinarily arbitrary, at least in the sense that the value judgments underlying them are clear and consistent.

Second, if one rejects RCBA and opts for the intuition and discourse of normal democratic channels (as do Rolston, MacLean, Dreyfus, and others), very likely he forgets the constraints on real-world decisionmaking in the sense that he probably fails to realize that the conditions for participatory democracy are rarely, if ever, met. Recent US history has exhibited many calls for participatory democracy, calls which (almost without exception) seem appropriate. Now presumably Rolston, MacLean, and others who issue such calls believe that the participation of members of an agreement group confers moral acceptability upon the policy that results. But if one examines this belief more closely, it becomes apparent that there are *conditions* under which participation confers moral acceptability upon resulting policies. As Norman Care pointed out in an excellent article in *Ethics*, these conditions include:

1. that all the *participants* be: noncoerced; rational; accepting of the terms of the procedure by which they seek agreement; disinterested; committed to community self-interestedness and to joint agreement; willing to accept only universal solutions; and possessed of equal and full information;
2. that the *policy* agreed to prescribe something which is both possible and non risky, in the sense that parties are assured that it will be followed through; and finally
3. that the means used to gain aggrement be ones in which all participants are able to register their considered opinion and ones in which all are allowed a voice.[56]

Once one considers these conditions, several facts become apparent. Perhaps the overarching one is that circumstances seldom permit the full satisfaction of these conditions for procedural moral acceptability. This being so, it is not obvious that proponents of the deficiency argument, especially those who favor its Kantian variants, are justified in believing that democratic procedure *alone* will produce policy which is more morally acceptable than that achieved by democratic procedure informed by RCBA or by some other well-defined analytic method.

A second consideration (derived from Care's excellent analysis of the constraints on the moral acceptability of procedural democracy) is that, if anything, the moral acceptability of the democratic process seems to require that analytic methods, so far as they are available, be employed. Care's second and eighth conditions, for example, require that participants in the democratic policymaking process be rational and in possession of equal and full information. Since considering the various risks, costs, and benefits of a project, and assigning numerical values to them or to preference orderings about them, is surely one way of being rational about policy-making, then use of RCBA would seem to enhance fulfillment of the rationality condition. Moreover, if precedurally moral decisionmaking requires that all participants possess equal and full information, then *not* to consider the type of information afforded by RCBA seems to put a constraint on fulfillment of this condition.

Without some well-defined, analytic input to democratic decisionmaking, it is obvious neither that the conditions on rationality and full information will be met as well as they might, nor that 'wisdom' and 'expertise' will characterize the policymaking process, as Dreyfus, Lovins, and other proponents of the deficiency argument suggest. Back scratching, payoffs, bribes, and ignorance could just as well take over, except that these moves would be harder to detect than within a clear and systematic decisionmaking approach. More importantly, however, there is no decision procedure for recognizing correct intuitions, expertise, or wisdom. This being so, use of an analytic technique, amenable to democratic understanding and control, might be preferable to reliance on some unspecified form of intuition exercised through current democratic channels. In other words, given no non-arbitrary decision procedure for identifying either intuitively correct or wise policies, we might do well to employ a policy technique which, however deficient, is clear, explicit, and systematic, and therefore amenable to democratic modification. Given no philosopher king and no benevolent scientist-dictator, opponents of RCBA need to explain why they believe that society has the luxury of *not* opting for some analytic procedure such as RCBA in democratic decisionmaking. Until there are clear criteria for expert policy-making, the most rational approach might be not to beg the question by eliminating consideration of any methods, such as RCBA. If opponents of RCBA maintain that reliance on existing democratic processes is preferable to employing RCBA within those processes, then they ought not to beg the question of the desirability of RCBA. At a minimum, they ought to explain why RCBA is unlikely to aid policymakers in satisfying the conditions for procedural democracy, just as proponents of the technique ought to argue why RCBA is likely to enable decisionmakers to meet those conditions.

In the absence of arguments about how to meet the conditions for procedural democracy, why are proponents and opponents likely to beg the question of the desirability of RCBA? Very likely they are so wedded to the belief that RCBA can or cannot take rights into account, for example, that (for them) there is no consideration which might make them change their minds. But if there is no possible way, either to prove the case against RCBA or to modify it so that it is acceptable, then these persons have begged the question. For such persons, the world must, of necessity, be viewed in one particular way. This means that their beliefs (about rights, for example) have become a metaphysics in the worst sense of the term. The way to avoid begging this question is not, for example, to ask now one generates models of preference orderings under the constraint that scalar comparisons cannot be made, but to ask whether, in a practical sense, such scalar comparisons are desirable, and what kind of world would permit them.[57] Otherwise one merely assumes that RCBA can or cannot be of help in rational policymaking.

3.5. *The Assumption That Many of the Constraints on Real-World Decisionmaking May Be Ignored*

A fifth reason why the inference, that RCBA deficiencies alone provide sufficient grounds for rejecting the technique, is questionable is that it fails to take account of many of the constraints on real-world decisionmaking. Once these constraints are recognized, a good case can be made for the claim that thinking of social-choice situations in RCBA terms may help one to make reasonable policies.

Realistic policymaking often requires that one make a decision, even though it is not clear that the factors involved are commensurable or that he can adequately take account of rights. In rejecting RCBA, either because it fails to give a phenomenologically 'true' account of all situations of human choice, or because it allegedly fails to take account of certain Kantian truths about ethics, proponents of the deficiency argument often forget that public policymakers are not like pure scientists; they do not have the luxury of demanding that truth be the only type of value that should be taken into account. There are other values, pragmatic ones, that also have to be considered, and this is precisely where RCBA might play a role. My earlier remarks on using the technique for social-cost pricing provide a case in point. Moreover, expressing the choice parameters in terms of risks, costs, and benefits forces one to make financial decisions in terms of competing claims upon his time or money.

For example, a police chief might want to equip all his patrolmen with imported, 4-cylinder cars as cruisers, since they are much cheaper than larger ones; he might reason that the money saved could be used to raise the officers' salaries. Once the chief realized, however, that patrolmen were more than twice as likely to survive a serious crash if they were in a large, heavy, 8-cylinder car, than in a smaller, lighter compact, he might decide that the risks of the smaller cars were not worth the benefits in gas and capital costs. Expressing his choice parameters in terms of risks, costs, and benefits thus would force him to make a decision in terms of realistic, competing claims (e.g., safety) upon his money, and not just in terms of only one consideration. In other words, an analytic scheme at least provides explicit rules for evaluating different decision alternatives relative to each other. Without such rules, realism is not served, because real-world constraints mean that one ought to make nonarbitrary, consistent decisions in the light of finite dollars and achievable policy alternatives.

The usefulness of an analytic scheme like RCBA is clear if one considers an example. Suppose citizens were debating whether to build a newer, larger airport in one section of their city. Suppose, too, that a RCBA was completed for the project and that the benefits were said to outweigh the costs of the project by $1 million per year, when nonmarket, qualitative costs were not taken into account. It would be very difficult for citizens to decide whether the qualitative costs of the project were offset by the $1 million margin of benefits. But if one *hypothetically* assumed that 50 thousand families in the vicinity of the airport each suffered qualitative costs of aircraft noise, traffic congestion, and increased loss of life worth $20 per year per family, then the decision would be easier to make. It would be much easier to ask whether it was worth $20 per family per year to be rid of the noise, congestion, and auto fatalities associated with a proposed airport, than it would be to ask whether the airport pros outweighed the cons, or what a wise or expert person would say about the situation.[58] Formulating the problem in terms of hypothetical monetary parameters, and using an analytic scheme forcing one to make either a cardinal estimate or an ordinal ranking of his preferences, appears to make this particular problem of social choice more tractable. Moreover, one need not believe that the hypothetical dollars assigned to certain costs are objective in any clear sense in order to benefit from RCBA. In most cases, e.g., the airport example, obviously they are not. One need believe only that RCBA preference ordering, including the assignment of numbers to these preferences, is a useful device for formulating the alternatives posed by social choice problems.

Admittedly one could always respond: "How is this $20-per-family 'gimmick' any illustration of the merits of RCBA? Couldn't one just claim that the airport proposal ought to be vetoed because its implementation would bring too much noise, congestion, and lowered quality of life?"[59] Yes, one could argue this way, but the obvious problem is that one's opponents could always charge that the benefits for the larger community were not given adequate weight, and that one's views on the issue were egoistic and myopic. How would one know or prove to others, for example, when congestion was 'too much', or when noise was 'excessive'?

Although the $20-per-family scheme is only an arbitrary thought experiment, it illustrates very well that the airport's margin of benefits is not very great in comparison to its costs. If nearby families were each willing to pay $20 to avoid the airport, then this too would provide further illustration of the fact that the benefit margin for society would have to be much larger, in order for citizens surrounding the airport to tolerate the noise. The RCBA example also captures an important fact about public policymaking. This is that, all things being equal, alleged public policies are more or less acceptable, respectively, depending on whether the ratio of benefits to costs is greater or smaller. (Of course, a high benefit-cost ratio cannot justify violations of rights.) Estimating consequent gains and losses often can provide a useful framework for thinking realistically about a project. Without such a framework, policymakers and the public might erroneously demand safety or benefits appropriate only to some ideal world with infinite dollars.

Despite all its problems related to the freedom of choice of the poor and the diminishing marginal utility of money, using RCBA to examine what individuals will 'trade off' for safety or for amenities does yield some insights about preferences and therefore about how to allocate societal dollars, as the airport example shows. Such examinations are no more attempts to *value lives* and other incommensurables than are examinations of the frequency with which people take vitamins. Rather, assigning RCBA numbers to preference orderings is aimed at gaining insight on how to spend money. It is not an attempt to *assign worth* but to elucidate choices and preferences, by numbers, once they have already been made. When philosophers have stewed over this issue for centuries, why should anyone expect that welfare economists (through RCBA) have discovered the ultimate principles of social choice? They haven't and they don't claim to have done so. RCBA can tell us about constraints on choice; it can't (and economists don't claim that it can) tell us about processes of valuation which guide these choices. The goal of RCBA is to provide relevant information (about people's preferences) which is useful for rational, democratic debate. Hence any decisive argument against RCBA cannot merely fault the technique for its admitted inability to enlighten us about the values which guide our choices. A good argument ought to show why some alternative to RCBA is better able to take account of the constraints on real-world decisionmaking, e.g., social-cost pricing and making choices limited by finite dollars.

Perhaps one reason why proponents of the deficiency argument leap too quickly from criticism of RCBA to the inference that it ought to be rejected is that they view policymakers as arguing that analytic methods (apart from the democratic process) are alone sufficient for rational, ethical, social choice. Harvard law professor Laurence Tribe, for example, claims that proponents of RCBA "have frequently supposed that their necessarily instrumental techniques encompass the *whole* of rationality".[60] Surely, however, this is not the case. I doubt that anyone knowledgeable in risk-cost-benefit or decision-theoretic methods has ever seriously thought that these techniques provided an ultimate justification for ethical choice or action.[61] Proponents of RCBA have a much more modest goal: to illuminate the various means of achieving policy ends. As Tribe pointed out astutely, methods like RCBA are not adequate for selecting ultimate societal *ends*, but merely for helping to choose the best *means* to those ends; they are merely "instrumental techniques".[62] Moreover, when RCBA parameters are quantified according to some cardinal or ordinal measurement scheme, of course there are numerous nonquantitative criteria (e.g., equity, aesthetics), at work in providing the basis for particular numerical assignments. Quantification is not a substitute for use of these other criteria. It is only a device for clarifying the way in which various policy options and ethical *ends* appear more or less reasonable as different *means*, different assumptions, are employed to generate various cardinal assignments or alternative preference orderings, each with different numerical measures.

4. THE FIRST PREMISE OF THE KANTIAN ARGUMENT: RCBA IS UTILITARIAN

But if RCBA methods are not meant to prescribe various ethical *ends*, but only to illuminate their consequences and alternative *means* for achieving them, then how is it that MacIntyre and other proponents of the Kantian variant of the deficiency argument can allege that RCBA is a "normative form of argument" which "reproduces the argumentative forms of utilitarianism"?[63] This brings us back to the issue of whether the first premise of the Kantian variant of the deficiency argument is true. Does RCBA have a utilitarian structure? Earlier I claimed (see section 2 of this chapter) that I would show that, because this first premise is highly doubtful, there is reason to question the conclusion of the deficiency argument, viz., that RCBA should be rejected.

Why are there grounds for doubting that RCBA has a utilitarian structure? Are there any ways in which it might be thought of as utilitarian? RCBA is indeed utilitarian in one crucial respect: the optimal choice is always determined by some function of the utilities attached to the *consequences* of all the options considered. Reasoning in RCBA is hence essentially consequentialist. It is not obvious, however, that it is *solely* consequentialist. For one thing, before one can apply RCBA strategies, he must first specify alternative courses of action open to the decisionmaker. But since one can consider only a finite set of options, the decisionmaker must make a value judgment that the eliminated alternatives would not be the best, were they left in the calculation. But a utilitarian value judgment will not suffice for reducing the set of options, because it would lead back to the original infinity of logically possible alternatives. In other words, an infinite set of options cannot be reduced by means of a utilitarian value judgment, because it would presuppose knowing the utilities attached to the consequences of an infinity of options. But it is impossible to know the utilities attached to the consequences of an infinity of options, both because they are infinite and because the only utilitarian grounds for reducing the options is to carry out the very calculations which cannot be accomplished until the options are reduced. Thus any application of RCBA principles presupposes that one makes some value judgments which cannot be justified by utilitarian principles alone.[64] One might decide, for example, that any options likely to result in serious violations of rights ought to be eliminated from the set to be subjected to RCBA calculations. In this case, use of the allegedly utilitarian RCBA techniques would presuppose a deontological value judgment. The key point is that some nonutilitarian value judgment is *necessary* if RCBA is to be used at all.

RCBA is also nonutilitarian in that it includes many presuppositions which can be justified by utilitarian principles only if one engages in an infinite regress. Hence these must be justified by some nonutilitarian principles. Some of the presuppositions are that each option has morally relevant consequences; that there is a cardinal or ordinal scale in terms of which the consequences may be assigned some number; that a particular discount rate may be used; and that certain values may be assigned to given consequences.[65] Moreover, as Ralph Keeney, a student of Howard Raiffa's, pointed out, these assignments could be made in such a way that RCBA was able to account for nonutilitarian considerations. For instance, one could always assign the value of negative infinity to consequences alleged to be the result of an action which violated some deontological principle(s).[66] Thus if Gewirth, for example, wants to proscribe policy actions which result in cancer's being inflicted on people, then he can satisfy this allegedly nonutilitarian principle by assigning a value of negative infinity to this consequence.[67] The point, of course, is that RCBA not only *allows* for such nonutilitarian assignments, but also *requires* nonutilitarian justifications, under pain of avoiding the infinite regress of utilitarian considerations.

Other arguments also support the claim that employment of RCBA does not necessarily commit one to a utilitarian ethical theory. Patrick Suppes, for example, argues that RCBA is not necessarily utilitarian because "the theory could in principle be adopted without change to a calculus of obligation and a theory of expected obligation". From the standpoint of moral philosophy, says Suppes, this material indifference means that RCBA has an incomplete theory of rationality.[68] For Suppes, RCBA is a formal calculus which can be interpreted in a variety of ways. Instead of interpreting only market parameters as costs, for example, one could conceivably interpret RCBA costs to include considerations such as violations of rights and distributive inequities. Dasgupta and Heal, for example, show that the social welfare function of RCBA can be interpreted according to at least three different moral frameworks, that of egalitarianism and intuitionism as well as of utilitarianism.[69]

If RCBA is amenable to many sorts of ethical systems, why is it that proponents of the deficiency argument, as well as a variety of other thinkers, tie RCBA and utilitarianism together?[70] Rosenberg claims that the misunderstanding is an example of commission of the genetic fallacy. He maintains that the person who weds economic techniques to capitalist utilitarianism simply forgets that the origins of the methods bear no necessary relationship to their logical and ethical underpinnings.[71] MacPherson makes much the same point and argues astutely that there is no necessary incompatibility between maximizing utilities and maximizing some nonutilitarian value. Rather, he claims that one *interpretation* of RCBA is what is opposed to nonutilitarian values. This interpretation is that market incentives, including the right of unlimited individual appropriation, is the only way to maximize utilities. According to MacPherson, only the market-society interpretation of maximizing utilities is incompatible with nonutilitarian values because its implementation is likely to lead to disproportionate holdings and therefore to political inequalities.[72]

If my arguments, as well as those of Suppes, Rosenberg, and others are correct, then RCBA is nonutilitarian in at least three senses: it presupposes nonutilitarian value judgments prior to the beginning of the calculations; it provides for weighting the consequences of the options considered (perhaps according to nonutilitarian ethical criteria); and, since the calculus is formal, it allows one to count virtually any ethical consideration or consequence as a risk, cost, or benefit. This means that there are at least three possible grounds for arguing that the first premise of the Kantian version of the deficiency argument is untrue. And if it is untrue, then there is reason to doubt its conclusion, that RCBA ought to be rejected as a means of policy analysis.

5. CONCLUSION

Where does all of this leave us, with respect to the status of RCBA, the deficiency argument used against it, and possible means of either dispelling or responding to various deficiencies? I have attempted to argue for at least seven conclusions. The first six deal with problems inherent in the main inference of the deficiency argument. In addressing the status of the deficiency argument, these six conclusions make an attempt to spell out what sort of reasoning ought and ought not to be counted as decisive against RCBA. The seventh conclusion summarizes my arguments about the first premise of Kantian variants of the deficiency argument. These conclusions are:

1. Arguments focusing merely on the alleged deficiencies of analytic approaches such as RCBA are necessary but not sufficient for proving that RCBA ought not be used to make policy decisions. Arguments which are sufficient must establish the relative superiority or inferiority of available approaches.
2. Arguments against RCBA or any other analytic approach are not decisive if their acceptance would preclude *all* rational, nonarbitrary forms of decisionmaking.
3. Arguments against RCBA or any other analytic method are not decisive if they ignore the fact that all decision theories have some residual unaccounted for by the theory. Decisive arguments provide criteria for assessing the relative importance of these residuals.
4. Arguments based on the disanalogies between *individual* decisionmaking (as in driving an automobile) and analytical methods of decisionmaking do not provide decisive reasons for rejecting use of analytical methods of *societal* decisionmaking.
5. Arguments based on appeals to intuition, wisdom, and expertise, in lieu of reliance on analytical decisionmaking, can provide decisive grounds for rejecting RCBA and other analytical methods, provided that they specify acceptable criteria for correct intuition, wisdom, and expertise – a difficult, if not impossible, task.

6. Arguments which ignore the real-world importance of making decisions among finite alternatives with finite resources do not provide decisive grounds for accepting or rejecting RCBA or other analytic methods.
7. RCBA is not wedded, in any essential way, to the deficiencies of classical utilitarianism.

Admittedly, those who affirm the utilitarian aspects of RCBA have an important point. Regardless of whether it can be interpreted in nonutilitarian ways, many (if not most) policymakers have used RCBA in a classical utilitarian manner. In this respect, proponents of the deficiency argument have a point and a highly significant one. It raises the question of what can be done to offset the tendency of RCBA to be used and misused according to only one of a variety of possible ethical frameworks. In Chapter Eight, I will discuss a way to weight RCBA by means of alternative ethical systems.

NOTES

[1] Quoted by Paul A. Samuelson, *Economics*, McGraw-Hill New York, 1964, p. 3.

[2] S. Hampshire, 'Morality and Pessimism', in *Public and Private Morality* (ed. by S. Hampshire), Cambridge University Press, Cambridge, 1978, pp. 5–6.

[3] S. Levine, 'Panel: Use of Risk Assessment . . . ', in *Symposium/Workshop . . . Risk Assessment and Governmental Decision Making* (ed. by Mitre Corporation), Mitre Corporation, McLean, Virginia, 1979, p. 634; hereafter cited as: Levine, Panel, and Mitre, *Symposium.*

[4] When an individual attaches a numerical quantity of desirability to alternative outcomes, a *benefit* becomes an outcome for which the individual exhibits a preference, and a *cost* is an outcome for which the individual exhibits a negative preference. In the nineteenth century, economists assumed that this quantity of desirability (*utility*) could be measured cardinally. While cardinalists still exist, the more common view among practitioners of risk-cost-benefit analysis is that utility is capable only of ordinal measurement, such that it can be said merely that utility is greater or less in one situation than in another. (A. K. Dasgupta and D. W. Pearce, *Cost-Benefit Analysis*, Barnes and Noble, New York, 1972, pp. 24–25. See also Amartya Sen, *On Economic Inequality*, Norton, New York, 1973, pp. 2–3; hereafter cited as: Sen, *OEI.*

One of the central tasks of RCBA is to use a social welfare function, a collective choice rule, for deriving a social ordering of preferences from individual orderings. More recently, there has been a trend back to cardinal utility, particularly when the purpose has been to explain behavior under uncertainty as did Von Neumann and Morgenstern and Friedman and Savage (See K. Basu, *Revealed Preference of Government*, Cambridge University Press, Cambridge, 1980, pp. 65 ff. and Y. K. Ng, *Welfare Economics*, John Wiley, New York, 1980). In defending RCBA, this essay presupposes adherence to no particular theory of measurement, whether cardinal or ordinal. Clearly some sorts of objections may be brought only against *cardinal* accounts, while others may be brought only against *ordinal* theories. In sidestepping these particular objections, my remarks address the difficulties common to all variants of RCBA, independent of the measurement scheme employed.

[5] L. J. Carter, 'Dispute Over Cancer Risk Quantification', *Science* 203 (4387), (March 30, 1979), 1324–1325; hereafter cited as: Carter, Dispute. See C. Starr and C. Whipple, 'Risks of Risk Decisions', *Science* 208 (4448) (June 6, 1980), 1118; hereafter cited as: Starr and Whipple, Risks.

[6] See I. Barbour, *Technology, Environment, and Human Values*, Praeger, New York, 1980, pp. 163–164.

[7] A. B. Lovins, 'Cost-Risk-Benefit Assessments in Energy Policy', *George Washington Law Review* 45 (5), (August 1977), 912; hereafter cited as: Lovins: CRBA.

[8] Fred Hapgood, 'Risk-Benefit Analysis', *The Atlantic* 243 (1), (January 1979), 35–37; hereafter cited as: RBA.

[9] Kenneth J. Arrow (*Social Choice and Individual Values*, Wiley, New York, 1951) argues that one cannot rationally aggregate individual preference structures into a single joint preference structure. His 'Impossibility Theorem', generally acknowledged as the outstanding problem in philosophy of economics, is that, given certain assumptions, there can be no ideally rational aggregation device. Since his proof is valid, all discussions of his theorem rest on his four conditions or assumptions which (he claims) any aggregation device must meet:

1. unrestricted scope
2. Pareto principle
3. nondictatorship and
4. independence of irrelevant alternatives.

Many studies relevant to technology-related decisionmaking avoid Arrow's problem either by assuming that the choice is eventually made by a single person or by directing attention solely at the decisionmaking process. (See Edith Stokey and Richard Zeckhauser, *A Primer for Policy Analysis*, Norton, New York, 1978.) Scholars more concerned with the theory behind Arrow's Theorem often avoid his conclusion by arguing that there are relevant grounds for modifying one or more of the four conditions which Arrow postulated. H. F. MacKay (*Arrow's Theorem: The Paradox of Social Choice*, Yale University Press, New Haven, 1980), for example, argues that, if one modifies the condition on unrestricted scope, then one can avoid the impossibility described by Arrow. For other discussions of Arrow's famous theorem, see P. K. Pattanaik, *Voting and Collective Choice*, Cambridge University Press, Cambridge, 1971; A. K. Sen, *Collective Choice and Social Welfare*, Holden-Day, San Francisco, 1970; Wulf Gaertner, 'An Analysis and Comparison of Several Necessary and Sufficient Conditions for Transitivity Under the Majority Decision Rule', in *Aggregation and Revelation of Preferences* (ed. by J. J. Laffont), Elsevier, New York, 1979, pp. 91–112; hereafter cited as: Gaertner, Analysis, and Laffont, *Aggregation*. See also E. A. Pozner, 'Equity, Nonfeasible Alternatives and Social Choice: A Reconsideration of the Concept of Social Welfare', in Laffont, *Aggregation*, pp. 161–173; C. W. Churchman, 'On the Intercomparison of Utilities', in *The Structure of Economic Science* (ed. by S. R. Krupp), Prentice-Hall, Englewood Cliffs, 1966, p. 255; Gordon Tullock, 'Public Choice in America', in *Collective Decision Making* (ed. by C. S. Russell), Johns Hopkins Press, Baltimore, 1979, pp. 27–45; and E. F. McClennen, 'Constitutional Choice: Rawls *versus* Harsanyi', in *Philosophy in Economics* (ed. J. C. Pitt), Reidel, Boston, 1981.

[10] Gage calls this the "greatest inadequacy" of RCBA. He claims that government decisions concerning risk, health, technology, and the environment are essentially "economic decisions involving optimization of the allocation of resources", but are based on "political, ethical, moral, and social attitudes" (S. Gage, 'Risk Assessment in Governmental Decision Making', in *Symposium/Workshop . . . Risk Assessment and Governmental Decision Making* (ed. by Mitre Corporation), Mitre Corporation, McLean, Virginia, p. 13; hereafter cited as Gage, Risk, and Mitre, *Symposium*). While Gage and others are correct in affirming that politics often controls policy, they are incorrect in assuming that RCBA proponents deny this fact. Rather than denying the power of politics, subjectivity, or values, proponents of RCBA maintain that following this technique puts political, subjective, or other assumptions "into prominence" (F. Farmer, 'Panel: Accident Risk Assessment', in Mitre, *Symposium*, p. 426), so that most major factors in an assessment are "openly laid out" and not merely "implied" (W. Lowrance, 'Discussion', in Mitre, *Symposium*, p. 152). Or, as one person put it, RCBA forces one to 'face up' to all kinds of difficulties (L. Lave, 'Discussion', in Mitre, *Symposium*, p. 190; hereafter cited as: Lave, DMC), rather than to deal with those difficulties in an *ad hoc* or secretive way. In other words, proponents of RCBA do not believe that their theories are substitutes "for the way we do business in this democratic society" (Gage, Risk, p. 14), but they do believe that RCBA forces one to clarify situations and to specify key assumptions often left unrecognized and therefore uncriticized. If this is so, then RCBA can help to mitigate the effects of politics and to reveal when methodological assumptions are merely the vehicle for politics.

[11] The utilitarian underpinnings of RCBA, with its consequent neglect of distributive inequities, are perhaps the key problems for which RCBA has been criticized (see for example, S. Jellinek 'Risk Assessment at the EPA', in Mitre, *Symposium*, pp. 63–64; S. Samuels, 'Panel: Accident Risk Assessment', in Mitre, *Symposium*, pp. 391–392; hereafter cited as: ARA. See also K. Shrader-Frechette, 'Technology Assessment as Applied Philosophy of Science', *Science, Technology, and Human Values* 6 (33), (1980) 33–50; hereafter cited as: TA.

Numerous authors have discussed the problems inherent in attempting to reduce a variety of cost, benefit, and risk parameters to some common denominator, e.g., money. The key difficulty in this regard is expressing qualitative and nonmarket factors (e.g., aesthetic benefits, psychological costs) in terms of some common unit (see Samuels, ARA, p. 391). A human life, say some scholars, can hardly be represented according to a quantitative measure. Numerous risk assessors, however, maintain that specifying all factors according to some common system of measurement, while not necessarily accurate or even desirable, does help to avoid 'ad hocism' and the failure to be explicit about how decisions are made (M. Baram, 'Panel: Use of Risk Assessment', in Mitre, *Symposium*, p. 622). If we don't attempt to quantify parameters in order to use RCBA, says one analyst, then we will run the risk of spending $10 million for every life saved through kidney dialysis and only $500 for other lives; in other words, the politics of those interested in dialysis might control things (A. Hull, 'Panel: Public Perception of Risk', in Mitre, *Symposium*, p. 579). Not to attempt to quantify parameters means that one must stay with his 'hunches', instead of attempting to deal with actions that will raise or lower given risks, says Lave (CMC, p. 577). Moreover, says Whipple, people continually make decisions which implicitly tradeoff lives and dollars. If those decisions were made explicit, he claims, especially through RCBA quantification, then more lives might be saved (C. Whipple, 'Panel: Public Perceptions of Risk', in Mitre, *Symposium*, p. 576). Or, as O'Neill puts it, without expressing various parameters in terms of some common denominator, one "simply can't think about this problem" (J. O'Neill, 'Discussion', in Mitre, *Symposium*, p. 31; hereafter cited as: Discussion).

[12] Since accurate RCBA requires consideration of all risks, benefits, and costs, the dearth of epidemiological data on health risks (see O'Neill, Discussion, pp. 23–24), as well as synergistic and other effects (A. Brown, 'Plenary Session Report', in Mitre, *Symposium*, pp. 693–694), makes *complete* RCBA close to impossible. The issue, then, is whether an incomplete RCBA is better than none, and whether using RCBA will force society to gather the data necessary to evaluate its policies.

[13] See Gage, Risk, pp. 11–13.

[14] See Gage, Risk, pp. 11–13.

[15] S. E. Dreyfus, 'Formal Models vs. Human Situational Understanding: Inherent Limitations on the Modeling of Business Expertise', *Technology and People* 1 (1982), 133–165; hereafter cited as: Formal Models. See also Stuart Dreyfus, 'The Risks! and Benefits? of Risk-Benefit Analysis', unpublished paper, presented on March 24, 1983, in Berkeley, California, at the Western Division meeting of the American Philosophical Association; hereafter cited as: Risk. (Stuart Dreyfus, a Professor of Operations Research at the University of California, Berkeley, is the brother of Hubert Dreyfus, a Professor of Philosophy at the same institution. They share the beliefs attributed to Stuart in this paper and often coauthor essays defending this point of view. See, for example, S. Dreyfus, 'The Scope, Limits, and Training Implications of Three Models of . . . Behavior', ORC 79-2, Operations Research Center, University of California, Berkeley, February 1979.

[16] Alasdair MacIntyre, 'Utilitarianism and Cost-Benefit Analysis', in *Ethics and the Environment* (ed. by D. Scherer and Thomas Attig), Prentice-Hall, Englewood Cliffs, 1983, pp. 139–151; hereafter cited as: MacIntyre, UCBA, and Scherer and Attig, *EE*.

[17] Alan Gewirth, 'Human Rights and the Prevention of Cancer', in *EE* (ed. by Scherer and Attig), pp. 170–177; hereafter cited as: Rights.

[18] Ronald N. Giere, 'Technological Decision Making', in *Reason and Decision* (ed. by M. Bradie and K. Sayre), Bowling Green State University Press, Bowling Green, Ohio, 1981, Part III, also takes opponents of RCBA to be employing a simple, four-step argument similar to the one sketched here. Giere's article is hereafter cited as: TDM.

[19] Some of the economists and philosophers of economics who assume that what is rational for individual decisionmaking is rational for societal decisionmaking include Steven Strasnick, Peter Self, and P. H. Wicksteed. See, for example, Steven Strasnick, 'Neo-Utilitarian Ethics and the Ordinal Representation Assumption', in *Philosophy in Economics* (ed. by J. C. Pitt), Reidel, Boston, 1981; hereafter cited as: Strasnick, Ordinal, and Pitt, *PE*. See also Alexander Rosenberg, 'A Skeptical History of Microeconomic Theory', in Pitt, *PE*, p. 48; hereafter cited as: Rosenberg, Microeconomic Theory; and Peter Self, *Econocrats and the Policy Process: The Politics and Philosophy of Cost-Benefit Analysis*, Macmillan, London, 1975, pp. 70–75; hereafter cited as: Self, *PPCBA*.

[20] Self, *PPCBA*, pp. 70–75.

[21] Dreyfus, Formal Models, p. 161; L. H. Tribe, 'Technology Assessment and the Fourth Discontinuity', *Southern California Law Review* 46 (3), (June 1973), 659; hereafter cited as: TA.

[22] R. H. Socolow, 'Failures of Discourse', in Scherer and Attig, *EE*, p. 169; hereafter cited as: Socolow, Failures. Douglas MacLean, 'Understanding the Nuclear Power Controversy', in *Scientific Controversies* (ed. by A. L. Caplan and H. T. Engelhardt), Cambridge University Press, Cambridge, 1983, Part V; hereafter cited as: NP.

[23] See MacLean, NP, Part V; Socolow, Failures, pp. 152–166, and Dreyfus, Formal Modes, p. 163.

[24] Dreyfus, Formal Models, p. 161.

[25] Dreyfus, Risk, p. 2.

[26] Self, *PPCBA*, p. 70; MacIntyre, UCBA, pp. 143–145; A. Lovins, 'Cost-Risk-Benefit Assessment in Energy Policy', *George Washington Law Review* 45 (5), (August 1977), 913–916, 925–926; hereafter cited as Lovins, CRBA. See also R. Coburn, 'Technology Assessment, Human Good, and Freedom', in *Ethics and Problems of the 21st Century* (ed. by K. Goodpaster and K. Sayre), University of Notre Dame Press, Notre Dame, p. 108; hereafter cited as: Coburn, TA, in Goodpaster and Sayre, *Ethics*. See also E. Mishan, *Cost-Benefit Analysis*, Praeger, New York, 1976, pp. 160–161 (hereafter cited as: *CBA*), who says that not all RCBA parameters can be "measured with honesty". One of the most common objections to the subjective theory of value is that utility and disutility are not quantitative and cannot be measured. See Gunnar Myrdal, *The Political Element in the Development of Economic Theory* (trans. by Paul Steeten), Harvard University Press, Cambridge, 1955, p. 89. See also S. S. Stevens, 'Measurement, Psychophysics, and Utility', in *Measurement: Definitions and Theories* (ed. by C. W. Churchman and P. Ratoosh), John Wiley, New York, 1959, pp. 36–52.

[27] A. Radomysler, 'Welfare Economics and Economic Policy', in *Readings in Welfare Economics* (ed. by K. J. Arrow and T. Scitovsky), Irwin, Homewood, Illinois, 1969, p. 89.

[28] Dreyfus, Risk, pp. 9–11.

[29] MacLean, NP, Part V.

[30] Lovins, CRBA, pp. 929–930. Douglas MacLean, 'Quantified Risk Assessment and the Quality of Life', in *Uncertain Power* (ed. by D. Zinberg), Pergamon, New York, 1983, Part V; hereafter cited as: MacLean, QRA.

[31] Tribe, TA, pp. 628–629; MacLean, QRA, Parts V and VI; MacIntyre, UCBA, pp. 139–142; and Alan Gewirth, 'Human Rights and the Prevention of Cancer', in Scherer and Attig, *EE*, p. 177; hereafter cited as: Gewirth, Rights.

[32] Gewirth, Rights, p. 175.

[33] Scherer and Attig, *EE*, is the book in question.

[34] Mishan, *CBA*, p. 161, makes this same point.

[35] This point is emphasized by D. W. Pearce, 'Introduction', in *The Valuation of Social Cost* (ed. by D. W. Pearce), George Allen and Unwin, 1978; hereafter cited as: *VSC*. See also Self, *PPCBA*, p. 78.

[36] Mishan, *CBA*, p. 383, makes this same point.

[37] See Self, *PPCBA*, pp. 165–171 for a discussion of this alternative.

[38] Self, *PPCBA*, p. 171.

[39] See Self, *PPCBA*, and R. M. Hare, 'Contrasting Methods of Environmental Planning', in Goodpaster and Sayre, *Ethics*, pp. 64–68.

[40] Dreyfus, Risks, p. 1.

[41] MacIntyre, UCBA, p. 151.

[42] Socolow, Failures, p. 152.

[43] See note 26.
[44] See notes 21 and 22.
[45] For a discussion of various simplifying assumptions in policy models, see Sergio Koreisha and Robert Stobaugh, 'Appendix: Limits to Models', in *Energy Future* (ed. by Robert Stobaugh and Daniel Yergin), Random House, New York, 1979, pp. 237–240.
[46] See Coburn, TA.
[47] A. K. Sen, 'Rawls Versus Bentham: An Axiomatic Examination of the Pure Distribution Problem', in *Reading Rawls* (ed. by Norman Daniels), Basic Books, New York, 1981, pp. 283–292; hereafter cited as: San, Rawls, and Daniels, *Rawls*.
[48] See notes 30 and 31.
[49] See B. R. Barber, 'Justifying Justice: Problems of Psychology, Politics, and Measurement in Rawls', in Daniels, *Rawls*, pp. 292–318.
[50] E. F. McClennen, 'Constitutional Choice: Rawls Vs. Harsanyi', in Pitt, *PE*, p. 95 claims that Rawls made such an argument in the Appendix to his 1978 Stanford Lectures. Since the lectures have not been published, I have been unable to verify his claim.
[51] MacLean, QRA, Part IV.
[52] M. Mandelbaum, 'Societal Facts', *The British Journal of Sociology* 6 (4), (December 1955), 312.
[53] D. M. Hausman, 'Are General Equilibrium Theories Explanatory?', in Pitt, *PE*, pp. 26–28, and E. J. Green, 'On the Role of Fundamental Theory in Positive Economics, in Pitt, *PE*, p. 14, make similar points.
[54] Kenneth Boulding, *Economics as a Science*, McGraw-Hill, New York, 1970, pp. 120–121; hereafter cited as: *EAAS*.
[55] I am grateful to Holmes Rolston for suggesting this analogy.
[56] Norman S. Care, 'Participation and Policy', *Ethics* 88 (1), (July 1978), 316–337.
[57] A similar point is made by C. W. Churchman, On the Intercomparison of Ultilities', in *The Structure of Economic Science* (ed. by S. R. Krupp), Prentice-Hall, Englewood Cliffs, 1966, p. 256.
[58] For a similar example, see Mishan, *CBA*, p. 161.
[59] This is Holmes Rolston's objection, raised in response to an earlier version of this essay.
[60] Tribe, TA, p. 636.
[61] Giere, TDM, Part III, agrees.
[62] Tribe, TA, p. 635.
[63] MacIntyre, UCBA, pp. 140–141.
[64] Giere, TDM, Part III, employs a similar argument.
[65] See the previous note.
[66] Keeney mentioned this in a private conversation at Berkeley in January, 1983.
[67] See Gewirth, Rights.
[68] Patrick Suppes, 'Decision Theory', in *Encyclopedia of Philosophy* (ed. by Paul Edwards), Vol. 1–2, Collier-Macmillan, New York, 1967, p. 311.
[69] P. S. Dasgupta and G. M. Heal, *Economic Theory and Exhaustible Resources*, Cambridge University Press, Cambridge, 1979, pp. 269–281.
[70] One person who ties utilitarianism and welfare economics together is Nicholas Rescher, *Distributive Justice*, Bobbs-Merrill, Indianapolis, 1966, pp. 11–12. See also Amartya Sen, *On Economic Inequality*, Norton, New York, 1973, p. 23.

[71] A. Rosenberg, *Microeconomic Laws*, University of Pittsburgh Press, Pittsburgh, 1976, p. 203.
[72] C. B. MacPherson, 'Democratic Theory', in *Philosophy and Technology* (ed. by C. Mitcham and R. Mackey), Free Press, New York, 1972, pp. 167–168.

PART II

GENERAL METHODOLOGICAL PROBLEMS

CHAPTER THREE

THE RETREAT FROM ETHICAL ANALYSIS

1. INTRODUCTION

There are many theoretical ghosts haunting supposedly factual research programs. One of these specters, positivism, was allegedly put to rest some twenty years ago by Hanson, Kuhn, Polanyi, Toulmin, and others.[1] It has reappeared recently, in the work of some social scientists, engineers, and lawmakers who do technology assessments, environmental-impact analyses, and studies of science policy. The positivistic doctrine which some of these scholars promote is that technology assessments (TA's), environmental-impact analyses (EIA's), and other science-related studies ought to be wholly neutral and objective descriptions of the facts, and ought not to include any normative, evaluative, or theoretical components. As such, this position really comes down to two theses: (1) that wholly neutral and objective technology assessments and environmental-impact analyses are possible, and (2) that they are desirable. This latter thesis is particularly disturbing, since it amounts to proscribing the role of the normative scholar in TA and EIA and to condemning the attempts of applied philosophers to come to grips with some of the issues of ethics and scientific methodology which are at the heart of policy questions.

Let us examine both of these widely held theses. I will argue that TA's, EIA's, and studies of science policy neither can nor ought to be wholly objective and neutral; that such requirements of objectivity and neutrality are based on a number of doubtful presuppositions, such as the pure science ideal and the fact-value dichotomy; and that acceptance of this position would lead to serious public policy consequences.

2. THE PRINCIPLE OF COMPLETE NEUTRALITY

The belief that TA, EIA, and science policy studies ought to exclude normative (ethical and methodological) components is based on the principle of complete neutrality. This is the assumption that TA's and EIA's can and should be wholly value-free, entirely neutral and objective, and that any foray into applied ethics or methodological criticism represents a lapse into advocacy and subjectivity. Representative statements of this principle have been given by various authors. The National Academy of Engineering says that "technology assessments should be produced in an environment free from . . . predetermined bias".[2] The US Office of Technology Assessment stresses that "all OTA assessments must . . . be . . . free from advocacy or ideological bias" and must be "objective".[3]

Numerous engineers and assessors have repeated these injunctions; they have claimed that TA's, EIA's, and science studies ought to be wholly 'objective' and completely devoid of any partisan stances or evaluative judgments.[4] As Dorothy Nelkin put it, in a classic formulation: "the role of social and philosophical studies of science and technology" is not "to reveal the rights and wrongs of policy choice", but only "to understand . . . to interpret . . . to draw a coherent picture of what is going on".[5]

So great is many social scientists' faith in the possibility and desirability of allegedly objective, value-free assessments, that some have even argued that different technology assessment teams, analyzing the same problem but working for different agencies of interest groups, ought to "come to the same conclusion".[6]

Such claims on behalf of the principle of complete neutrality appear reasonable, in large part, because they make an implicit appeal to what historian George H. Daniels calls 'the pure-science ideal'.[7] Pure science is disinterested, dispassionate, and allegedly has no value component. When the principle of complete neutrality is interpreted in the light of the pure-science ideal to mean that one ought to aim at unbiased analysis, it is surely correct. Virtually any good researcher or lawmaker would laud the consistency, impartiality, and even-handed treatment of both data and people sanctioned by the principle of complete neutrality. As such, the principle appears to be both the cornerstone of empiricist philosophy and the key to elevating science- and technology-related studies from prejudice and superstition to the heights of systematic, rational investigation. Adherence to the principle of complete neutrality, whatever else can be said about it, represents a noble aim. Without it, disinterested, empirical scholoarship and enlightened politics would be impossible.

Necessary as it is to protect technology-related studies from deliberate bias, the principle of complete neutrality is in reality a complex thesis admitting of at least three interpretations. When its proponents maintain that TA's, EIA's, and other technology-related studies can and ought to be value free, they often mean to assert one or more of the following claims, not all of which are obviously true:

1. Studies of science and technology can and ought to be free of *bias*

values (deliberate misinterpretations and omissions to serve one's own purposes).

2. Studies of science and technology can and ought to be free of *constitutive values* (adherence to values underlying particular rules of scientific method).
3. Studies of science and technology can and ought to be free of *contextual values* (personal, social, cultural, or philosophical emphases and goals).

Following Longino's analysis of constitutive and contextual values,[8] I will argue in this section that, while it is possible to avoid bias values, it is *in principle* impossible for any scholarly pursuits, even pure science, to avoid constitutive values, and that it is *in practice* impossible to avoid contextual values. In later sections, I will argue that, because it is impossible to avoid constitutive values, philosophers of science and others ought to engage in methodological criticism in TA, EIA, and other science-related studies. Likewise, because it is impossible to avoid contextual values, I will argue that ethicians and others ought to engage in normative analyses in these studies. First, however, let me show why it is impossible to avoid constitutive and contextual values.

2.1. *One Case Against Complete Neutrality: Constitutive Values*

When proponents of the principle of complete neutrality assert that TA's and EIA's ought to be value free, they often assume that if the TA's or EIA's are free of *bias* values (falsification of data, interpreting data to support one's own prejudices, etc.), then they are value free altogether. What they forget is that nothing, not even pure science purged of one's own prejudicial interpretations of data, is completely value free or neutral. Science is laden with values and normative constraints generated by the goals of scientific activity; these values are a function of considerations of what counts as a good explanation of a phenomenon. For example, explanations are said to be good on the basis of their problem-solving capacity, their simplicity, their predictive power, or other *constitutive values*, values that are the source of rules determining what constitutes acceptable scientific practice or method.[9]

Although everyone ought to attempt to avoid *bias values* in his work, there is no research, even in science, which is free from *constitutive values*. All facts are theory-laden, as can be seen from several considerations. For one thing, even purely scientific research cannot be achieved merely be collecting data; it requires some assumptions about the behavior being studied, so that the researcher knows what data to collect and how to interpret it. Moreover, perception does not provide us with pure facts; knowledge, beliefs, values, and theories we already hold play a key part in determining what we perceive. No researcher, and especially no scientific one, simply records everything he or she observes. Rather, he records only those things which the theories (he holds) suggest are significant. The high-energy physicist, for example, does not count all the marks on his cloud chamber photographs as observations of pions but only those streaks which his *theories* indicate are pions.

Another indication, that all alleged facts are unalterably laden with constitutive values, is that even in the supposedly clearest case of an observation contradicting a theory, the practicing scientist need not reject his theory. Numerous classic examples from the history of science illustrate this point. For example:

1. In Aristarchus' time, failure to observe the parallax counted as a counter-instance causing the rejection of the theory of the moving earth. By the eighteenth century, however, this 'counter-instance' was treated instead as a problem to be solved by the currently accepted theory of the moving earth.
2. From the time of the ancients to that of Kepler, the controlling astronomical theory was that heavenly motions were circular. During this time, all observations to the contrary were treated not as counter-instances but as problems to be solved.
3. When beta decay was seen early in this century, the observation was not taken as a counter-instance to the theory of conservation of enerty and momentum. Rather, the decay was interpreted as a problem for which an explanation was sought.[10]

These examples and many others suggest that, although scientists ought to attempt to be as objective and unbiased as possible, they cannot begin with pure facts and then use them, without constitutive values, to confirm or reject theories. Rather, accepted values necessarily guide research and determine how scientists deal with observed phenomena. Or, as Kuhn would say, 'normal science' operates such that a fundamental theory organizes and structures experimentation, helps to determine the meaning of observations, which ones are relevant, how problems are to be attacked, and what counts as solutions to them. Even if one could somehow obtain some unbiased, wholly atheoretical 'facts', one would not have objective research, but merely an array of useless information in need of some guiding theory to organize it, make it intelligible, and employ it for problem solving. In other words, all research is unalterably theory-laden in requiring both a definition of the research problem and a criterion for relevant evidence. Use of such definitions and criteria mean that research can never be neutral in the sense of being free from constitutive values.

2.2. *Another Case Against Neutrality: Contextual Values*

Other values arise, even in allegedly pure science, as a result of implicit or explicit *contextual values*. These are personal, social, philosophical, or cultural constraints or influences, and they include factors such as trends in areas of inquiry and emphases created because of the goals of certain funding agencies. If one decides to follow an atomistic, rather than a field-theoretic, paradigm in his high-energy physics research, for example, he might be subscribing to a particular philosophical or personal contextual value. According to the Mertonian school of history and sociology of science, even before the questionable connection between science and goals external to the pursuit of science, social and cultural values had an impact on the kinds of scientific research pursued and the way it was pursued. On this view, the kinds of questions thought important to investigate are determined as much by contextual values as by constitutive ones.

It is impossible to avoid contextual values, both in scientific work and generally in any research because all scholarly pursuits take place in a context. That context includes personal beliefs, financial influences, cultural baggage, and social norms. Although none of these influences need be negative or detrimental to the research at hand, they influence the way it is conceived, accomplished, and received.

One area permeated by contextual values has been the study of human interferon. As Longino points out, industrial microbiology, as practiced by small firms of biochemists, has been heavily influenced by cultural and financial values determining the way research results are tested and announced. Because of interferon's alleged therapeutic value in treating cancer, the actual scientific activity surrounding it has been heavily influenced by commercial values such as the profit motive.[11]

Another area heavily influenced by contextual values has been that of developing oral contraceptives. Korenbrot showed that the selection of health risks to be considered was often determined by the extra-scientific values of those performing the testing. Some of these contextual values included the desire to check population growth. Because of the researchers' commitment to this social value, they tended to overestimate the medical benefits of using an oral contraceptive and to underestimate its health risks.

In spite of available data on its hazards, many researchers emphasized the drug's minimal therapeutic properties and downplayed its enormous medical dangers. This illustrates, not that scientists working on oral contraceptives were deliberately dishonest, but that their goals and personal values influenced the way they determined what effects to test for and what significance to accord them.[12]

Contextual (social and cultural) values also influence research because they often are part of the assumptions required to mediate between hypotheses and theories or observations and experiments. A good example of the mediating influence of contextual values occurs in the case of using differential distribution of hormones between males and females to explain behavior differences between the sexes. Since testosterone has been observed to bring about aggressive behavior in laboratory animals, many scholars jump from this observation to the theory that testosterone can explain male-female behavioral and status differences. Often mediating the gap between the *observation* of laboratory animals and the *theory* about humans, however, is the cultural or *contextual* norm that aggressivity is a feature of male rather than female behavior, that aggressivity is biologically determined in humans, or that male social dominance is natural and inevitable.[13] Although such contextual values frequently play a role in the allegedly scientific explanation of behavior, often scientists fail to recognize their nonempirical character.

One reason why contextual values have played such a large role in many areas of scientific activity, such as the interferon, oral contraceptive, and sexual differences cases and, indeed, why contextual values seem in practice unavoidable, is that they often fill the gap left by limited knowledge. Since any research is hampered by incomplete information in some respect, there is usually room for contextual values to come into play. Where one does not know enough about a phenomenon either to explain it or to predict its behavior, or even to choose appropriate methods for prediction or explanation, there is a gap in understanding. This gap provides an opportunity for the determination of scientific procedures by contextual (social and moral) values which have little to do with the factual adequacy of these procedures. In other words, ignorance about a phenomenon frees research from the ordinary constraints imposed by *constitutive* norms and leaves it vulnerable to *contextual* norms.[14]

2.3. *The Fact-Value Dichotomy*

At the heart of the rhetoric surrounding the principle of complete neutrality and its denial of the role played by constitutive and contextual values in guiding factual research is adherence to a famous tenet of positivism. This is the fact-value dichotomy, the belief that facts and values are completely separable, and that there are facts which are not value-laden (see note 1). Applied to science-related scholarship, this claim is that TA, EIA, and policy studies ought to consist of *factual* and neutral presentations of alternative courses of action, although the policy *decisions* dependent on them, and made as a consequence of them, may be *evaluative*. Many assessors, scientists, and even the US Office of Technology Assessment (OTA) affirm that one can and ought to obtain hard 'data' or 'information'. They claim that one can engage in purely descriptive studies of science and technology, and that one may take an evaluative stance only after the TA or EIA is finished.[15]

Despite the fact that scholars such as J. S. Mill, Lionel Robbins, Milton Friedman, and others have affirmed the fact-value dichotomy,[16] there are a number of reasons for doubting both it and the pure-science ideal associated with it. For one thing, belief in the dichotomy is incompatible with formulation of any scientific *theory* or analysis to explain causal connections among phenomena. Second, presenting alternative accounts of one's options can never be a purely descriptive or *factual* enterprise. This is because one has to use *evaluative* criteria to select which options to present, and these normative criteria are outside the scope of allowable inquiry.[17] Moreover, to subscribe to the fact-value dichotomy is to subscribe to the belief that there can be pure facts and presuppositionless research. As the earlier discussion of constitutive and contextual values revealed (sections 2.1 and 2.2), there is no presuppositionless research, although everyone ought to attempt to avoid bias in his work. All facts are laden, in principle, with constitutive values and, in practice, with contextual values. Hence even pure science, if there is such a thing, cannot be achieved merely by collecting data, since there are no pure or brute facts.

The traditional empiricist motivation behind the belief that there are brute facts is a noble and important one. Many empiricists believe that only value-free observations or facts guarantee the objectivity of one's research. This conclusion follows, however, only if values *alone* determine the facts. A great many philosophers of science maintain that both our values and the action of the external world on our senses are responsible for our perceptions, observations, and facts (see note 1). In other words, they believe neither that perception is passive observation, nor that its objects are created out of nothing, but that perception is both structured by the world outside us, yet susceptible to the imprint of our own presuppositions, values, and theories. Because our observations or facts may be seen in *different* ways, it does not follow that they may be seen in *any* way. *Just because observations or facts are value-laden, this does not mean that there is no sufficient reason for accepting one theory over another. One theory may have more explanatory or predictive power, or unify more facts, or be more coherent.* One need only think of quark theory for a dramatic example of this familiar point. Hence the value-ladenness of one's 'facts' (whether in science or the humanities) implies relativism and lack of objectivity *only if* one accepts the positivist presupposition that only the observation of value-free data provides reasons for accepting one theory rather than another. But if there are no wholly value-free data, then either this presupposition is false, or there are no reasons to accept one theory rather than another. Since I have just mentioned some reasons commonly used for rationally accepting one theory over another, the positivist presupposition must be wrong. Reasons other than the observation of value-free data can and do provide grounds for accepting one theory over another. Hence belief in the fact-value dichotomy is not necessary to good science or trustworthy research.

3. THE IMPOSSIBILITY OF WHOLLY OBJECTIVE TECHNOLOGY ASSESSMENT AND ENVIRONMENTAL-IMPACT ANALYSES

Apart from the theoretical and conceptual difficulties just discussed with the fact-value dichotomy, there are numerous concrete reasons why, contrary to the OTA's belief, technology assessments, environmental-impact analyses, and science policy studies, in particular, have not been and cannot be completely neutral, nonpartisan, and objective.[18] If constitutive and contextual values prevent science from being completely neutral and purely objective, then this is doubly the case for TA and EIA. These activities are far more practical and applied than is purely scientific work. In fact, the appeal to objectivity and neutrality works best in situations where one is able to deal uniformly with logical abstractions from reality, and worst in situations where one must handle practical problems, as indeed all questions of TA, EIA, and science policy are. It simply is not possible to have an activity which is *both* wholly objective ('positive') *and* about the real world. As soon as allegedly neutral symbols are applied to actual facts, that application itself becomes inexact and value-laden.[19] This being so, it may well be that, like the ideology of free enterprise, the TA–EIA rhetoric of purity becomes louder as the connections with the practical world become stronger.

An Engineers Joint Council report recently admitted that, although perhaps not intentionally biased, nearly all government assessments have been performed by those who were "usually pro technology". They have been done under the specific, limited, and often purely political directives of the person or group (Congressman, committee, or industry) commissioning the study, and they have been influenced heavily by mission-oriented agencies seeking to maximize only short-term goals or to obtain a return on sometimes questionable R and D investments.[20] Admittedly, however, such difficulties only can be said to have characterized many technology assessments and environmental-impact analyses in the past. Are there any reasons for believing that unavoidable factors will limit the objectivity and neutrality of such studies in the future?

It seems to me that, in every technology-related assessment, as Biderman puts it, there is "a structure of concepts that is largely implicit and poorly understood".[21] Each structure carries with it various value-laden assumptions. Although we can clarify these assumptions, we cannot rid ourselves of all of them and hence cannot attain wholly neutral or objective TA's and EIA's. Let us see why this is so.

3.1. *Methodological Assumptions in Technology Assessment*

Consider first the case of methodological assumptions. If an assessor wishes to model the real world, in order to compare various environmental and technological options, it is inevitable that he make certain simplifying presuppositions, or what Koreisha and Stobaugh call "red flags". These include assumptions about: *exclusion* (of empirical evidence); *aggregation* (of data); *range* (of experience treated); *reversibility* (of elasticities used to make forecasts); and *time lag* (for estimated effects to occur).[22] In the case of assumptions regarding *range*, for example, every assessment must have some value-laden resolution of the problem of bounding the study; the existence of higher-order impacts makes particularly clear the fact that this is an open-ended question.[23].

Although some methodological assumptions are necessitated by the fact that "there are no agreed-upon techniques for carrying out a technology assessment",[24] other value judgments arise from the particular methodology adopted for use in a specific technology assessment or environmental-impact analysis.

Consider, for example, the methodological assumptions imbedded in cost-benefit analysis.[25] It has been touted widely as "the cornerstone of these new techniques" of "objective and reliable modes of public policy decisionmaking concerning science and technology".[26] Despite its alleged objectivity, however, its practitioners must make numerous evaluative judgments. Some of these include: whether something (e.g., cheap, abundant energy) is an authentic benefit or a cost; how to assess an unknown parameter (e.g., risk of nuclear-plant sabotage); which, if any, discount rate to use for future costs; whether to compare average or marginal costs; how to quantify nonmarket parameters; and how to correct market prices for imperfections.[27] Because of intrinsic problems such as these, cost-benefit analysis is inherently value-laden; although preeminently useful, it is closer to systematized intuition than to discovered 'facts'. Moreover, this value-ladeness is unavoidable. Without using some assumptions about the commensurability of different factors, cost-benefit analysis would have no way of comparing disparate values.

The positivist view of cost-benefit analysis, of course, is that it can be objective, and that values need not enter the calculations until *after* we weight costs and benefits differently; they maintain that only *after* this evaluative weighting ought it to be possible to obtain alternative answers as to what policy is most desirable.[28] However, all the controversial methodological assumptions (previously noted) enter the analysis *before* this 'weighting' stage at which all of the values are supposed to enter.

In the case of evaluating the costs and benefits of a ban on chemical pesticides, values clearly enter the analysis. This happens not only when weights are added to the costs and benefits already computed, but also earlier, when the assessor decides whether or not to include a given cost. In one recent study, for example, the evaluator erroneously assumed that agricultural adjustments after a pesticide ban would be limited only to factors of production, when in reality they could include changing the crops or the locations of their production. In the same study, the assessor ignored the resultant modification in the amount of agricultural crops produced, as well as the adjustments that the economy was likely to make in response to changing production costs.[29] Because of both these types of omissions, it is very likely that the assessors' own evaluative assumptions contributed to an overestimation of the costs of a pesticide ban.

The effects of value-laden, cost-benefit assumptions are also particularly evident in two recent studies of energy technologies. Kneese, *et al.* showed that, when normal discounting methodologies were used, the benefits of nuclear-generated electricity outweighed the costs, but when all costs and benefits were treated equally, the costs of atomic energy outweighed the benefits.[30] Likewise I showed recently that when one follows one set of methodological assumptions (which place high priority on *both* present and future costs), coal-generated electricity can be shown to be more cost-effective than nuclear, but that when other presuppositions (which place priority *only* on present costs) are followed, then nuclear can be shown to be more cost-effective than coal-generated electricity.[31] Both these examples illustrate, not only that there are numerous unavoidable methodological assumptions in cost-benefit analysis, but also that variations in key presuppostions can change allegedly objective conclusions about the desirability of alternative policies or programs.

Similar methodological and evaluative assumptions also occur in risk-benefit analysis and render its results highly normative. Many of its leading practitioners, however, fail to recognize this fact and continue to subscribe to the positivistic principle of complete neutrality, to the fact-value dichotomy, and to application of the pure science ideal to risk assessment. Okrent, for example, is convinced that this quantitative technique may be called "objective".[32] Likewise, both Lowrance and Häfele maintain that the calculated probability, $p_i(x_i)$, of a given hazard's occurring "is an objective thing".[33]

In adhering to this positivistic stance, assessors fail to recognize that even a mathematical probability is a product, in part, of a specific methodology and particular human judgments. Moreover, for the low-probability, high-consequence events, e.g., liquefied-natural-gas accidents typically treated in risk-benefit analysis, there is little hard data on frequency of hazard occurrence. This means that many allegedly objective probabilities are the products of highly variable methodological assumptions used in particular theoretical approaches. This is why, for example, the US Environmental Protection Agency (EPA) criticized the US Nuclear Regulatory Commission (NRC) for its calculation of nuclear-accident probabilities in the government document, WASH-1400; they charged that the latter agency ignored, among other things, the contribution of sabotage and terrorism to risk.[34] Criticisms such as these reinforce the point that risk assessors inevitably have to make evaluative judgments when they choose what issues to consider, what hazard paths are relevant, and what methods to use.

One of the most questionable evaluative assumptions often made by risk-benefit analysts is that societal risk may be equated with annual fatalities, or that aversion to a societal risk is directly proportional to the number of deaths the hazard is likely to cause. Numerous assessors, such as Starr,

Whipple, Okrent, Maxey, Cohen, and Lee make this assumptions.[35] Yet, if several psychometric studies are correct, then aversion to a given risk often has far more to do with *values*, such as whether the risk is equitably distributed, than with the allegedly objective probability that a number of fatalities will occur.[36] This means that positivistically inclined assessors not only make unrecognized value judgments about how to *define* their key concept, risk, but also fail to recognize the value components in societal judgments of risk.

Besides cost-benefit and risk analysis, numerical indicators are another methodological source of the normative judgments present in TA and EIA. Kenneth Boulding, Ezra Mishan, and others have written widely publicized accounts of the deficiencies of the GNP as an allegedly objective measure of economic health.[37] I will not repeat those criticisms here. Even social indicators, often proposed in lieu of economic measures such as the GNP, have their evaluative components. For example, whenever one uses regular and continuing trend statistics (e.g., on unemployment) as social indicators, one is limited to comparing only the phenomena which are important for evaluating a wide range of institutions, which can be anticipated, and which, (being part of a *series*) do not include unique, noncomparable events.[38]

Since the information requirements for evaluating most technologies are not limited to regular series, assessors must gather *ad hoc* data. With this comes the problem of deciding precisely which data are relevant and of anticipating events that might be observed in order to gather them. Biderman outlines at least four inherent weaknesses of obtaining *ad hoc* data, all of which result in assessors' incorporating judgments into their findings. They tend, for example, to rely on *post hoc* reasoning methods without *ante hoc* observations and to emphasize only the data from sources, e.g., newspapers, that leave indelible records.[39]

Inevitably, use of any social indicators forces assessors to take the aspects of reality that they are best able to observe or to comprehend as those which are most important, or those in terms of which technologies may be evaluated. As a consequence, they are often able to obtain precise figures for representing specific concepts. Their figures, however, have a 'specious accuracy'. They yield somewhat reliable but useless data, e.g., in the case where one uses *price* to represent *value*. This means that attempts to avoid normative assumptions are likely to result in obtaining accurate but irrelevant data, while attempts to obtain useful data are likely to result in obtaining evaluative but highly relevant information.[40] For this reason, use of any meaningful social indicators necessitates use of nonneutral evaluative assumptions.

3.2. *Evaluative Assumptions in the Light of Inadequate Data*

Many normative assumptions in TA and EIA also arise, not merely from the methodology chosen, but from the necessity to proceed in a situation in which there is a lack of data. (Recall that earlier (section 2.2), it was argued that contextual values often play a large role in science when information is incomplete.) One prominent student of technology has observed, in fact, that of the 45 representative, technology-related case studies which he investigated, the information required for decisionmaking was sufficient in only 9 instances.[41] When the necessary information is inadequate, it is inevitable that evaluative assumptions are made regarding the status of what is unknown.

For example, in the case of estimating the risks, costs, and benefits of ingesting various levels of pesticide residues, assessors clearly use statistical extrapolations from high-dose data, from animal research, and from relatively small sample spaces, in order to determine the risks, costs, and benefits to humans. As a consequence, "much of the data presently available on pesticide use" is not "inclusive, valid, quantifiable" or "reliable".[42] To assume that it is, and to use it as if it were, is to make a number of highly theoretical assumptions about pesticide technology.[43] The Food and Drug Administration (FDA), for example, used the lack of definitive knowledge about the effects of the chemical, DES, as a basis for allowing it to stay on the market.[44] As a matter of fact, it is generally the case that, so long as available technical data are ambiguous or uncertain, political criteria for assessment of a technology are likely to prevail.[45] This is exactly what has happened in the case of data regarding risks from low-level radiation. Some scholars support current US NRC radiation standards, while others allege that they are too lenient by at least a factor of 10. As a consequence, the standards have been set politically. The US government requires that radiation emissions be "as low as is reasonably achievable" by industry.[46]

Likewise some authors support current government assessments of nuclear safety, while others fault their methodology. At least part of this controversy arises because of the uncertainty of the scientific data and because of the evaluative character of the methodological assumptions made in the face of this uncertainty, e.g., about fault-tree analysis.[47] Public policy experts estimate, for example, that *reliable* data on nuclear risk probabilities differs by a factor of 1000.[48] In such a situation it is obvious that nuclear proponents are likely to focus on the upper limits of this range, while their opponents are sure to emphasize the 1000-fold lower limit. Whichever group controls a given nuclear technology assessment, for example, will determine whether the upper or lower risk probabilities are likely to be assumed to be accurate. As one scholar put it: "Technology assessment is best used to justify political actions rather than to provide rational appraisals . . . participants [in TA] use the products of assessment to support and explain their partisan positions".[49]

3.3. *Political Assumptions in TA and EIA*

The situation of partisan technology assessment and environmental-impact analysis occurs in part because wholly factual/objective assessments are impossible (for all the reasons already given) and in part because "most of the organizations that do TA's are heavily funded by federal agencies with promotional or regulatory interests in the technology, or alternatively, by industries with economic interests".[50] Although these funding patterns do not mean either that the assessments are intentionally biased or that they are erroneous, they do indicate that there is a discrepancy between the constituency of the sponsor [of a TA or EIA] and society in general. The problem is not that, for example, a given federal regulatory agency (e.g., the NRC) is anti-society, but that its goals (e.g., promoting nuclear energy) are different than those of society generally (e.g., finding the safest and cheapest source of energy). As a consequence, an allegedly objective TA or EIA, done under contract to a given sponsor, is more likely than not to reflect implicitly the goals, and therefore the normative assumptions, of the sponsor.

In the face of the facts that presuppositionless research is impossible, and that numerous methodological and politically-determined assumptions apparently characterize TA and EIA, many persons are probably willing to give up the principle of complete neutrality. They are likely to assert that, even though *purely descriptive* analyses are not possible, they ought to be sought nevertheless as a goal or ideal, in order to guide research and to keep it as objective as possible. In other words, as one social scientist put it, even if we cannot attain presuppositionless research, we ought to try to do so. We ought not to talk explicitly about the ethical or methodological "rights and wrongs" of policy regarding science and technology; we ought not to do applied ethics and applied philosophy of science.[51] Instead we ought to pursue the *ideal* of complete neutrality.

4. THE IDEAL OF COMPLETE NEUTRALITY

Such a condemnation of applied ethics and methodological criticism in TA and EIA is misguided for at least two reasons. *First*, it rests on a highly doubtful epistemological presupposition about the nature of objectivity. *Second*, if accepted, it would lead to a number of dangerous consequences.

4.1. *The Presupposition That Objectivity = Neutrality*

The main problematic presupposition behind the condemnation of applied ethics and applied philosophy of science is that the only objective studies of science and technology are those which are neutral, those in which all policy alternatives are represented as being equally plausible/desirable. Proponents of this presupposition maintain that if the weight, either of the evidence or of the evaluator's analysis implicitly suggests or explicitly affirms that one policy alternative is better than another, then that TA or EIA can be faulted for taking a position of advocacy or of criticism.[52]

It is not clear, however, that the ideal TA or EIA presents all policy options as equally plausible or desirable. For one thing, to assume that a position of evaluation or criticism ought never be taken in a TA or EIA is to assume that an assessor ought never to analyze or criticize the constitutive and contextual values (see sections 2.1 and 2.2 earlier in this chapter) imbedded in the data or methodology which he and other researchers use. If indeed these values cannot be avoided (although, as was pointed out earlier, deliberate bias can and ought to be avoided), then not to analyze them is merely to provide a TA, EIA, or policy study which *sanctions* those values, regardless of the errors to which they may lead. Succinctly put: if one does not engage in ethical and methodological analysis of the values imbedded in his data, he implicitly sanctions those values. Hence, if objectivity is identified with failure to engage in normative analysis, then objectivity is therefore identified with sanctioning received or status-quo values. But if objectivity is identified with sanctioning a particular set of values, then those who argue for pursuing the ideal of objectivity, by avoiding normative analysis, are inconsistent. This is why, once one admits that constitutive and contextual values are unavoidable (as was argued in 2.1 and 2.2), one cannot consistently argue that neutrality and objectivity are served by refusing to engage in ethical and methodological analyses of those values, insofar as they are implicit in, or presupposed by, the data and methods which researchers use.

Another reason why complete *neutrality*, in the sense of refusing to engage in ethical or methodological evaluation, ought not to be identified with *objectivity* in TA or EIA is that such an identification presupposes that one position on a given issue is as good as another. Obviously it is not. If all policy alternatives were *in fact* equally desirable in all relevant respects, there would be no reason to have an assessment. Presumably one only undertakes an assessment of a variety of options in order to help adjudicate the question of which option is best, or which option is better, relative to another. For example, if one were assessing the use of x-ray technology to determine whether shoes fit properly, then surely one ought not to claim that any one opinion on this issue were as good as any other. It would not be true to say that using and not using the technology for this purpose were equally reasonable positions.

Far from contributing to objectivity, it seems to me that describing widely diverse policy options in an equally plausible way would, in many instances, serve anything but the interests of objectivity. Consider the case in which largely political (but logically and technically unjustfied) reasons were widely publicized by special-interest groups, in order to support a particular technological alternative. To present such an alternative, without ethical and methodological criticism of the contextual values implicit in it, would not serve the interests of objectivity. One need only think of the various energy-policy positions in order to grasp this point. In particular, consider the alternative 'expert' assessments of future energy demand. The utilities' projections for needed electricity are extraordinarily high, considering the recent downward trends in consumption, while environmentalists' statistics are relatively low, even in light of lowered rates of electricity use caused by higher prices. The think-tank, foundation-supported, and government calculations of future energy demand usually lie somewhere in between the high projections of the utilities and the low estimates of the environmentalists.[53]

If proponents of the ideal of complete neutrality are to be believed, then apart from what the facts are (e.g., regarding elasticity of energy demand, continuation of past trends, etc.), each of these alternative positions (i.e., regarding planning for a high-, medium-, or low-energy-use scenario) is to be treated as equally plausible, and no position is to be criticized or advocated on methodological or ethical grounds. Yet, some scientific and technical assumptions made within specific demand scenarios are certainly more plausible than others, just as some contextual assumptions are more plausible than others. Obviously, for example, it is questionable to assume that utilities and environmentalist groups have the same context for projection, just as it is questionable to assume that electricity demand, during times of greatly increasing electricity costs (in real dollars), will be the same as demand during times of decreasing electricity costs. Nevertheless this latter assumption is central to many of the high-demand scenarios.[54] Not to criticize it is unwise, both because someone else might fail to do so, and because accepting it could lead to disastrous public and economic policies. The most *objective* thing to do, in the presence of a fallacious assumption, is often to be *critical* of it.

Likewise, the most *responsible* thing to do is often to be critical of it. One assessor claims that a TA which is critical is not 'responsible'. Yet, in the face of great or unrecognized errors or evils, perhaps only a critical press is a responsible one. This is why Skolimowski argues that all genuine assessment must terminate in value terms, and why he says that the 'real expertise' in TA is social and moral, not technical.[55]

Of course, the problem arises when someone is unfairly or unjustifiably critical, or when he falsely judges an assumption to be erroneous or dangerous. That someone can err in criticism or advocacy hardly justifies the thesis that one ought never to take a position of criticism or evaluation. All activity of any kind would be proscribed if a necessary condition for engaging in it were that one would never err by doing so. Hence, it appears implausible to assume that the likelihood of error justifies the prohibition against TA's or EIA's which include critical or normative analyses.

Admittedly, allowing for criticism or advocacy in TA will probably encourage the *initial* emergence of numerous dogmatic, irrational, or misguided positions. At the same time, however, explicit provision for advocacy and criticism, especially within the context of adversary assessment and public scrutiny of alternative evaluations, is likely to provide an *ultimate* policy-making framework in which various normative positions can be recognized, exposed for what they are, and decided upon. The antidote for various moralistic excesses or absolutist pronouncements (such as those of the Moral Majority, for example) is not to condemn stances of advocacy or criticism. Rather it is to provide an assessment framework within which alternative normative analyses are developed, compared, and criticized. (The purpose of Chapter Nine will be to outline such an adversary framework.) In this way society could subject the claims of all would-be critics and advocates to the rigors of open inquiry.

It might also be argued that one ought not to engage in advocacy or criticism in technology assessment, because the decisionmaking in such matters should be left to government or to the public. Although I agree that policy ought not be made by technology assessors, philosophers' engaging in ethical and methodological analyses in TA and EIA does not preclude decision-making by the public. If anything, including normative analyses in technology-

and environment-related studies helps to guarantee that certain critical issues are discussed in a scholarly way and brought to a public forum. Providing such analyses helps to insure that the often implicit evaluative assumptions of assessors are made explicit and, therefore, available for public scrutiny. The purpose of normative analyses in TA and EIA ought never to be to remove decisionmaking power from those democratically charged to do it. Rather it is to provide a basis for democratic choice, a basis for value judgments by the public. Seen in this light, ethical and methodological evaluation in TA and EIA elucidates various means for the achievement of goals; it clarifies the presuppositions built into them and the consequences following from pursuing some goals rather than others; and it points out conflicts among goals thought to be equally desirable. Most importantly, ethical and methodological criticism in TA and EIA enables policymakers and the public to grasp the corollaries of their choices and therefore to determine if they can live with them.[56]

A variant of the objection that evaluation ought to be left to the public and to policymakers and, therefore, that TA and EIA should not include normative analysis, is that assessors ought to serve the public interest. Any sort of ethical or methodological criticism, goes the objection, serves a particular interest and not the public interest.

There are at least two reasons why this 'public interest' objection is not plausible. Most obviously, as was mentioned earlier, it fails because it ignores the fact that refusing to engage in normative analysis does not serve the public interest, but serves whatever interests are represented by the constitutive and contextual values implicitly built into the TA/EIA data and methodology. Hence there is no basis for the presupposition that refraining from criticism or evaluation somehow serves the public interest.

A second problem with this public-interest objection is that it ignores the reality that public policy is made in a *political* environment.[57] Policy is not made by some hypothetical policymaker pursuing the 'public interest'. It is made by interaction among a plurality of partisans.[58] And each of those partisans would probably claim that his own conception of 'the public interest' is the correct one. On this analysis, there is no abstract 'public interest', devoid of partisan values, just as (see Sections 2.1 and 2.2 of this chapter) there are no abstract 'facts', devoid of constitutive or contextual values. This being so, the practitioner of TA or EIA serves the public interest not by alleged neutrality but by effective ethical and methodological criticism of all reasonable TA/EIA policy options.

To attempt to be 'neutral' in waters already muddied by unavoidable politics is simply to serve the implicit or most powerful political values already present. Hence, the 'public-interest' objection to normative analysis would work only if there were data, methods, or policy options *not* affected by politics and values. Since they are so affected, the best way to deal with them is to render them explicit, spell out the presuppositions they embody, and the consequences to which their adoption would likely lead.

4.2. *The Presupposition That There Is No Place for Philosophical Evaluation in Technology Assessment and Environmental-Impact Analysis*

Given the impossibility of avoiding constitutive and contextual values in TA and EIA, the real question is not whether value issues ought merely to be 'left to the public', but whether particular assessors have in fact raised them in a way which gives proponents of various positions an opportunity for a real hearing. This point is totally missed in positivist critiques of applied philosophy. They condemn all normative discourse, not merely certain *examples* of philosophical criticism because it specifically violates given standards of procedural justice or fair play. Such positivist condemnations are typified by the recent statement, already noted, of a prominent social scientist. She maintains that "the role of social and philosophical studies of science and technology" is not "to reveal the rights and wrongs of policy choice", but only "to understand ... to interpret ... to draw a coherent picture of what is going on".[59]

In excluding criticism or analysis directed at *methodological evaluation* of TA's and EIA's and *ethical evaluation* of science policies, this scholar has advocated a classic positivistic epistemology regarding the role of philosophical studies (see note 1). All those who claim that philosophers and social scientists ought not to analyze the 'rights and wrongs' of science policy, however, are inconsistent. In advocating this position, they themselves are promoting an allegedly 'right' policy for studies of technology and science. Not only is their advocacy of such a position inconsistent on their own terms, but also it is questionable because it is unsubstantiated. Presumably what is most damaging about certain evaluations of 'right' and 'wrong' policy is that they are subjective, biased, or inadequately argued. If so, then whenever they give no reasons for holding positivist presuppositions, those who condemn normative analysis have fallen victim to what is allegedly most unconscionable about the prescriptive studies they condemn. They appear to have leapt prematurely to the conclusion that the inapplicability of the positivist model of rationality, to TA or EIA, establishes the arationality (and therefore the purely relative character) of philosophical analyses of science and technology. Obviously, however, if one accepts a different model of rationality, such as that of Hanson (see note 1), then logical, methodological, and ethical studies of science and technology do not necessarily have an arational and relativistic character. In any case, it is inconsistent for critics of applied philosophy to denounce all forms of ethical and methodological advocacy in TA and EIA, but to adopt an unsubstantiated position of advocacy for the positivist presupposition that one can and ought to be purely descriptive and objective in his analyses (see note 52).

The real issue regarding applied philosophy is *not* whether philosophers ought to speak normatively about science, technology, and public policy, since theories, norms, and values unavoidably structure our science and research (see sections 2.1 and 2.2 of this chapter). The real issue is whether a given philosophical account is normative in a way that is misleading, incoherent, incomplete, question-begging, or implausible. In other words, the naive epistemological presuppositions of positivist practitioners of TA and EIA lead them to criticize work simply because it contains normative conclusions, rather than because those conclusions are not reasonable or not substantiated. They ask the wrong question. They fail to recognize that the point of ethical analysis should not be blind advocacy, but often to show that, *if* policymakers accept certain assumptions about methods of evaluating technology, *then* certain consequences are likely to follow. For example, *if* one uses standard cost-benefit analysis, *then* policy based on these analyses is likely to take no account of distributive variations in the costs and benefits. In arguing that failure to take account of distributive variations does not serve equity, philosophers have maintained that serving equity is desirable. They have argued that the failure to serve equity usually results in discrimination against the poor, in violations of minority rights, and in discrimination in favor of "the most powerful and articulate groups" in society.[60] If something is wrong with their position, then it would have to be established that their arguments are deficient somehow, not merely alleged that philosophers ought not to have any normative arguments, or ought not to talk about "rights and wrongs".

The real point of tracing the consequences to justice that result from given modes of public policy and technology assessment, however, is not that philosophers are somehow advocating a particular ethic, but that assessors often do not recognize that their analyses fail to serve justice. That most policymakers would probably claim to subscribe to an ethic of justice or equity, and that they sometimes sanction assessment methods insensitive to considerations of justice or equity, is the real point. If, in analyzing these assessment methods, the applied philosopher cannot substantiate *ethical* claims about what actions ought to be done, or *epistemological* claims about what methodology ought to be preferred, or *logical* claims about what arguments are valid – in short, if one cannot speak prescriptively and theoretically – then indeed positivist presuppositions will have banished not only applied ethics and philosophy from TA and EIA, but also all evaluative thought and language as well. And if these presuppositions preclude critical analyses, then they undercut the most important reason for engaging in TA/EIA in the first place.

4.3. *Consequences of Condemning Applied Ethics and Philosophy of Science*

Even were there no basic epistemological reasons for believing that it is impossible to avoid implicitly evaluative research, however, there would be practical reasons for continuing explicitly normative studies. It makes as much sense for positivist practitioners of TA and EIA to tell science scholars to avoid employing prescriptive conclusions as it does for the AMA to tell the public to avoid making value judgments about the practices of given medical doctors. In both cases, uncriticized ethical and methodological values lead to great abuses. To suggest that those who study either science policy or medical policy ought to content themselves with a nonevaluative description of 'what is. . . going on'[61] is both unwise and inconsistent. It is inconsistent, because allegedly 'descriptive' studies of policy always contain constitutive and contextual values and hence implicitly sanction the status quo. To allow only allegedly descriptive studies is also unwise, because it deprives the public of a major forum for scholarly debate about the rights and wrongs of public policy. Moreover, if rights and wrongs are not subject to intelligent public debate, then it is more likely that undesirable, simplistic, uncritical, authoritarian, or repressive views will hold sway. In fact, positivistic condemnations of applied philosophy appear likely to lead to at least four undesirable consequences: (1) sanctioning ethical relativism; (2) accepting the status quo; (3) masking central evaluational assumptions; and (4) deemphasizing nontechnological policy solutions. As I suggested previously, one of the most dangerous of these consequences is relativism.

4.3.1. *Sanctioning Ethical Relativism*. Basically, the positivists' line of thought is that, since only facts are objective, and since values are merely subjective, then there are no rational grounds for deciding questions of value. Hence, they maintain, only a position of relativism is possible.[62]

As a methodological assumption, ethical relativism is held by a number of social scientists; it is an explicit axiom of research of the Columbia University school of anthropology, for example, a starting point of much psychoanalytic practice, and the underlying ethical framework of much current work in the social sciences.[63] Many scientists appear to be relativists by virtue of their belief that, because values may be seen in *different* ways, they may be viewed in *any* way. (See section 2.3 of this essay.) Apart from the epistemological problems with such an assumption, ethical relativism leads to dangerous consequences because the social sciences are often applied to help resolve questions of social, political, legal, and economic policy, as in technology assessment and environmental-impact analysis. It could be disastrous to adopt a stance of ethical relativism in a science-policy situation having potentially catastrophic consequences, e.g., use of nuclear weapons. Ethical relativism, in any policy context, is highly questionable.

One noted social scientist, for example, famous for studies of technology-related controversy, condemns ethical analyses of public policy on the grounds that ethics is relative. She claims that the values on which philosophical evaluations are based are arbitrary; that ethical analyses could be based on love, justice, or brotherhood; and that there is nothing special about justice, as opposed to other values.[64]

For my part, I find it inconceivable that justice not be central to ethical analyses of science policy, since policy is in large part built on regulation and law, and good regulation and law is based on *nonpartiality* and *consistency*, i.e., on justice. To say that justice is no more important than other public-policy values or, more generally, to sanction ethical relativism is to allow acknowledged evils to continue uncriticized. This is exactly what happened during World War II, when some Columbia-University anthropologists were asked about their position on the Nazis' actions. As a response, they affirmed a position of relativism. They said that "in this conflict of value systems", they had to "take a professional stand of cultural relativity" and write themselves "down as skeptics".[65] Were everyone to take such a position in similar situations, consequences of inestimable harm would likely follow.

One of the biggest problems with adopting the thesis of relativism is that it could lead one to remain silent while abuse was showered on innocent people. Perhaps this is why Dante says that the worst places in hell are reserved for those who are neutral in a time of crisis. Moreover, in condoning such silence, by virtue of their refusal to allow ethical and methodological analyses – evaluative studies – relativists are themselves adopting an evaluative position, viz., that values are relative.

Another difficulty with the stance of relativism is that it forces one to assume that progress is meaningless, in spite of the fact that there are many good reasons for believing in the concept of progress, and that it is intuitively accepted by virtually everyone. 'Progress' only makes sense when one can make judgments concerning the desirability of one position relative to another. Such judgments have no place in a situation in which all positions are defined as equally desirable, as they must be for a consistent relativist.

Another problem is that relativism is built on the questionable assumption that, just because an action *is* done or a particular ethic is practiced, therefore it *ought* to be. Obviously, however, the mere occurrence of certain behavior does not endow it with normal acceptability. If it did, for example, then the occurrence of wartime crimes against gypsies, Jews, and leftists would have to be said to be morally acceptable. Yet, most people do not agree that these actions were tolerable. Moreover, if ethics were relative, then the Nuremburg War Trials and other trans-cultural normative undertakings would have to be said to have been conducted with no authentic ethical basis. Since there appear to be many good reasons for affirming the legitimacy of the trials, if one is consistent, then he ought to recognize that ethics cannot be wholly relative.

Admittedly it is often impossible to specify when a given action is ethical and when it is not. All the opponent of relativism need do, however, is to establish that there are some acts which, across cultures, are worthy of moral approval or disapproval. Given at least one good counterexample to relativism, it follows that discourse about the 'rights and wrongs' of policy choice cannot necessarily be dismissed merely as subjective. This means, in turn, that when positivists dismiss applied ethics and the attempt to articulate, justify, and criticize certain policies, they are running the risk of encouraging unjustified abuses of human beings merely because these aberrations are sanctioned by some relativistic ethic, no matter how nonsensical.

4.3.2. *Accepting the Status Quo*. Another likely consequence of prohibiting talk about policy rights and wrongs in technology assessment and environmental-impact analysis is that such a position is almost certain to constitute a normative judgment, viz., sanctioning the status quo. If one claims to be engaged in wholly neutral analysis, then there is indeed a problem with work which gives implicit assent to a particular situation or position.

Far from implicitly condoning the status quo because of our failure to criticize it, there often seems to be a moral obligation not to be completely neutral, or silent, about a given policy. If assessors were evaluating a position on a certain technology-related issue which could be shown to promote racism, sexism, or violations of civil liberties, then surely the policy ought to be criticized on these grounds.

I have often wondered why, when Albert Einstein condemned Hitler's violations of civil liberties in 1933, the Prussian Academy of Sciences, in turn, denounced Einstein.[66] Was it that Einstein's colleagues believed that they ought to 'get the facts' but to leave evaluative and political judgments to the people or to national policymakers? Did they ever consider the fact that, by attempting to avoid all normative judgments, and by condemning Einstein because he made a normative judgment, they were in fact making a strong normative judgment in favor of the status quo? Albert Camus phrased my point well:

> We are guilty of treason in the eyes of history if we do not denounce what deserves to be denounced. The conspiracy of silence is our condemnation in the eyes of those who come after us.[67]

Abraham Lincoln said much the same thing: "silence makes men cowards".[68]

Apart from the moral obligation to beneficence,[69] which suggests that, in certain situations, one ought not refrain from normative judgments, there is also a very practical reason for sometimes including explicitly evaluative judgments in TA's and EIA's and thereby avoiding sanctioning the ideal of complete neutrality and the status quo. If assessors do not provide a range of explicit, normative, alternative recommendations, then purely political pressures could force policymakers to ignore the assessment results. As Selwyn Enzer, Associate Director for Futures Research, Graduate School of Business, University of Southern California, put it: "Technology assessors must take a stand, make value judgments and specific recommendations. Otherwise, this has the effect of weakening the significance of the assessment results", and contributing to its lack of utility and vulnerability to criticism.[70]

The rationale behind such an imperative is that, if assessors do not present critical remarks about a range of policy alternatives, and if they attempt to be neutral, then their silence about alleged errors, evils, or misrepresentations will simply help to legitimate whatever policy is currently being followed. In this sense, acceptance of the ideal of complete objectivity can serve explicitly political ends.[71] In legitimating the status quo, particularly the technological status quo, it obscures the unavoidably political components of technology assessment and environmental-impact analysis.

4.3.3. *Masking Central Evaluational Assumptions*

In other words, if one explicitly avoids either advocating or criticizing the ethical and methodological presuppositions of a given policy position, and he claims that his assessment is neutral, then he ignores the fact that his alleged neutrality actually serves the constitutive and contextual values (see sections 2.21 and 2.22) in the status quo. Moreover, to the extent that one believes that avoiding advocacy or criticism guarantees neutrality, then to that same degree is acceptance of the ideal of complete neutrality likely to lead one to ignore important theoretical/evaluative assumptions implicit in the TA or EIA. If presuppositionless research is impossible, and if one fails to realize this, then provided that an assessment contains no *explicit* criticism or advocacy, the *implicit* methodological/ethical assumptions are likely to be ignored. Obviously one is likely to miss a normative component of one kind (e.g., the effects of given quantitative presuppositions) if he erroneously believes that normative statements only take the form either of explicit advocacy or criticism.

One social scientist, for example, claims that scholars ought to avoid *explicit* evaluation of the "rights and wrongs" of science policy. Yet, at the same time, she sanctions their *implicit* value judgments about such policy because she says they must "discover order in the complex social systems of science and technology" and "interpret the political stakes involved".[72] Once one gets into the business of "discovering order" and "interpreting", however, he is in some sense coloring the facts (perhaps unintentionally) with his own evaluative theories, assumptions, and his own connotations of right and wrong. If so, then it is incoherent to *approve* such interpretations but to *disapprove* normative discourse. They are both of a piece. This means that many proponents of the ideal of complete neutrality are in the questionable position of condemning *explicit* normative argumentation by those trained to do it, while sanctioning the *implicit* (and usually unsubstantiated) normative judgments involved in interpretation. Apart from whether this stance is consistent, it is not coherent, especially if the objective of the earlier condemnation is to avoid subjectivity. The worst sort of subjective research, that posing as objective and unbiased, would seem to occur, not when a philosopher argues normatively, as he is wont to do, but when he erroneously believes that he is avoiding evaluative or normative judgments by following the directive "to interpret" or "to draw a coherent picture of what is in fact going on". Interpretations or pictures of the facts are less overtly, and therefore more dangerously, normative.

The best safeguard of authentic objectivity in a TA or EIA would therefore appear to be the realization that no wholly objective stance is possible. Because of this recognition, one would be forced to investigate alternative normative assumptions and to attempt to recognize them when they occurred.

Unfortunately, there is little evidence that assessors realize that their analyses cannot be wholly objective. This means that not only do they often miss evaluational assumptions in their work, but also they fail to see how they figure in that of others. For example, the well-known risk assessor, W. Häfele, in an assessment of nuclear technology, makes a number of normative assumptions about human ability to contain and control radioactive waste in perpetuity. Not recognizing these assumptions, and viewing himself as a scientist who 'has the facts', he dismisses the opinions of those who are critical about nuclear technology because of the radwaste problem. He says simply that fears about radioactive waste come from an ignorant public who doesn't understand the facts of the situation.[73]

Starr, Whipple, Maxey, Cohen, and Lee all take much the same position. Failing to recognize the normative and methodological assumptions in their work, they miss alternative, but perhaps equally plausible, evaluative presuppositions in the work of their opponents. Since they subscribe to both the principle and the ideal of complete neutrality and the fact-value dichotomy, they place too great an emphasis on obtaining the facts. They assume that their opponents disagree with them only because they don't know the facts.[74] For example, they give evaluatively determined probabilities for technological hazards such as LNG (liquefied natural gas) or nuclear accidents then, when other experts or representatives of public-interest groups register opposition to particular uses of either of the two technologies, they dismiss this opposition by saying that the public does not understand the probabilities.[75] In reality, the disagreement is over the *evaluative assumptions* used to estimate the probabilities, although many risk assessors see the controversy as caused by mathematical ignorance. This means that acceptance of the principle and the ideal of complete neutrality is likely to encourage the heavy-handedness of experts who often dismiss the importance of the value judgments made by the public or by those who disagree with them. Apparently they make such a dismissal because they see themselves as the technical experts dealing with a purely technical situation.[76]

Another problem with experts' dismissal of the importance of normative assumptions is that it often leads them to a pro-technology bias. Assessments of technologies of the future, usually high technologies, rely much more on prediction than on experience. Because of this, and because decisionmakers are more reluctant to prevent risks that are uncertain than to approve projects whose benefits are speculative, over-emphasis on allegedly objective, scientific information is often likely to create a bias in favor of a given future technology.[77] If this is true, and if social scientists cannot get away from value judgments, then the danger comes, not from making them, but from being unaware of the nature of the values that are operating.[78]

5. DEEMPHASIZING NONTECHNICAL POLICY SOLUTIONS

Lack of awareness of the values operating in a given TA or EIA often leads one to overemphasize the purely technical aspects of public policy, as the example of LNG and nuclear assessments suggests. But if these technical aspects are overemphasized, then this can only mean that the evaluative aspects are underemphasized, with the consequence that the *real sources* of policy controversy are often not recognized. And if they are not recognized, then there is little chance that the debates they fuel will be settled, and little chance that political, social, or normative, as well as purely technical, solutions will be proposed. As a consequence, public policy is likely to be far less rational and successful than it might be.

Langdon Winner argued that America has for too long substituted technical solutions for problems that were either political or moral in nature.[79] If he is correct, then part of what might be done in order to redress this wrong is to avoid both the principle and the ideal of neutrality, since it apparently leads to our underemphasizing the importance of nontechnical policy solutions.

6. ALTERNATIVES TO THE PRINCIPLE AND THE IDEAL OF COMPLETE NEUTRALITY

Perhaps what needs to be done most to correct this situation is to provide a brief account of a revised method of technology assessment and environmental-impact analysis, a method which rests upon neither implicit nor explicit acceptance of the principle and the ideal of complete neutrality. I can think of at least four components that ought to be included in such a new method: (1) explicit admission of the methodological, ethical, factual, and theoretical assumptions upon which the TA or EIA conclusions are contingent; (2) use of much broader and much more varied social indices; (3) employment of an adversary system of technology assessment and environmental-impact analysis; and (4) evaluation of alternative philosophical positions on various policy options. I will briefly explain these four suggestions here. Later, in Chapters Eight and Nine, I will give a fuller treatment of my two most important proposals, (3) and (4).

Component (1), explicit admission of the methodological, ethical, factual, and theoretical assumptions, is a response to the long-standing call of Nobelist Gunnar Myrdal for the inclusion of explicit value premisses in social inquiry.[80] Such a component is important because it is rare to find candid admissions of uncertainty in TA's and EIA's.[81] Without this hallmark of the scientific method, assessments are likely to be interpreted as more objective than they really are. Moreover, to require that normative assumptions be made overtly and explicitly is to increase the possibility that societal debate over given technologies and environmental impacts focuses on the real issues that often divide people. Although it is unrealistic to expect that explicit admission of assumptions is likely to resolve controversies over technology, this admission is usually a necessary condition for real communication among those who disagree.

By increasing the variety of components indexed and the number of social indices, one is also likely to help avoid giving the impression either that a given TA, EIA, or index is wholly neutral or that there are no alternatives to it. In a classic article on social-systems accounting, M. B. Gross gave some suggestions for broadening social indicators, so as to avoid being trapped by the narrow methodology of only one index. He urges that input-output matrices, national balance sheets, and analyses of wealth and income distribution be developed to evaluate factors such as the nation's educational, artistic, recreational, and humanistic capacities, as well as its crime and delinquency, its mobility, its stratification, its recognition of human rights, and so on.[82] By employing indicators for a variety of factors, one helps to focus on alternative values underlying various measures of well-being. Moreover, by using a number of different indices, one helps to minimize both the inaccuracy stemming from use of only one statistical method, and the tendency to overestimate the importance of the results determined from only one index.[83]

Component (3), employment of an adversary system of TA and EIA, is perhaps the major means for putting the principle and the ideal of complete neutrality to rest. (I will devote Chapter Nine to this suggestion. Suffice it here merely to characterize briefly my proposal.) The heart of the adversary system is the assumption that, since there is no one, wholly objective TA or EIA for a given project, then alternative assessments (based on facts interpreted in the light of different assumptions) must be obtained. Another assumption, perhaps also an insight, of the adversary method is that the key problem in building an assessment system is not how to provide an objective analysis of data, but how to obtain an analysis of the data which provides equal time, opportunity, and attention for articulating all major positions of advocacy and criticism regarding a given project. In other words, the real problem is to be sure that all the important issues surface and receive attention, not to seek some unattainable standard of objective truth on those issues. This is why Amory Lovins, Harold Green, and others who support an adversary system often have emphasized that they believe that the *process* by which an assessment is done is more important than its factual correctness.[84] Since no one can guarantee *correct facts*, the best we can do is to attempt to provide a *correct method* for handling all views of the facts.

What might such a method be like? Probably it would be based on procedural justice. It would provide for all views of the facts, even criticism or advocacy, to come to the forefront and to be debated in the context of adversary assessment. In other words, since there is no such thing as strict objectivity, an assessment mechanism would do well if it encouraged all biases, values, perspectives and assumptions on various sides of a given issue to be made explicit, so that all alternative positions in turn could be evaluated by the public.

Numerous objections to such an adversary system have already been raised. Where they do not rest on explicit appeal to the ideal of objectivity,[85] most of these complaints focus on the inefficiency of turning every technological or environmental dispute over to the delay and legal wrangling involved in adversary proceedings. Such an objection, however, seems to me to rest on the erroneous assumption that, in all situations, efficiency is more important than democracy. Clearly, to continue the present system of TA and EIA prejudices policymakers in favor of both the status quo and the methodological assumptions of a small group of assessors who, although they do not make policy decisions, do create the value-laden measures and conceptions of public welfare in terms of which policy is made. Although the adversary system is less efficient than the current method of TA and EIA, it is clearly the more desirable, insofar as democratic freedom and procedural justice are concerned.[86] There is no other system of TA and EIA which explicitly provides for all major positions on an issue to receive a hearing.

Another common objection to the adversary system is that it will stifle progress. While this might be true, there are at least two reasons for giving the adversary system a trial use, in order to evaluate its effects. *First*, the proponents of a given technology or environment-related development usually have much greater financial resources, and therefore have potentially greater access (to the government and to the public) to present their case, than do their opponents. If this is true, then far from stifling progress, the adversary system may well help to equalize the opportunities of all persons to receive a hearing on a given issue. *Second*, and more important, 'progress' itself is a notoriously normative term. Moreover there is no clear reason why citizens of a democracy should be forced to accept a given notion of progress (e.g., one car per person; the urbanization of all areas) that it might not want to accept.[87] Hence any arguments that an adversary system of TA or EIA would thwart progress have to be examined very closely. Chapter 9 will consider such arguments in detail and will spell out the particulars of one such system.

6.1. *The Importance of Including Philosophical Analysis*

In addition to the adversary system, it is also important to include component (4), evaluation of alternative philosophical positions on various policy options. (Chapter 8 will be devoted entirely to a specific proposal in this regard.) This requirement for inclusion of a philosophical component in assessment is based on the importance of several key premisses. One is practical, the other, somewhat theoretical.

From a practical point of view, if particular value positions and methodological presuppositions are left *implicit* in a TA or EIA, rather than explicitly analyzed and evaluated, then there is less chance that the public and policymakers will be able to deal with them clearly and make an informed judgment concerning them. If they are critically evaluated, however, then decisionmakers are more likely to recognize the normative component in their assessments and to avert the dangerous consequences likely to come about as a result of sanctioning particular values.

From a theoretical point of view, the basic reason for including normative (ethical, methodological) analyses in TA and EIA is that only philosophical investigations can deal properly with questions such as whether given *ends* (e.g., accelerating use of fossil fuels) are wise or desirable. Apart from whether a certain technology or environmental impact is necessary as a *means* to a particular end,[88] rational and democratic policymaking requires that assessors also enable the public to evaluate the *ends* of public policies. In other words, only properly philosophical reflection deals with the acceptability of the various *goals* or alternative *final causes* that technology and its impacts have been assumed to serve.

6.2. *The Positivistic Attack on Philosophy*

Of course, the traditional positivist response to the claim that philosophical analysis is necessary to certain TA and EIA enterprises, e.g., evaluating the ends or goals of policy, has been to deny the importance and uniqueness of this role of philosophy. Part of the positivistic heritage, embraced by many social scientists, is that methodological and ethical, i.e., normative or philosophical, analyses, require no special skills and have no place in TA or EIA. One famous social scientist, for example, criticizes the notion "that skill in philosophical analysis and command of ethical arguments provides ... a specialized form of expertise".[89]

In assuming that philosophical and ethical evaluation requires no special expertise, this positivist appears to be saying that, contrary to what Plato, Aristotle, and others believed, anybody can do methodological and ethical analysis. But if anybody can do methodological and ethical evaluation (a premise which is obviously false, but whose falseness is not my point here), and no special expertise is needed, then anybody's analysis must be as good as anybody else's. But if one believes that any philosophical or ethical analysis is as good as any other, then surely one is a relativist. As I already pointed out, there are numerous reasons for rejecting the position of relativism (see section 4.3.1 of this chapter).

Besides their implicitly sanctioning a relativistic position on ethics, another problem with traditional positivistic condemnations of applied philosophy is that they are misdirected. For example, one social scientist claims that "to rely on moral theory to understand the quandries of science and technology is to assume that people act out of abstract principles that are divorced from their social and political settings".[90] Here, however, the attack is on a straw man. I know of no person doing applied ethics who claims (or even assumes) either that moral theory *alone* is sufficient for understanding scientific controversies or that people act *only* out of abstract moral principles. Both assumptions are simplistic. Obviously many types of analyses (e.g., social, economic, ethical, and biological) are necessary to understand these controversies, and obviously people act for a great many reasons. Because one believes in using ethical theory to do science policy studies does not mean that he believes in using ethical theory *alone* to do science policy studies. The argument is that ethical analysis is *necessary* to a complete study of public policy controversies, not that it is *sufficient*. Hence the charge against applied philosophy is misdirected; it confuses necessary and sufficient conditions.

Other criticisms of applied philosophy rest on this same confusion of necessary with sufficient conditions. "To argue", says one science-policy expert, "that philosophical analysis is especially useful in understanding policy issues . . . is to attribute such endemic problems as the neglect of social costs to false principles, methodological lapses, or inadequate concepts".[91] Here the expert erroneously assumes that because applied philosophers believe that methodological analyses are *necessary* to understanding the policy issues in TA and EIA, they believe these analyses are *sufficient*. I know of no one who believes or who has argued that this is the case. Rather, the point is that questions of technological and environmental policy are many-faceted. The methodological and ethical facets of TA and EIA can no more be ignored than can the biological or economic components. Numerous authors have shown how given methodological assumptions can lead to erroneous policy regarding technology. This is one of the main points of the recent Harvard Business School study of energy policy. Misleading methodological and conceptual assumptions, such as the authenticity of market price, have "helped to create some of the impasses and stalemates in US energy policy"; methodological errors in computing the price of oil, say the authors, ultimately caused excessive promotion of petroleum, a weakening of the dollar, and a failure to provide incentives for conservation and the use of solar technology.[92]

The claim that methodological and conceptual errors in technology assessment and environmental-impact analysis cause no technological controversy or no policy problems is dangerous, however, as well as false. It is dangerous precisely because it is *anti-intellectual* in ascribing problems to 'politics' rather than also to methodological assumptions which are sometimes the vehicle of politics.[93] If one assumes that methodological issues never play a role in technological controversy, then he ignores one of the most subtle, and therefore potentially the most dangerous, manifestations of politics. Moreover, to assume automatically that political clout causes technological controversy is to prejudge the issues that often divide reasonable persons. It is also anti-intellectual, because methodological and conceptual disagreements (e.g., as to the accuracy of distributed methods of cost-benefit analysis) often are at issue, apart from whether political disagreements (e.g., as to whether it is in industry's interest to compute social costs) are present. All this is ignored when positivist scientists misrepresent the claims of those who do ethical and methodological analysis and then criticize them on the basis of their misrepresentations.

7. CONCLUSION

If ethical and methodological analysis is as important as I have argued it is, then philosophical components clearly have a place in technology assessment and environmental-impact analysis. Admittedly, philosophers and others who study value theory are "normally not a component of most interdisciplinary [TA or EIA] teams of any kind".[94] Admittedly, too, the scientists on interdisciplinary teams are often looked upon as the experts, even on issues which involve value judgments.[95] This suggests that the positivistic principle and ideal of complete neutrality is extremely pervasive. As a consequence, all nonscientific, or allegedly evaluative, analyses are viewed as purely subjective and therefore unimportant. If my remarks in this essay are correct, however, then there are numerous reasons for doubting both the principle and the ideal of complete neutrality.

Why, then, does the myth of complete neutrality persist? In part, it is because it gives us some reassurance, even if erroneous, that we can 'get at' the truth in a simple way. Without this belief, it is far more complex to attempt to safeguard the nonbiased and empirical quality of important TA and EIA work.

Another reason for the persistence of the myth was suggested by Kenneth Boulding. He himself never seems to suffer from the disease he diagnosed as *agoraphobia*, the fear of open spaces, the fear of research which cannot be closed, clean, tight and objective. Because we fear open spaces, especially conceptually open ones, said Boulding, we tend "to retreat into the cosy closed spaces of limited agendas and responsibilities".[96] If Boulding is right, then perhaps the biggest problem with technology assessment and environmental-impact analysis is not how to make them completely neutral, but how to protect ourselves from those who want us to believe that they can be.

228640

NOTES

[1] See Norwood Russell Hanson, *Patterns of Discovery*, Cambridge University Press, Cambridge, 1958; Thomas S. Kuhn, *The Structure of Scientific Revolutions*, University of Chicago Press, Chicago, 1962, 1979; Michael Polanyi, *Personal Knowledge*, Harper and Row, New York, 1958, 1964; and Stephen Toulmin, *Foresight and Understanding*, Harper and Row, New York, 1961. According to N. Abbagnano, 'Positivism', in *The Encyclopedia of Philosophy* (ed. by Paul Edwards), Macmillan, New York, 1967, vol. 6, p. 414, "the characteristic theses of positivism are that science is the only valid knowledge and facts the only possible objects of knowledge; that philosophy does not possess a method different from science. ... Positivism, consequently ... opposes any ...

procedure of investigation that is not reducible to scientific method". As Laudan points out, because they have disallowed talk about developments in "metaphysics, logic [and] ethics ... 'positivist' [sociologists,] philosophers and historians of science who see the progress of science entirely in empirical terms have completely missed the huge significance of these developments for science as well as for philosophy". (*Progress and Its Problems: Toward a Theory of Scientific Growth*, University of California Press, Berkeley, 1977, pp. 61–62.)

[2] Quoted by C. V. Kidd, 'Technology Assessment in the Executive Office of the President', in *Technology Assessment: Understanding the Social Consequences of Technological Applications* (ed. by R. G. Kasper), Praeger, New York, 1972, p. 131.

[3] US Congress, Office of Technology Assessment, *Annual Report to the Congress for 1976*, US Government Printing Office, Washington, D.C., 1976, p. 63; hereafter cited as: *AR 76*. US Congress, Office of Technology Assessment, *Technology Assessment in Business and Government*, US Government Printing Office, Washington, D.C., 1977, p. 9; hereafter cited as *TA* in *BG*.

[4] L. H. Mayo, 'The Management of Technology Assessment', in Kasper, *op. cit.*, p. 107, and R. A. Carpenter, 'Technology Assessment and the Congress', in Kasper, *op. cit.*, p. 40.

[5] Dorothy Nelkin, 'Wisdom, Expertise, and the Application of Ethics', *Science, Technology, and Human Values* 6 (34), (Spring (1981), 16–17; hereafter cited as: Nelkin, Ethics. Inasmuch as Nelkin condemns philosophical discussion of the "rights and wrongs" of science policy, she disallows talk about the ethics of science policy. Hence her disallowing this talk about ethics is tantamount to subscribing to a positivistic model of the role of philosophy and of studies about science and technology.

[6] Carpenter, *op. cit.*, p. 42.

[7] M. D. Reagan, *Science and the Federal Patron*, Oxford University Press, New York, 1969, p. 9.

[8] See Helen Longino, 'Beyond "Bad Science": Skeptical Reflections on the Value-Freedom of Scientific Inquiry', unpublished essay, March 1982, done with the assistance of National Science Foundation Grant OSS 8018095. Hereafter cited as: Longino, Science.

[9] See Longino, Science, esp. pp. 2–3.

[10] For discussion of these three examples from the history of science, see Harold I. Brown, *Perception, Theory and Commitment*, University of Chicago Press, Chicago, 1977, pp. 97–100, 147, and K. S. Shrader-Frechette, 'Recent Changes in the Concept of Matter: How Does "Elementary Particle" Mean?', in *Philosophy of Science Association 1980*, vol. 1 (ed. by P. D. Asquith and R. N. Giere), Philosophy of Science Association, East Lansing, 1980, pp. 302 ff.

[11] Longino, Science, pp. 6–9.

[12] Longino, Science, pp. 10–12.

[13] Longino, Science, pp. 16–19.

[14] Longino, Science, p. 25.

[15] W. K. Foell, 'Assessment of Energy/Environment Systems', in *Environmental Assessment of Socioeconomic Systems* (ed. by D. F. Burkhardt and W. H. Ittelson), Plenum, New York, 1978, p. 196; US Congress, *Technology Assessment Activities in the Industrial, Academic, and Governmental Communities*, Hearings Before the Technology Assessment Board of the Office of Technology Assessment, 94th Congress, Second

Session, June 8–10 and 14, 1976, US Government Printing Office, Washington, D.C., 1976, pp. 66, 200, 220; hereafter cited as: *TA* in *IAG*. See A. B. Lovins, 'Cost-Risk-Benefit Assessments in Energy Policy', *George Washington Law Review* 5 (45), (August 1977), 940.

[16] See M. C. Tool, *The Discretionary Economy: A Normative Theory of Political Economy*, Goodyear, Santa Monica, Ca, 1979, p. 279. See R. M. Hare, 'Contrasting Methods of Environmental Planning', in *Ethics and the Problems of the 21st Century* (ed. by K. E. Goodpaster and K. M. Sayre), University of Notre Dame Press, Notre Dame, In., 1979, p. 65.

[17] See Tool, *op. cit.*, p. 280.

[18] For statements of the OTA's belief, see US Congress, Office of Technology Assessment, *Annual Report to the Congress for 1976*, US Government Printing Office, Washington, D.C., 1976, p. 4; and US Congress, Office of Technology Assessment, *Annual Report to the Congress for 1978*, US Government Printing Office, Washington, D.C., 1978, p. 7; hereafter cited as: *AR76* and *AR78*, respectively.

[19] A. L. Macfie, 'Welfare in Economic Theory', *The Philosophical Quarterly* 3 (10), (January 1953), 59, makes this same point.

[20] H. P. Green, 'The Adversary Process in Technology Assessment', in Kasper, *op. cit.*, pp. 51, 52, 55, 60, 61; hereafter cited as: Green, Adversary; and H. Fox, Chair, *Technology Assessment: State of the Field*, Second Report of the Technology Assessment Panel of the Engineers Joint Council, Engineers Joint Council, New York, 1976, pp. 3–5; and S. G. Burns, 'Congress and the Office of Technology Assessment', *George Washington Law Review* 5 (45), (August 1977), 1146.

[21] A. D. Biderman, 'Social Indicators and Goals', in *Social Indicators* (ed. by R. A. Bauer), MIT Press, Cambridge Ma., 1966, p. 101.

[22] Sergio Koreisha and Robert Stobaugh, 'Appendix: Limits to Models', in *Energy Future: Report of the Energy Project at the Harvard Business School* (ed. by Robert Stobaugh and Daniel Yergin), Random House, New York, 1979, pp. 237–240.

[23] Mr. Selwyn Enzer, speaking before the committee, as quoted in Congress, OTA, *TA in IAG*, p. 225, makes this same point.

[24] The OTA itself admits this. See Congress, OTA, *TA in BG*, p. 13; Congress, OTA, *AR76*, pp. 63, 66.

[25] For discussion of this topic, see K. S. Shrader-Frechette, 'Technology Assessment as Applied Philosophy of Science', *Science, Technology, and Human Values* 6 (33), (Fall 1980), 33–50; hereafter cited as: Technology Assessment.

[26] H. Green, 'Cost-Risk-Benefit Assessment and the Law: Introduction and Perspective', *George Washington Law Review* 5 (45), (August, 1977), 908; hereafter cited as: Cost. See also T. Means, 'The Concorde Calculus', *George Washington Law Review* 5 (45), (August 1977), 1037.

[27] See Lovins, *op. cit.*, pp. 913–937; W. D. Rowe, *An Anatomy of Risk*, John Wiley, New York, 1977, pp. 145–147, 225, 243; and A. L. Porter, F. A. Rossini, S. R. Carpenter, and A. T. Roper, *A Guidebook for Technology Assessment and Impact Assessment*, North Holland, New York, 1980, pp. 266–267, all of whom discuss these points.

[28] For an example of someone who holds this position, see C. Starr, *Current Issues in Energy*, Pergamon, New York, 1979, p. 10; hereafter cited as: *CIE*.

[29] D. J. Epp, *et al.*, *Identification and Specification of Inputs for Benefit-Cost Modeling*

of Pesticide Use, EPA-600/5-77-012, US Environmental Protection Agency, Washington, D.C., 1977, p. 26, makes a similar point.

[30] A. V. Kneese, S. Ben-David, and W. D. Schulze, 'The Ethical Foundations of Benefit-Cost Analysis Techniques', in *Energy and the Future* (ed. by D. MacLean and P. G. Brown), Rowman and Littlefield, Totowa, N.J., 1982, pp. 59–73.

[31] K. Shrader-Frechette, 'Economic Analyses of Energy Options: A Critical Assessment of Some Recent Studies', in *Energy and Ecological Modelling* (ed. by W. Mitsch, W. Bosserman, and J. Klopatek), Elsevier, New York, 1981, pp. 773–778.

[32] D. Okrent, 'A General Evaluation Approach to Risk-Benefit . . .', UCLA-ENG-7777, UCLA School of Eng., Los Angeles, 1977, pp. 1–9.

[33] Rowe, *op. cit.*, p. 3, and W. Häfele, 'Benefit-Risk Tradeoffs in Nuclear Power Generation', in *Energy and the Environment: A Risk-Benefit Approach* (ed. by H. Ashley, R. Rudman, and C. Whipple), Pergamon, New York, 1976, p. 181.

[34] US Nuclear Regulatory Commission, *Reactor Safety Study: An Assessment of Accident Risks in US Commercial Nuclear Power Plants* (WASH-1400), US Government Printing Office, Washington, D.C., 1975, Appendix XI, p.2–2; hereafter cited as: WASH-1400.

[35] See, for example, C. Starr, 'Benefit-Cost Studies in Sociotechnical Systems', in *Perspectives on Benefit-Risk Decision Making* (ed. by the Committee on Public Engineering Policy), National Academy of Engineering, Washington, D.C., 1972, pp. 26–27.

[36] See, for example, B. Fischhoff, P. Slovic, S. Lichtenstein, S. Read, and B. Combs, 'How Safe is Safe Enough?', *Policy Sciences* 9 (2), (1978), 150, and P. Slovic, B. Fischhoff, and S. Lichtenstein, 'Facts and Fears: Understanding Perceived Risk', in *Societal Risk Assessment* (ed. by R. Schwing and W. Albers), Plenum, New York, 1980, pp. 190–192.

[37] See K. Boulding, *Economics as a Science*, McGraw-Hill, New York, 1970, and E. J. Mishan, 'Whatever Happened to Progress?' *Journal of Economic Issues* 2 (12), (1978), 405–425. See also K. Shrader-Frechette, *Environmental Ethics*, Boxwood, Pacific Grove, 1981, pp. 135–36, 212–216.

[38] A. D. Biderman, 'Anticipatory Studies and Stand-by Research Capabilities', in Bauer, *op. cit.*, pp. 272, makes these same points.

[39] Biderman, *op. cit.*, pp. 273–274.

[40] Biderman, 'Social Indicators and Goals', in Bauer, *op. cit.*, p. 97, makes the same point, as does Oskar Morgenstern, *On the Accuracy of Economic Observations*, Princeton University Press, Princeton, 1963, pp. 26, 35–37, 62, 194–107.

[41] E. Lawless, *Technology and Social Shock*, Rutgers University Press, New Brunswick, N.J., 1977, pp. 497–498.

[42] Epp *et al.*, pp. 111, 73–78.

[43] One problem with most assessments of carcinogens, for example, is that a test result of 0 tumors in 100 animals is statistically consistent with a true risk of 4.5% when an assurance level of 99% is employed. (See Epp *et al.*, p. 55.) Another area in which similar statistical problems and difficulties with extrapolation occur is radiation hazards. Dose-response coefficients are determined by making a *theoretical* assumption (generally, that dose-response is linear) about the validity of extrapolating on the basis of high doses (See US Atomic Energy Commission, *Comparative Risk-Cost-Benefit Study of Alternative Sources of Electrical Energy*, WASH-1224, US Government Printing Office, Washington, D.C., 1974, pp. 4–13, to 4–15.

[44] See. S. G. Hadden, 'DES and the Assignment of Risk', in *Controversy: Politics of Technical Decisions* (ed. by D. Nelkin), Sage, Beverly Hills, pp. 118–119.
[45] This point is documented by S. Hadden, *op. cit.*, pp. 122–123.
[46] US Environmental Protection Agency, *Proceedings of a Public Forum on Environmental Protection Criteria for Radioactive Waste*, ORP/CSD-78-2, US EPA, Washington, D.C., 1978, p. 121, letter from Dr. Thomas Mancuso. See also pp. 122–123, and I. Bross, Director of Biostatistics at Roswell Park Memorial Institute, in Committee on Commerce, Science, and Transportation, *Radiation Health and Safety*, US Senate, 95th Congress, First Session, No. 95-49, US Government Printing Office, Washington, D.C., 1977, pp. 176–177. For information on US radiation standards, see K. S. Shrader-Frechette, *Nuclear Power and Public Policy*, Reidel, Boston, 1983, Chapter 2.
[47] G. J. Lieberman, 'Fault-Tree Analysis as an Example of Risk Methodology', in Ashley, *et al.*, *op. cit.*, pp. 247–276, also discusses this point.
[48] R. Zeckhauser, 'Procedures for Valuing Lives', *Public Policy* 4 (23), (Fall 1975), 444.
[49] Burns, *op. cit.*, p. 1150.
[50] Dr. Don E. Kash, Director of the Science and Public Policy Program, University of Oklahoma, in Congress, OTA, *TA in IAG.*
[51] See note 5. H. Skolimowski, 'Technology Assessment as a Critique of Civilization', in *PSA 1974* (ed. by R. S. Cohen, *et al.*), D. Reidel, Boston, 1976, p. 461, points to a similar condemnation of applied ethics and normative policy analysis. He cites an article by Genevieve J. Knezo, 'Technology Assessment: A Bibliographic Review', which was published in the first issue of the periodical, *Technology Assessment.* He says that Knezo condemns all normative literature as "emotional, neoluddite, and polemic" in nature, and "designed to arouse and mold mass public opinion". At the same time, says Skolimowski, Knezo praises all allegedly "neutral" work, and says it "usefully serves inform the public, through a responsible press, of the pros and cons of a public issue of national importance".
[52] See Nelkin, Ethics, pp. 16–17, for an example of someone who holds this position. See note 51.
[53] For two quite diverse views of future energy consumption, see Ford Foundation, Final Report, Energy Policy Project, *A Time to Choose*, Ballinger, Cambridge, Massachusetts, 1974, and W. G. Dupree and J. S. Corsentino, *United States Energy through the Year 2000* (revised), US Department of the Interior, Bureau of Mines, Washington, D.C., 1975.
[54] See Dupree and Corsentino, *op. cit.*
[55] See note 45. See also Skolimowski, *op. cit.*, p. 461.
[56] C. H. Weiss and M. J. Bucuvalas, *Social Science Research and Decision Making*, Columbia University Press, New York, 1980, p. 26, makes this same point.
[57] For another discussion of this same point, see E. S. Quade, *Analysis for Public Decisions*, American Elsevier, New York, 1975, pp. 269 ff.
[58] This same point is also made by C. E. Lindblom and D. K. Cohen, *Usable Knowledge: Social Science and Social Problem Solving*, Yale University Press, New Haven, 1979, p. 64.
[59] See notes 5 and 51.
[60] See, for example, Shrader-Frechette, Technology Assessment, pp. 35–38.
[61] Nelkin, Ethics, p. 17, argues that scholars ought to engage in nonevaluative description. See also Knezo (note 51).

[62] See R. M. Hare, 'Contrasting Methods of Environmental Planning', in Goodpaster and Sayre, *op. cit.*, p. 76, who outlines this rationale of positivists.
[63] See John Caiazzo, 'Analyzing the Social "Scientist" ', *The Intercollegiate Review* 2 (16), (Spring/Summer 1981), 96.
[64] Nelkin, Ethics, p. 17.
[65] Ruth Benedict, cited in Caiazzo, *op. cit.*, p. 96.
[66] For an account of this episode, see Albert Einstein, *Ideas and Opinions* (trans. by S. Bergmann), Crown, New York, 1954, pp. 205–210; Albert Einstein, *The World As I See It* (trans. by A. Harris), Philosophical Library, New York, 1949, pp. 81–89; and Philipp Frank, *Einstein: His Life and Times* (trans. by G. Rosen), Alfred A. Knopf, New York, 1947, pp. 234–235.
[67] *Notebooks* (trans. by J. O'Brien), Knopf, New York, 1965, p. 146.
[68] Quoted by J. Primack and F. von Hippel, *Advice and Dissent: Scientists in the Political Arena*, Basic, New York, 1974.
[69] See, for example, W. D. Ross, *The Right and the Good*, Clarendon Press, Oxford, 1930, Chapter 2.
[70] Congress, OTA, *TA in IAG*, pp. 233–234.
[71] D. Dickson, *The Politics of Alternative Technology*, Universe Books, New York, 1975, p. 189, makes a similar point: "the stress placed on the cultural importance of abstract science legitimates the ideology of scientism, yet disguises not only the exploitative way in which science is put to practical use through technology, but also the very fact that the existence of contemporary science – in terms of support for R and D – results directly from this practical use. A further aspect of scientism is that it promotes a passive acceptance of an existing state of affairs. . . . It dismisses as irrational or unscientific any attempts to challenge our contemporary situation in terms of the class interests which it maintains". See also pp. 186–195.
[72] Nelkin, Ethics, pp. 16–17.
[73] W. Häfele, 'Energy', in *Science, Technology, and the Human Prospect* (ed. by C. Starr and P. Ritterbush), Pergamon, New York, 1979, p. 139.
[74] For arguments to this effect, see K. Shrader-Frechette, 'Economics, Risk-Cost-Benefit Analysis, and the Linearity Assumption', in *PSA 1982*, Volume 1 (ed. by P. D. Asquith and T. Nickles), Edwards, Ann Arbor, Michigan, pp. 219–220; hereafter cited as: Economics.
[75] For arguments to support this claim, see K. Shrader-Frechette, Economics, pp. 219–223.
[76] R. Kasper, 'Perceptions of Risk and their Effects on Decision Making', in *Societal Risk Assessment* (ed. by R. Schwing and W. Albers), Plenum, New York, 1980, p. 77, makes the same point.
[77] H. Green, Cost, p. 901, makes this same point.
[78] M. Lutz and K. Lux, *The Challenge of Humanistic Economics*, Benjamin/Cummings, London, 1979, p. 3, makes the same point.
[79] L. Winner, *Autonomous Technology*, The MIT Press, Cambridge, Massachusetts, 1977, pp. 10–11.
[80] Cited by Tool, *op. cit.*, p. xvi.
[81] R. Andrews, 'Substantive Guidelines for Environmental Impact Assessments', in *Environmental Impact Analysis* (ed. by R. Jain and B. Hutchings), University of Illinois Press, Urbana, 1978, p. 40, substantiates this point, as does B. Gross, 'The State of the Nation: Social Systems Accounting', in Bauer, *op. cit.*, p. 165.

[82] Gross, *op. cit.*, pp. 268–271.
[83] Gross, *op. cit.*, pp. 266, also holds this position.
[84] Lovins, *op. cit.*, pp. 941–942; Green, Cost, p. 910, and Green, Adversary. See also S. Enzer, Associate Director, Center for Futures Research, Graduate School of Business, University of Southern California, *op. cit.*, p. 235 (note 17); Biderman, *op. cit.*, p. 134; Congress, OTA, *TA in IAG*, p. 198, statement by D. Kash, Director of the Science and Public Policy Program, University of Oklahoma; and A. Kantrowitz, 'Democracy and Technology', in Starr and Ritterbush, *op. cit.*, pp. 199–211.
[85] M. Bauser, 'The Atomic Energy Commission's ECCS Rule-Making', *Atomic Energy Law Review* 1 (16), (Spring 1974), 74, for example, says that the "adversary system of trial . . . is not compatible with an objective presentation and scholarly, detached evaluation of technical information".
[86] H. Green, Adversary, p. 58, defends this point.
[87] Green, Adversary, pp. 58–59, makes many of these same points.
[88] A similar point is made by R. Cohen, 'Ethics and Science', in *For Dirk Struik*, Boston Studies in the Philosophy of Science, XV (ed. by R. Cohen, J. J. Stachel, and M. W. Wartofsky), D. Reidel, Boston, 1974, p. 310. See also Hare, *op. cit.*, p. 65, and Skolimowski, *op. cit.*, p. 459, as well as Porter *et al, op. cit.*, p. 255; Lovins, *op. cit.*, p. 936; and J. R. Ravetz, *Scientific Knowledge and Its Social Problems*, Clarendon Press, Oxford, 1971, pp. 400–401, 431–432.
[89] Nelkin, Ethics, p. 16.
[90] Nelkin, Ethics, p. 16.
[91] Nelkin, Ethics, p. 16.
[92] Robert Stobaugh and Daniel Yergin, 'The End of Easy Oil' in *Energy Future: Report of the Energy Project at the Harvard Business School* (ed. by R. Stobaugh and D. Yergin), Random House, New York, 1979, pp. 4, 6, 11. See also 'Conclusion: Toward a Balanced Energy Program' in Stobaugh and Yergin, *ibid.*, p. 227. Also in the same collection, see the essay by M. A. Maidique, 'Solar America', p. 211. See also Laudan, *op. cit.*, (Note 1), pp. 59–61.
[93] Nelkin, Ethics, p. 16, ascribes problems to politics rather than to faulty methodology.
[94] G. E. Brown, California, member of the Technology Assessment Board, OTA, in Congress, OTA, *TA in IAG*, p. 201, has emphasized this point.
[95] Dr. D. Kash, *op. cit.*, p. 202, agrees with this point.
[96] K. E. Boulding, quoted by B. M. Gross, 'Preface', in Bauer, *op. cit.*, p. xvii.

CHAPTER FOUR

THE FALLACY OF UNFINISHED BUSINESS

1. INTRODUCTION

It is a truism that the wise person is not necessarily known by the answers he gives, but by the way he formulates the questions. Perhaps this is because even the best of us often cannot resolve a problem when it has been formulated in terms of the wrong questions.

For the last several decades, at least, we have intently formulated our science- and environment-related problems in terms of technological questions about purely technological solutions. We have asked, for example, whether we ought to store radioactive waste in salt mines or in deep-drilled wells, in solidified ceramic form or as a liquid inside double-walled steel tanks.[1] Not surprisingly, we have been getting technological answers, ones which respond to the questions asked, but which fail to resolve the more difficult problems which generated the original inquiry. In the case of the environmental hazard posed by radioactive wastes, the really intractable problems are not the technical ones of what storage techniques to adopt, but the ethical and social ones, such as what risk we can impose on future generations and how we ought to determine the acceptability of a given risk. We have not answered questions such as these in part because we have been asking, not wrong questions, but incomplete ones. We have been asking questions that are epistemologically loaded, questions that presuppose a definition of a given problem for which only an answer in terms of the technological status quo counts as a solution.

Many current technology assessments (TA's), environmental-impact analyses (EIA's), and science policy studies implicitly ask questions that presuppose narrowly technical definitions of the problems they address. As a consequence, they arrive at conclusions which, being narrowly technical, are less useful than they might be. In defining the problems they address in such a circumscribed way, practitioners of TA and EIA often fall victim to an ethical and methodological assumption that Keniston termed 'the fallacy of unfinished business'.[2] Related to one version of the naturalistic fallacy, this assumption is that technological and environmental problems have only technical, but not social, ethical, or political solutions. Proponents of this assumption believe that all that is necessary to solve our technological and environmental problems is more 'business as usual', more and better technology. After using several TA's and EIA's to illustrate the policy consequences of the fallacy of unfinished business, I suggest how it might be overcome. I present three standard arguments, repeatedly used in technology assessments and environmental-impact analyses, by those who defend and subscribe to this fallacy. I briefly examine the logical, consequentialist, and historical reasons for rejecting all three arguments. If my suggestions in this chapter are correct, then TA and EIA is not only a matter of discovering how to finish our technological business, but also a question of learning how to recognize the ethical and epistemological dimensions of our assessment tasks.

2. SUBOPTIMIZATION AND THE STATUS QUO

A recent Environmental Protection Agency (EPA) study provides a clear example of how assessors' subscribing to the fallacy of unfinished business contributes to substantial weaknesses in their TA conclusions. The question addressed in the study was simple: ought society to develop coal or nuclear fission in the nine-state Ohio River basin area, in order to meet electricity demand between now and the year 2000?[3] The TA team's allowable technological answers to this question – four different 'mixes' of coal and nuclear power – were not very satisfying, owing both the the well-known problems besetting nuclear utilities since 1974 and to the atmospheric and geographical characteristics (air inversions and cascading effects) unique to the Ohio River valley. Precisely because their formulation of this technological question admitted only of a purely technological answer, in terms either of coal or nuclear fission, the EPA assessment team ignored alternative ethical, social, and political solutions to the problem of meeting electricity needs. Members of the assessment team did not consider that (as the authors of the recent Harvard Business School study pointed out) – up to the first ten-million barrels per day of oil equivalent – conservation and some low-technology forms of solar power are all cheaper than any conventional energy source, including coal and nuclear power.[4] Quite typically, the team was willing to look at all the *technological options* that might render coal or nuclear power more cost effective, but members were not willing to examine all the ethical, social, and political parameters, e.g., peak-hour prices, tax incentives, regulations, consumer education, that could make conservation or on-site solar more cost effective than conventional energy sources. For this reason, the purely technical formulation of the main assessment question gave the team a purely technical and, to that extent, incomplete and unsatisfying answer.

The same sorts of unsatisfying answers appear in many environmental-impact analyses and technology assessments, and it can be shown that their incompleteness stems from a common methodological assumption, the fallacy of unfinished business. In a recent OTA (Office of Technology Assessment) study of automobile technology, for example, the authors calculated existent and predicted rates of pollution caused by automobiles and analyzed the effectiveness of various pollution-control devices. Because they investigated largely *technological solutions* (e.g., better pollution-control equipment) to the environmental problems caused by automobiles, the assessors failed to consider adequately the role of social-political solutions, such as regulatory incentives for mass transit. As a consequence, they sanctioned continuation of the status quo: increasing use of private auto transport, with no future increase in employment of mass transit. By their own admission, the price paid for this conclusion will be that, by the year 2000, approximately half of all persons in the US will be regularly exposed to dangerous levels of toxic chemicals from automobile emissions.[5]

What is reprehensible about both the EPA energy study and the OTA auto assessment is not that they sanctioned hard rather than soft technology. Nor is it that they sanctioned the status quo rather than innovative solutions to public-policy problems. The big failing of both assessments is their not considering key options in the set of solutions to policy problems. In considering only technological solutions, both groups of assessors have no firm basis for believing either that their conclusions are methodologically sound or that they serve authentic human interests. Their commission of the fallacy of unfinished business has precluded their providing *any* general answer – whether in terms of hard or soft technology – to the policy problems they have addressed.

As the EPA energy study and the OTA auto assessment illustrate, even though assessment teams working under US government contract are specifically directed to examine alternative means of solving impact problems and attaining given goals,[6] many EIA's and TA's fail to evaluate nontechnological or non-status-quo options. This situation seems to present a classic case of what Boulding called suboptimization. He defined suboptimization as finding the best ways to do things which might better be left undone.[7] In assessing automobile technology, apparently the OTA authors fell victim to suboptimization. They studied the auto's impacts and found the best way to continue to use something, conventionally designed automobiles for private travel, even though this mode of transport may not itself be a truly optimal alternative. Likewise, in assessing energy technology, apparently policymakers found the best and the cheapest way to do something, use up nonrenewable resources and import fifty percent of this country's oil, even though such an extensive import policy should not have been followed at all. The policy weakened the dollar, discouraged conservation, and threatened political security.[8]

If these examples about "suboptimal" technology and environmental-impact assessments are representative, and I think they are, then they lead one to question the methodological and ethical assumptions undergirding many assessment results. (1) Do evaluators merely find a narrowly technical solution to a problem which is prematurely and perhaps erroneously defined purely as technical? Or (2) do they also generate a number of possible ethical, social, and political solutions whose complexity requires more than merely technical analyses? A review of representative, recent TA's and EIA's suggests to me that most studies fit the paradigm of (1) and not (2).[9] Because of the way most impact questions are formulated, it is rarely the case that key ethical, social, or political parameters are either recognized or analyzed. This is particularly true if solution scenarios in which they might appear are built on atypical presuppositions, such as that more technology is not necessarily better technology,[10] or that need for a technology or commodity is not accurately defined by demand for it.[11]

3. THE FALLACY OF UNFINISHED BUSINESS AND THE NATURALISTIC FALLACY

If, as has been suggested, committing the fallacy of unfinished business leads to suboptimal results, why do the authors of many TA's and EIA's often continue to assume that more of the status quo and more technology (whether contraceptives for India or broad-spectrum pesticides for Malaya), are all that is needed to cure our technological, environmental, and social malaise? Why do assessors often believe that 'real' answers are technical or scientific, but not ethical or social? Why has it been so hard to learn, for example, that India's population problem is not a consequence of a dearth of technology – contraceptives – but a result of a *social* situation in which parents have many children to insure that some will survive to care for them in old age? Why have assessors been so slow to learn that Malayan pest infestation has worsened in the face of chemically induced pest immunity, and that more chemical technology is unlikely to solve the problem? If it is so clear that providing more of the status-quo technology is not necessarily the answer to the problems addressed in TA and EIA, then why do persons subscribe to the fallacy of unfinished business?

Perhaps, in attempting to reduce complex ethical, social, and political problems to purely technological ones, assessors hope to avoid the morass of allegedly subjective normative judgments and to arrive at more objective, scientific ones. Perhaps they hope that their reduction (of ethical to technical factors) will help achieve closure on the problem at hand. If so, then it is likely that the fallacy of unfinished business arises for some of the same reasons as does one version of the famous naturalistic fallacy. A variant of the latter error occurs, according to Moore, whenever one attempts to give scientific reasons, alone, as a justification for ethical beliefs.[12] (For example, if one argued that imposing a certain technological or environmental risk on society were moral, solely because it had a low probability of causing catastrophe, then one would commit this error. The problem here would be the assumption that a purely scientific property, i.e., a low probability, constitutes a sufficient condition for judging a risk as ethically acceptable. Obviously, however, considerations such as distributive equity, rights, and duties, also play a necessary role in such judgments.[13]) As Moore pointed out, empirical, inductive, or statistical considerations represent only a part of what must be addressed in making ethical judgments.[14]

Insofar as they subscribe to the fallacy of unfinished business, a number of assessors appear to assume that empirical, statistical, or technical considerations are sufficient for making judgments about environmental and technological impacts. Admittedly, most assessors would not claim to be making ethical judgments by means of their technical considerations. In fact, most impact studies do avoid overtly normative statements. Nevertheless they fall into implicit normative assumptions by virtue of what they are willing to count as relevant to the problem at hand. For example, consider the EPA ORBES assessment mentioned earlier. Certainly the project members, who were directed to find the best way of meeting energy demand in the basin, would not claim to have made any ethical judgments about what ought to be done. Yet, by virtue of their own decision that coal and nuclear fission were the only viable options to be assessed, they did make an implicit evaluative judgment about whether society ought to consider alternatives such as on-site solar or nuclear fusion.[15] In this sense, assessors' subscribing to the fallacy of unfinished business does seem to have reduced their (implicitly ethical) impact studies to purely scientific considerations.

Although some variants of the naturalistic fallacy have been quite controversial – some have claimed that they are not fallacies at all – there is widespread agreement that reducing an ethical problem to a purely scientific one ought to be avoided.[16] This can be seen easily in the case of using such a reduction in TA or EIA. If one reduced environmental or technological problems to purely scientific ones, then a number of obviously undesirable consequences could be shown to follow. Perhaps most importantly, policymakers would have to equate the criteria for the ethical acceptability of an action, e.g., using broad-spectrum pesticides, with the criteria for assessing the purely scientific or technical impacts of that action, e.g., increasing crop yield. By definition, considerations such as (1) whether a given action is accepted voluntarily or involuntarily, (2) whether the costs and benefits of the action are distributed equitably, and (3) whether the action is 'worth it' would be excluded from consideration. Hence the attempt to reduce policy considerations to purely scientific analyses obviously ignores a number of relevant, and perhaps central, parameters.

One of the most important questions about environmental impacts is not whether to choose polluting technology *A* or polluting technology *B* as a *means* to attain some end, *C*, but whether to choose *C* or some other *end*. But, if assessments deal with purely technological questions, then of course they deal solely with choosing *means* to an assumed (usually status-quo) end, rather than with also evaluating alternative *ends*.

One desirable result of avoiding the fallacy of unfinished business is thus likely to be that once ethical, social, and political solutions to given problems are considered, then the *ends* of policy actions, and not just the *means* to them, will be assessed. In other words, environmental-impact analyses and technology assessments would likely be addressed not merely at the impacts of various technical *means* to some presupposed end (e.g., generating more electricity). Instead, they would also investigate the impacts of alternative *ends* (e.g., generating more/less/the same amounts of electricity) as well. Only when assessments and impact analyses make the consequences of the broadest scope of public choices apparent, can those choices be rationally evaluated. Otherwise TA's and EIA's merely beg the questions they are designed to answer.

4. ONE SOLUTION: BROADENING THE SCOPE OF TA AND EIA

Practically speaking, avoiding the ethical and methodological assumption known as the fallacy of unfinished business and learning to assess ends as well as means might be accomplished by broadening the scope of one of the criteria expressly used for technology assessment and environmental-impact analysis in the US. In the third of its eleven criteria, the OTA enjoins assessors to determine the costs and benefits of "various policy options regarding a given technology".[17] If this criterion could be interpreted, such that the concept of 'policy options' were understood in the widest possible sense, then environmental-impact analysis and technology-assessment methodology might move a long way in the right direction.

'Various policy options' could be defined so as to include alternatives that might be viable within different ethical, legislative, fiscal, regulatory, social, or political frameworks. With this new understanding of the options, policymakers would have before them alternatives which differed in ends as well as in means. They would have various ethical alternatives, some of which challenged, and others of which supported, existing assumptions about, for example, the necessity of growth, the environmental price of progress, and the value of particular pollution controls.

By redefining what we mean by 'policy options', and broadening it to include nontechnological solutions, we could expand the range of choices open to a policymaker confronted with a given problem. One such range of choices was exhibited recently by Ben-David, Schulze, and Kneese, who illustrated how different social, ethical, and political assumptions could be used to generate alternative environmental/technological policies, each of which could be said to be cost-effective depending on the assumptions used. They concluded, for example, that on the basis of Nietzschean, Bethamite, and Golden-Rule ethics, strict standards for automobile emissions are not cost-effective, but that they are cost-effective when considered in the light of Rawlsian ethics.[18]

Such a proliferation of policy alternatives is likely to increase both the freedom and the power of those who make public decisions, Freedom is, after all not only a function of the *number* of alternatives that one has, but also a function of the number of distinct and *different* options he has. This is why using various ethical, social, and political assumptions to weight the costs and benefits being assessed, as Kneese *et al*, have done, clearly expands the range of public choices in environmental and technological decisionmaking. (This weighting technique will be discussed in greater detail in Chapter Eight.)

5. OBJECTIONS TO BROADENING THE SCOPE OF TA AND EIA SO AS TO AVOID THE FALLACY OF UNFINISHED BUSINESS

To this proposal for expanding the concept of 'policy options', the experienced practitioner of TA and EIA has several clear responses. (1) Such an expansion would require every impact-analysis team to spend much more time and money examining policy alternatives having only a small probability of ever coming to be. (2) There would be no way to obtain 'hard data' on new, untested, or alternative technologies or policy options. And (3) predicting impacts on the basis of alternative ethical, social, or political, e.g., institutional and governmental, frameworks, is risky and problematic, and would provide only a questionable basis for policymaking.

While each of these objections raises a central and important point, a number of considerations suggest that they are not as devastating as might appear. Objection (1) is a particularly compelling one because it focuses on the practical impossibility of indefinitely expanding the list of policy alternatives. Although it is true that all possible alternatives could never be assessed, practically speaking, the force of this objection could be mitigated by two considerations. *First*, if a study only considers highly probable alternatives under the technological status quo, then the assessment itself wrongly encourages a self-fulfilling prophecy and implicitly sanctions 'business as usual', regardless of how praiseworthy or blameworthy it may be. Only by consideration of less probable, more complex, but perhaps more desirable, policy options, is it likely either that those desirable options will become reality, or that the question-begging character of many assessments will be avoided. *Secondly*, although a study realistically cannot assess all possible policy options, an achievable goal might be simply to evaluate, not a greater number of policy options, but a greater variety of alternatives.

In the EPA energy study discussed earlier, for example, the project members considered only four possible energy scenarios for the year 2000: 100 percent nuclear, 100 percent coal, 80 percent nuclear/20 percent coal, and 20 percent nuclear/80 percent coal, based on BOM (Bureau of Mines) statistics regarding energy demand.[19] Obviously the team could have narrowed the number of alternatives considered by eliminating some of these scenarios as unrealistic for the year 2000 (e.g., the 100 percent nuclear option). At the same time, they could have provided an evaluation of more kinds of options, e.g., one with twenty-percent, on-site solar, or one with twenty-percent cogeneration. They could also have helped to present a varied policy list by assessing some options based on Ford Foundation energy demand statistics, rather than BOM data, which project an extremely high energy demand. In other words, increasing the *variety* of policy options does not necessarily entail increasing the overall *number* of alternatives assessed.

Objection (2) also has force because it rightly points out that there is little hard data on new, untested, or alternative technologies and policies. This is the obvious problem with anything new. If one refrained from analyzing new policies or technologies because there was little hard data to assess, however, then more date would never be generated. Thus new policies and technologies would never be studied, and environmental policy, environmental goods, and technology all would stagnate. When carried to its logical consequences, objection (2) clearly leads to false and undesirable results.

It is also not apparent that there is in fact as little data on new options as is often alleged. In this regard, US assessment teams have much to learn from their colleagues abroad. Canadians, for example, have done outstanding work on social and political alternatives (legal, regulatory) for avoiding the negative impacts of rail accidents and railroad technology. Their studies have enabled them to cut their rail accidents, per train mile traveled, to half that of the US.[20] The varied policy alternatives represented in their technology assessments and environmental-impact analyses have much to teach US teams. Likewise China, which feeds its own people (and exports rice as well) has much to tell us about the policy option of nonchemical pest control, a low-technology alternative not widely practiced in this country and adequately considered by US assessment teams.[21] The Swedes, too, have had much success in cutting automobile pollution and traffic accidents by means of regulatory incentives for mass transit.[22] To my knowledge, their experiences have not been used to assess US policy options in the same area. In fact, the latest US auto assessment is recalcitrantly pro auto and pro status quo.[23] Finally, we have much to learn from the English, who appear to have done a commendable job of pricing the social costs (externalities) of some environmental impacts.[24] Because of this economic-assessment technique, they have provided an alternative policy option for responding to the problems posed by numerous technological developments. US teams, on the other hand, are only beginning to assess such pricing mechanisms as possible legal and non-technical solutions for holding down social costs.

Although the cultural, economic, social, and political differences among the US and countries such as Canada, China, Sweden, and England, are great, there might be some profitable data (admittedly assumption-laden and context-dependent) which could help US assessment teams broaden the list and types of policy options they typically consider. More probably, the widened scope of assessment methodology sometimes used in other countries could very likely be modified for use in the US.

Objection (3) focuses on the well-known difficulty that any forecasting is tricky business, but predicting consequences of policy options under alternative (perhaps nonexistent) regulatory, price, political, ethical, or social frameworks is doubly problematic because both the future and the effects of an untried framework are unknown. While this objection focuses on unarguably correct points, it sanctions a position which is both questionable in itself as well as inconsistent with current practice.

It is questionable in itself because its wholesale acceptance could lead to devastating consequences; it ultimately leads to the conservative and paralyzing position of sanctioning only the status quo and only allegedly known and understood policy options. If objection (3) were accepted, it would be difficult, if not impossible, for any sort of technological progress to occur.

Moreover, acceptance of objection (3) is obviously inconsistent with current practices and policies regarding technology. The impacts of many conventional technological policies are not fully known and understood prior to the policies' being adopted. Consider, for example, the extent to which the effects of a nuclear (fission) core melt are understood.[25] It is also false that policies are considered and adopted only after their long-term consequences have been, or can be, spelled out. Current US policy allows the annual generation of exponentially increasing amounts of nuclear waste, for example, and yet we have as yet to discover whether it can be safely stored in perpetuity.[26] US society is likewise now using nonrenewable natural resources at an exponential rate, while we have yet to determine what the effects of such actions will be. Clearly the claim, 'but I don't know the environmental impacts of the future effects of the technology', has not stopped numerous policies from being implemented. Hence, the unknowability of the consequences of alternative-technology scenarios does not appear, in itself, to be *a priori* grounds for failing to include an analysis of them in assessments.

Perhaps the controlling reason why this is the case is that the ability of a given scenario to yield good predictions is less important than the value the society or the assessors place on realization of the scenario in question. Assessment scenarios involving solar energy, for example, are often seen as incapable of yielding hard, predictive data because the technology has not been used as extensively as others. Yet, as a recent government study indicated, the impacts of solar technology are for *easier* to predict than those of conventional energy sources, since the latter are complicated by the possibility of cartels and other international political-legal maneuvers.[27] Many assessors, nevertheless, resolutely pursue predictions based on conventional energy sources, but describe solar predictions as too difficult to assess. What just might be going on here is an implicit value judgment about the relative merits of conventional versus nonconventional energy sources. Perhaps this judgment has far more to do with accepting the fallacy of unfinished business than with looking for hard data.

6. CONCLUSION

If these brief outlines of possible responses to objections are partially correct, then perhaps there are no compelling methodological reasons against expanding the concept of 'policy options'. If not, then such an expansion (via OTA assessment criteria) may well provide a basis for avoiding the fallacy of unfinished business. Perhaps then we will move closer to assessing environmental and technological assets and liabilities in a way that is less an *apologia* for the status quo and more a challenge for complete and non-question-begging policy analysis.

NOTES

[1] See J. M. Deutsch and the Interagency Review Group on Nuclear Waste Management, *Report to the President*, TID-2817, National Technical Information Service, Springfield, Va., October, 1978; hereafter cited as: IRG, *Report.*

[2] K. Keniston, 'Toward a More Human Society', in *Contemporary Moral Issues* (ed. H. K. Girvetz), Wadsworth, Belmont, Calif., 1974, pp. 401–402.

[3] See J. J. Stukel and B. R. Keenan, *Ohio River Basin Energy Study Report*, vol. 1, Research Grant R804848-01, US Environmental Protection Agency, Washington, D.C., 1977; hereafter cited as: Stukel and Keenan, ORBES. The author served on this TA team and has first-hand knowledge of its work.

[4] Robert Stobaugh and Daniel Yergin, 'Conclusion: Toward a Balanced Energy Program', in *Energy Future* (ed. by Stobaugh and Yergin), Random House, New York, 1979, p. 277. See also Yergin, 'Conservation: The Key Energy Source', in *Energy Future* (ed. by Stobaugh and Yergin), pp. 136–182

[5] US Congress, OTA, *Technology Assessment of Changes in the Future Use and Characteristics of the Automobile Transportation System*, 2 vols., US Government Printing Office, Washington, D.C., 1979. See vol. 1, p. 16; hereafter cited as OTA, *Auto*.

[6] See note 17.

[7] K. E. Boulding, 'Fun and Games with the Gross National Product: the Role of Misleading Indicators in Social Policy', in *Environment and Society* (ed. by R. T. Roelofs, J. N. Crowley, and D. L. Hardesty, Prentice-Hall, Englewood Cliffs, N.J., 1974, p. 136.

[8] Stobaugh and Yergin, 'The End of Easy Oil', in *Energy Future* (ed. Stobaugh and Yergin), pp. 4–11. See also Stobaugh and Yergin, 'Conclusion', p. 227 (note 3).

[9] In addition to the automobile and energy studies already cited, see, for exmaple, US Congress, OTA, *A Technology Assessment of Coal Slurry Pipelines*, US Government Printing Office, Washington, D.C., 1978, where ethical, legal, and social parameters involved in use of slurries are disregarded; hereafter cited as OTA, *Coal*. See also US Congress, OTA, *Policy Implications of the Computed Tomography (CT) Scanner*, US Government Printing Office, Washington, D.C., 1978, which neglected to consider alternative social, regulatory, legal, political, and ethical frameworks within which the scanners might best be used and misuse avoided. The same failure to evaluate alternative social, political, and ethical frameworks within which a technology is used, or within

which environmental impacts occur, appears in nearly all assessments. See, for example, (1) US Atomic Energy Commission, *Comparative Risk-Cost-Benefit Study of the Alternative Sources of Electrical Energy*, WASH-1224, US Government Printing Office, Washington, D.C., 1974; hereafter cited as AEC, *Risk*; (2) Congress, OTA, *An Evaluation of Railroad Safety*, US Government Printing Office, Washington, D.C., 1978; hereafter cited as OTA, *Railroad*; (3) Congress, OTA, *Pest Management Strategies*, vol. 3, US Government Printing Office, Washington, D.C., 1979; hereafter cited as OTA, *Pest*. The one significant exception to this rule is US Congress, OTA, *Application of Solar Technology To Today's Energy Needs*, vol. 1, US Government Printing Office, Washington, D.C., 1975; hereafter cited as OTA, *Solar*, in which alternative social, political, and ethical frameworks are considered.

[10] In this regard, see K. S. Shrader-Frechett, *Environmental Ethics*, Boxwood Press, Pacific Grove, Calif., 1981, pp. 154–194.

[11] *Ibid.*, p. 162.

[12] G. E. Moore, *Principia Ethica*, Cambridge University Press, Cambridge, 1951, p. 40.

[13] A discussion of this error, as applied to assessment of nuclear fission, may be found in Shrader-Frechette, *Nuclear Power and Public Policy*, D. Reidel, Boston, 1983, pp. 136–137.

[14] Moore, *Principia Ethica*, pp. 23–24; see also p. 36. Although Moore argues that ethical judgments ought not be reduced to purely scientific ones, he does not deny that causal or empirical propositions are a part of ethics. In this regard, see F. Snare, 'Three Skeptical Theses in Ethics', *American Philosophical Quarterly* 14 (2), (1977), 129–130.

[15] For discussion of this point, see Stukel and Keenan, ORBES, IV (1978), pp. 50 ff.

[16] For example, one of the forms the naturalistic fallacy has taken is the attempt to derive 'ought' statements from 'is' statements. This definition of the *fallacy* is controversial, in large part, because it appears to presuppose a fact-value distinction. In this regard, see J. R. Searle, 'How to Derive "Ought" from "Is" ', *Philosophical Review* 73 (1), (1964): 43–58. See also Moore, *Principia Ethica*, pp. 73, 108. A number of philosophers (e.g., Bruening, Frankena, White, as well as Snare), however, do not believe that the naturalistic fallacy is committed whenever one attempts to derive an 'ought' from an 'is'. See, for example, L. Kohlberg, 'From Is to Ought: How to Commit the Naturalistic Fallacy and Get Away with It . . .', in *Cognitive Development and Epistemology*, (ed. T. Mischel), Academic Press, New York, 1971, p. 154. Failing to consider the open question has also been considered by authors such as Kohlberg and Giarelli to be another variant of the naturalistic fallacy. See J. M. Giarelli, 'Lawrence Kohlberg and G. E. Moore', *Educational Theory* 26 (4), (1976), 350. This variant is also controversial, however, because not all philosophers are willing to challenge the analyticity of definitions of good.

[17] US Congress, OTA, *Annual Report to the Congress for 1978*, US Government Printing Office, Washington, D.C., 1978, p. 73.

[18] Shaul Ben-David (Department of Economics, University of New Mexico), Allen V. Kneese, Resources for the Future, Washington, D.C.), and William D. Schulze (Department of Economics, University of Wyoming), 'A Study of the Ethical Foundations of Benefit-Cost Analysis Techniques', unpublished research done under NSF-EVIST funding, working paper, August 1979, p. 130.

[19] Stukel and Keenan, *ORBES*, I (1977).

[20] See US Congress, OTA, *Railroad Safety – US–Canadian Comparison*, US Govern-

ment Printing Office, Washington, D.C., 1979, pp. vii–viii. See also OTA, *Railroad*, esp. pp. 14, 37, 141, 156.

21 Robert Van Den Bosch, *The Pesticide Conspiracy*, Doubleday, Garden City, 1978, pp. 147–151; see also pp. 152–178. The late Van Den Bosch, a Berkeley entomologist, spent his life doing research on nonchemical forms of pest control.

22 See Goran Backstrand and Lars Ingelstam, 'Should We Put Limits on Consumption?' *The Futurist* 11 (3), (1977), pp. 157–162.

23 See OTA, *Auto*.

24 See M. R. McDowell and D. F. Cooper, 'Control Methodology of the UK Road Traffic System', in *Environmental Assessment of Socioeconomic Systems* (ed. D. Burkhardt and W. Ittelson), Plenum, New York, 1978, pp. 279–280. Cf. OTA, *Auto*, Vol. 1, pp. 16, 21–25, 31.

25 Core-melt probabilities were computed in the only allegedly complete study of nuclear-reactor safety, WASH-1400, known as the Rasmussen Report. Released in 1975, this study concluded that fission reactors presented only a minimal health threat to the public. Early in 1979, however, under growing knowledge of core-melt hazards, the Nuclear Regulatory Commission withdrew its support from WASH-1400. See US NRC, *Reactor Safety Study*, WASH-1400, NUREG-75/014, US Government Printing Office, Washington, D.C., 1975. See also Shrader-Frechette, *Nuclear Power*, pp. 3–4.

26 In fact, the US government says it will not know the answer to this question until well beyond the year 2000, since the first test-model storage facility will not be built until then. See IRG, Report, p. xxxiii.

27 OTA, *Solar*, vol. 1, p. 11.

PART III

PARTICULAR METHODOLOGICAL PROBLEMS

CHAPTER FIVE

RCBA AND THE AGGREGATION ASSUMPTION

1. INTRODUCTION

Just as Galilei's and Copernicus' views "resulted in a totally new conception of the physical universe and man's place within it", so also technological innovation has posed unheard-of questions for the present age; "as Thomasso Campanella poignantly expressed it, 'If Galilei's conclusions are right . . . we shall have to philosophize in a new way' ".[1] Technology assessment (TA) and environmental-impact analysis (EIA), two of the new modes of analysis triggered by technological innovation, represent the efforts of natural and social scientists and humanists to come to grips with the implications of their disciplines for the guidance of public policy and human behavior.

Two of the important policy questions addressed in TA and EIA are whether given developments would, in fact, enhance societal welfare and whether the risks and costs they impose are outweighed by their benefits. One of the most common ways of answering these questions is by means of risk-cost-benefit analysis (RCBA), which has been called "the final test of public policy".[2]

2. THE AGGREGATION ASSUMPTION

Perhaps the most basic methodological assumption of RCBA is that societal welfare may be measured as the algebraic sum of compensating variations (CV's). I call this "the aggregation assumption". According to RCBA theory, a CV for a given individual is

> the sum of money which, if received or paid after the economic (or technological) change in question, would make the individual no better or worse off than before the change. If for example, the price of a loaf of bread falls by 10 cents, the CV is the maximum sum a man would pay in order to be allowed to buy bread at this lower price. *Per contra*, if the loaf rises by 10 cents the CV is the minimum sum the man must receive if he is to continue to feel as well off as he was before the rise in price.[3]

When the CV's of the gainers (a positive sum) are added to the CV's of the losers (a negative sum), and the resulting algebraic sum is positive, then the gainers can compensate the losers and the economic change causing the gains and losses may be said to realize the Pareto concept.[4] Generally speaking, the Pareto Optimum is that "position from which it is not possible, by any reallocation of factors, to make anyone better off without making at least one person worse off".[5] More specifically, *potential* (potential because gains *can* be, but may not be distributed in this way) is measured in principle as the algebraic sum of all CV's. In other words, the Pareto criterion is met when, according to this definition, there is an excess of benefits over costs, where benefits and costs are understood in terms of compensating variations.[6] "An amount of money calculated at the given set of prices will suffice to measure the CV".[7] Since the concept of a 'compensating variation' is central to the Pareto criterion, and the Pareto criterion is the basis of risk-cost-benefit analysis, which in turn is widely regarded as one of the most important evaluative tools in a technology assessment, I wish to provide a brief analysis of the aggregation assumption, the notion that societal welfare may be measured as the algebraic sum of compensating variations.

3. METHODOLOGICAL PRESUPPOSITIONS IMPLICIT IN THE AGGREGATION ASSUMPTION

What is implicit in the aggregation assumption? There appear to be at least three basic presuppositions packed into it: (1) that compensating variation is a measure (when CV's are summed to equal a (potential) Pareto Improvement) of how gains can be so distributed as to make everyone in the community better off;[8] (2) that the criterion for whether one is 'better off' is how 'well off' he feels subjectively;[9] and (3) that one's feelings of being well off or better off are measured by a "sum of money" judged by the individual and "calculated at the given set of prices" on the market.[10] It will be instructive to analyze briefly the concept of 'compensating variation' by unpacking the (often unrecognized) *assumptions* underlying, and the *consequences* following from, these three notions.

3.1. *The Aggregation Assumption and the First Presupposition*

First, what assumptions are made under (1), by virtue of the fact the CV sums are used to measure everyone's being 'better off' according to the Pareto Criterion? Most obviously, this notion presupposes that gains and losses, costs and benefits, for every individual in every situation can be computed numerically in terms of CV's. This is something notoriously difficult to do.[11] In a famous passage in his *Manuel*, Pareto himself pointed out that, for 700 commodities and 100 persons, his theory required solving not less than 70,699 equations.[12] Hence the concept of compensating variation appears applicable only *in principle*. And, as one economic theorist has pointed out, neither Pareto nor his followers have even proved that, in principle, a solution for a CV could be computed "in a concrete case of application".[13]

A second important assumption built into the first presupposition of the aggregation assumption is that it is acceptable to employ an economic change to improve the community welfare, even though distributional effects of this change are ignored. In other words, the assumption is that a satisfactory criterion for community welfare permits economic change resulting in universal improvement in well-being, even though one subset of persons receives a disproportionately greater share of benefits than does another subset. Mishan expresses this point by noting that the use of CV sums according to Pareto theory "ignores the resulting change in the distribution of incomes".[14] Suppose, for example, that an economic change made a given set of individuals better off by a total of $10x$ dollars, at the expense of another set of individuals made worse off by a total of x dollars. This change would produce an excess gain of $9x$ dollars for the community as a whole. Nevertheless, even redistributive measures directed at compensating the second set of individuals (by a total of at least x dollars) would result in disproportionate benefits for the two sets of persons, despite the fact that total gains would exceed losses. As the example suggests, a consequence of accepting this assumption (viz., the permissibility of ignoring distributional effects in CV calculations) is that it is likely that those in the community who are made worse off, or less well off, are to be found primarily among lower-income groups.[15] This is probably one of the reasons why Biderman cautioned that the use of quantitative economic criteria of well-being ought not to be assumed to lead to greater equity and even-handedness in policy. Rather, because it allows economists to eschew evaluation of distributional effects, employment of the CV concept is likely to "reflect the dominant ideological orientations of the most powerful and articulate groups affected by the phenomena measured".[16]

3.2. *The Aggregation Assumption and the Second Presupposition*

Another presupposition built into the aggregation assumption also suggests that it has a particular (and perhaps unrecognized) ideological or normative component. This is the idea (2), identified previously, that the CV criterion for whether one is 'better off' is how he feels subjectively, as measured in quantitative terms. (Of course, in the preceding example of an economic change causing one set of persons to be better off by a total of 10*x* dollars, at the expense of another set who were made worse off by a total of *x* dollars, a member of the second set might not feel as well off as a member of the first set, particularly if he knew what the relative gains and losses of the two sets were. This suggests that Pareto-induced distributional inequities probably increase in proportion to the ignorance of members of the second set regarding their relative gains and losses.) Obviously built into this criterion is the goal of maximizing individual well-being, of questing for 'more'. This, however, involves the assumption that welfare is defined in terms of egoistic hedonism. Arrow admits as much in discussing compensating variations and Pareto Optimality.[17] What is important about the assumption, I think, is that whether this egoistic and hedonistic notion of well-being is right or wrong, it is *normative* to the core, a fact apparently ignored or denied by many economists (e.g., Friedman, Mises) who claim their discipline is value-free and objective.[18] One author has gone so far as to assert: "my charge against most economists is that they are ready to exclude other [than CV-based or Pareto-based] normative viewpoints as unscientific, while permitting this one to crawl under the fence".[19]

Also implicit in this aspect of the CV concept is the assumption that the individual is the best judge of what will make him happy. In other words, individual costs and benefits are understood by reference to what people prefer, as measured by their compensating variations. The difficulty, however, is that persons often prefer things that don't actually increase their well-being (e.g., smoking, a particular marriage partner).[20] The discrepancy between what is preferred and what actually increases well-being is important, since it seems rational to maximize benefits and minimize costs only if they are truly connected with human welfare. If one's best interests are assumed to be identical with what he prefers, however, then at least four undesirable consequences follow. *First*, the *quality* of one's choices is ignored.[21] *Second*, there is no recognition of the classical Platonic-Aristotelian doctrine that there is a difference between what makes men good/secures justice or happiness and what fulfills their wants – in other words, between needs and wants.[22] *Third*, the question of a distinction between *utility* or personal welfare and *morality* or moral principle is begged.[23] And *fourth*, group welfare is assumed to be merely the aggregate of individual choices or preferences, as expressed by summing all compensating variations.

Obviously, however, public well-being is not simply the aggregate of individual preferences. An individual's personal welfare might be maximized, for example, if he alone made a choice to use a disproportionate share of natural resources; the public welfare obviously would not be optimized if it were conceived of as the aggregate of CV's in which everyone made this choice. Moreover, in a rapidly changing situation, where leaders must act on the basis of likely future events, public welfare clearly is not merely the aggregate of individual preferences. This is because present decisions about costs and benefits in part determine future values and preferences. If one presently ascribes certain low costs and high benefits to transportation by private automobile (such a technology assessment will be discussed later), for example, then this will affect more than merely the future weight given to the use and value of the private auto. Most importantly, such a calculation of costs and benefits will also affect future values and preferences regarding location of residences and businesses, air quality, and highway construction. Hence it is very difficult to determine which *present* public policy is likely to maximize *future* benefits, particularly if that *social* policy is defined as merely the aggregate of *individual* costs and benefits as measured by CV's. Moreover, to assume that such a definition is adequate is to ignore the fact that good leaders attempt to reform public opinion; this means that, at least in some instances, authentic public welfare and the aggregate of citizens' choices are not identical.[24]

Because of these four consequences, it is not clear that the concept of a CV actually measures costs and benefits, gains and losses, or authentic well-being. Economists have answered this charge, however. They have appealed to the notion of "rational preference *ordering*", relative to compensating variation, as a substitute for relying merely on an individual's feelings about his well-being. This approach, nevertheless, has epistemological drawbacks. How could one obtain such an ordering except through a questionnaire? And if one is unclear in the beginning about how to maximize his authentic welfare, then how could such an ordering or a questionnaire remove these difficulties?[25] This 'solution', to the conceptual problems suggested by the four consequences, seems to serve only to transfer the lack of clarity from the relationship between 'preference' and 'welfare' to the relationship between 'rational *ordering*' and '*welfare*'. Just as one's preferences are no infallible guide to maximizing one's well-being, so also what is 'rational' according to the economists' sense of CV is not necessarily 'rational' in the best, or even ordinary, sense of the term.[26] What seems to be going on here is that, because of the way the CV concept is defined, welfare economists are employing a stipulative definition of 'welfare' by virtue of their technical understanding of the term. Provided the definition is correct, there is not necessarily a problem. What appears likely, however, is that this technical *economic-science sense* of the term, 'welfare', will be forgotten and then exported to other contexts (e.g., used in a technology assessment) where it will be used, in an *ordinary-language sense*, as an attempt to clarify specific ethical issues. (More will be said later about this likelihood, when I examine specific technology assessments.)

A final problem with the assumption (explicit in the CV concept), that the individual is the best judge of his well-being, as measured in quantitative terms, is that it appears to lead to the consequences that wealthy and poor individuals are not both able to make an equally desirable judgment based on monetary criteria. Consider, for example, a case in which an individual determines the maximum sum of money (the CV) he would pay in order to be allowed to 'buy' open heart surgery after its cost has decreased by $200. Obviously, assuming he has no insurance coverage, a rich man would be able to pay a much higher CV than would a poor man. In fact, the point of this example is well substantiated. It has been shown statistically that, as income increases, one is willing to pay a much higher CV for environmental quality,[27] as well as for medical care, improved life expectancy, transportation-safety equipment, home repairs, and job safety.[28]

As these statistics suggest, there are obvious discriminatory effects of measuring CV on the basis of what a person "is willing to pay, or to receive, for the estimated change of risk".[29] Nevertheless, subjective determination of CV (either through the questionnaire approach or observation)[30] is touted by cost-benefit analysts as "the only economically justifiable" means of applying the CV concept.[31] Some theorists have suggested that this problem (of CV computation leading to discrimination on the basis of wealth) could be solved by using an average value, computed from the CV's of both wealthy and poor individuals. This solution, however, is also arguably discriminatory. Some (probably wealthy) persons could claim that if CV's (using average values) were aggregated, this aggregation would not represent the real economic worth of individuals in their situation. Obviously a General-Motors executive, making $300,000 per year, could argue that if his CV (for a particular means of improving life expectancy) were lowered through the averaging process, then the lowered value would not represent an accurate economic compensation for him, given his higher earning power. This is probably why nearly all cost-benefit theorists insist on the notion of subjective (rather than average) determination of the amount of an individual's compensating variation.

Such a subjective interpretation of the CV concept, however, has several problems, in addition to those mentioned previously. For one thing, the consequences of discrimination on the basis of wealth might be contrary to

Rawls' notion of justice as fairness and, in fact, contrary to the egalitarian ethical framework we have come to regard as part of democratic institutions.[32] *Second*, the consequences of grounding the CV on a means of calculation resulting in discrimination is that the cost-benefit model of economic transactions does not maximize freedom as much as has been alleged by market proponents. Typically, the subjective determination of one's own price for a CV is what he would pay or be willing to take as compensation on the market, in exchange for some economic benefit or risk. And theoretically, the market is supposed to maximize freedom, since no external source controls exchanges or transactions.[33] Yet obviously a man is not free to pay a particular price, a CV, in order to obtain some benefit, e.g., reduced health risk, if his economic opportunities or sources of income are of lesser magnitude than those of another man. This means that, in affirming the adequacy of computing costs and benefits according to subjectively measured CVs, one is likewise affirming the adequacy of aggregating and comparing costs and benefits which were determined according to quite *different criteria*. That is, even though the quantities (CV's) will be aggregated via the Pareto theory, (which presupposes their being of a certain type or kind), each of them was obtained under quite different conditions; the conditions were based on the varying degrees of financial constraint operative on an individual's ability to pay a given amount for a certain good.

Employment of the subjectively determined CV points up, not merely problems with discrimination and lack of equal economic opportunity, but also difficulties with one of the classical simplifying assumptions employed in econometric and technological modeling, viz., aggregation. The problem with this assumption generally, and in this instance, is that it suggests wrongly that given data (e.g., CV's) are homogenous when in fact they are not. And, to the extent that they are not, conclusions regarding aggregates may be either misleading or false.[34] For all these reasons, it would be well for economists to discuss the degree of error possibly arising from using subjective valuations of compensating variation.

3.3. *The Aggregation Assumption and the Third Presupposition*

Another presupposition (mentioned previously) built into the concept of compensating variation is (3) that one's feelings of being well off or better off are measured by a "sum of money" judged by the individual and "calculated at the given set of prices" on the market.[35] What assumptions are made under (3), and what do they reveal about the aggregation assumption?

Probably the most basic presupposition implicit in (3) is that market prices may be used in computing CV's, and hence that market prices are measures of the value of goods. This is a significant assumption in that, until the eighteenth and nineteenth centuries, the distinction between *value* and *price* was clearly made. Aristotle, for example, distinguished between the "fair price" (value) and the market price. In the last century, however, the distinction has been abandoned, and economics has moved from a normative to a positive (in Comte's sense) emphasis. As one theorist put it, to the extent that there is a valid distinction between the price and the value of a thing, to that degree contemporary economics "could well use a modern Aristotle".[36]

Oscar Wilde is perhaps one of the great popularizers of a distinction between price and value, since his famous remark is often quoted; he accused the cynic of knowing the price of everything and the value of nothing. This suggests, of course, as Hobson was careful to point out, that value 'includes' something 'price' does not. Price takes account neither of "the intrinsic service [a thing] . . . is capable of yielding by its right use", nor of (what Seligman called) the ability to satisfy desire, not the "capacity for satisfying wholesome human wants". Obviously, then, Hobson and others believed that price was a function of the intensity of human wants, but not necessarily a function of the intensity of correct or desirable human wants. But if price is not based on authentically good wants, then price might not always be based on value, but on what Hobson calls 'illth'.[37]

Boulding expressed Hobson's point in a more specific way. The problem with saying price = value, he says, is that the concept of value (apart from price) always includes a 'residue', an 'integrative system' involving things like status, love, honor, community, and identity. Without these, says Boulding, no economic prices of any kind could develop, because it would presuppose trust and credibility. In distinguishing 'economic values' and 'human values' in this way, Boulding argues that the existence of human values is a necessary condition for the existence of economic values, which are much more restricted than the former. He terms the broader, integrative system (which is not based on price) the "grants economy". In this economy, unlike the market one, *quids* are given without any *quos*. Hence, Boulding would say, not only is "market price" or "economic value" too narrow to be identified with "human value", but also, economic value or price is far less important in the ultimate scheme of things. Obviously, for example, the value of goods is not *causally* determined by economic exchange, as Anderson noted, any more than the amount of water in a vessel is causally determined by one's measuring it; this suggests that using price as a measure of value gives price too great an ontological status.[38]

Apart from whether Hobson, Anderson, and Boulding are philosophically correct, however, there are some clear reasons, economically speaking, why market prices diverge from authentic values. Several of these are the following: (a) the distorting effects of monopoly;[39] (b) the failure to compute effects of externalities;[40] (c) speculative instabilities in the market;[41] and (d) the absence of monetary-term values for natural resources and for "free goods" (e.g., air) or "public goods", even though these items obviously have great utility to those who use them.[42] All these ill effects, visited on market-based economic statistics such as prices, are widely known and discussed among both classical and non-classical economists. What may not be recognized, however, is that whenever (and this is usual) market prices are used to determine CV's, then use of the concept of compensating variation automatically entails a *normative bias* in favor of the status quo. This is because, whenever one attempts to bring the economy toward an optimum (usually defined as the sum of CV's, according to Pareto theory) by using the existing set of prices, these prices themselves are a function of "the existing income distribution".[43] Hence, to the extent that application of the CV concept includes employment of market prices, it also involves use of the existing income distribution as a necessary condition for optimizing welfare. But if the existing income distribution is a necessary condition for optimizing welfare, then to the degree that this distribution is neither egalitarian nor socially just, then to a similar extent will the optimization achieved through calculation of CV's be neither egalitarian nor socially just. In other words, to the degree that the criteria (market price) for optimization of welfare via CV's are biased toward the *status quo* or are normative, then to a similar extent will the CV's themselves be biased. If this reasoning is correct, then perhaps the market and its associated mechanisms (e.g., price) do not promote social justice as much as has been thought by some classical economists.[44]

The discrepancy between market price and value, or market price and fair price, appears to illustrate a common difficulty with economic statistics. In fact, it has been called one of the three main sources of "false representation" of economic concepts. The problem is that the economist-observer selects what to include (e.g., price) in a particular concept, since complex phenomena (e.g., value) are never exhaustively describable.[45] This necessary selection process then results in a situation in which one applies economic concepts (e.g., 'price', 'compensating variation') having 'specious accuracy'. They are given this name because, despite their accuracy, they are so narrow as to be useless for many practical applications. Since the definition of the concept (e.g., 'price') is irrelevant to the *purpose* for which it was introduced (e.g., for calculating 'value'), the concept is one of 'specious accuracy'.[46] This is why economists have remarked that "only if the goals for which the evidence is sought are known can the right indicators be found".[47] Otherwise, particular econometric concepts are too wide or (as in this case with 'price' and 'value') too narrow of the mark. Without specification of the goals, for whose realization the concept is expected to be useful, the concept is not likely to measure what it purports.

Of course, the problem with assuming that price = value when computing CV's is not merely that price purports to measure value, but does not. Rather, it is that we are likely to think that price is important simply because we can measure it. As one theorist observed, "the aspects that we are best able to observe and comprehend seem to be those that become important".[48] In other words, our penchant for measurement and alleged objectivity leads us into the further epistemological error of treating a surrogate concept (e.g., price) as though it were identical with the concept (e.g., value) we wish to represent. In our dalliance with the surrogate, we forget the inferences we have made and we create a false aura of 'hardness' about our data.[49]

Logically speaking, however, perhaps the most serious problem (with employing 'price' to measure 'value' in calculating compensating variations) is that the most critical questions are simply begged. If human values are defined, via RCBA, in terms merely of choice (as measured quantitatively by price), then to claim that the uncontrolled market maximizes human welfare (as measured in summed CV's) is to state a tautology. This judgment about the way risks, costs, and benefits should be allocated is allegedly rationally evaluated by criteria that antecedently presuppose acceptance of the market system. Therefore use of the CV concept and Pareto Optimum beg the very questions they are designed to answer. What is required is to *test* the thesis that market allocations constitute maximum welfare, or social justice, and not merely to assume it.[50]

4. THEORETICAL STATUS OF THE AGGREGATION ASSUMPTION

If most of these points – about the assumptions built into, and the consequences following from, the three presuppositions included in the aggregation assumption – are reasonably correct, then several conclusions may be drawn regarding RCBA methodology. These conclusions can perhaps best be stated in terms of the extent to which the aggregation assumption provides a scientifically sound means for determining how to maximize human welfare according to Pareto economics. With respect to the *validity* of the aggregation assumption for this purpose, it is clear that it might help to provide a measure of welfare, but only of welfare as defined within a cooperative, egoistic, hedonistic framework. Likewise a CV might be a valid measure of an individual's well-being, so long as the individual is in fact the best judge of what will make him happy, i.e., so long as his rational preference orderings are indicators of his authentic welfare. The validity of CV as a measure of a person's well-being is undercut, however, by the fact that the sum of money associated with the CV is a function of the individual's wealth and of the economic constraints under which he must live, rather than only a measure of the value he would like to put on his welfare.

So far as *coverage* is concerned, the aggregation assumption is somewhat adequate in providing an overall estimate of societal well-being, but inadequate inasmuch as it takes no account of income distribution and inequities in the degree to which all persons are made better off. The *comprehensibility* of the assumption is also limited, because market prices are ordinarily used to measure CV (the value of a particular change in well-being), and these prices do not take account of, for example, spillover effects, authentic costs of natural resources and public goods, and distorting effects of market instabilities and monopolies. The comprehensibility of the aggregation assumption is also questionable to the degree that CV's are based on market prices which might not reflect the whole range of phenomena referred to as 'value'. The 'fit' between price and value has been assumed rather than established and, because of this fact, parameters (like CV) based on price cannot unequivocally be said to be measures of value or well-being. The *experimental utility* of the aggregation assumption is perhaps its weakest point. Not only is it difficult to compute CV's for nonmarket items, but also the summing of CV's (according to Pareto theory) is in practice impossible because of its complexity.

The constraints on the validity, coverage, comprehensibility, and experimental utility of the aggregation assumption point to a number of its epistemological properties. Most obviously, the assumption cannot be said to be "value-free" or "objective", a claim which Friedman and others have made on behalf of the whole of economics.[51] If this is the case, then it would be well to follow the suggestion of persons such as Marc Tool and Gunnar Myrdal, who argue that *explicitly* stated value premisses and error estimates be included in economic inquiry.[52] Besides, to the extent that problematic value premises (e.g., value = market price) are *implicit* in the aggregation assumption, then to that degree the widespread belief, that "there is consensus on what should be observed" and measured by economists,[53] is probably false.

5. APPLICATIONS OF THE AGGREGATION ASSUMPTION TO TECHNOLOGY ASSESSMENT AND ENVIRONMENTAL ANALYSIS

Although the difficulties with the aggregation assumption which I have just outlined have no clear theoretical resolution, it will be instructive to determine how the methodological analysis of the preceding sections (3–4) may be applied. This application can contribute, both to understanding the strengths, and to ameliorating the weaknesses of the RCBA methodology often uncritically employed in technology assessments and environmental-impact analyses.

When one considers a representative group of environmental-impact analyses and technology assessments, with regard to how they employ the aggregation assumption, they illustrate clearly the presuppositions implicit in this assumption and the consequences following from its use. Typically, the assessments: (1) ignore the distributional effects of costs and benefits; (2) use subjective feelings as criteria for assessing well-being; (3) assume that the individual is the best judge both of his and of the public welfare; and (4) build implicit acceptance of the market system into the calculations. As a consequence of (1)–(4), these assessments fall victim to a number of questionable policy moves. I will argue that, insofar as these assessments (1) ignore distributive effects, for example, they advocate policy sanctioning violations of minority rights in order to serve the majority and they neglect the negative effects of technological decisions on public policy and foreign affairs. Insofar as they (2) use subjective feelings as criteria for well-being, I maintain that these assessments support conclusions in which actual societal well-being often is not maximized and confuse needs for a given technology with demands for it. Likewise, I show that insofar as assessors (3) assume that the individual is the best judge both of his and the public welfare, the assessment is likely to allow discrimination against the poor and to place inconsistent valuations on the same items (e.g., human life) in similar situations. Finally, I argue that, insofar as assessors (4) accept the market system, they are likely to: calculate 'value' in terms of 'market price'; fail to include assessment of the social costs of technology; support assessment conclusions based on incomplete analysis of all costs and benefits; misrepresent the relative costs and benefits of various, often competing, technologies; and allow an apparent assessment bias in favor of technology, industry, and maintaining the status quo. Let us examine several key technology assessments in order to understand *how* the aggregation assumption has been employed and *why* its use has led to the consequences just noted.

5.1. *Consequences of Ignoring Distributive Impacts*

As a first example, consider the technology of coal-slurry pipelines, evaluated in a recent study by the US Office of Technology Assessment. This report was described by its authors as an analysis of "the costs and potential economic, social, and environmental impacts of coal-slurry pipelines".[54] Although the authors did not say so explicitly, they apparently employed the standard RCBA concept of Pareto Optimality, obtained by summing CV's, in order to arrive at their conclusion that "slurry pipelines can, according to this analysis, transport coal more economically than other modes [of transport, e.g., railway]".[55] The methodologically puzzling aspect of this conclusion, however, is that allegedly all costs and benefits of the proposed technology were considered, including "potential economic, social, and environmental impacts", yet at least one key social cost was apparently not calculated. This is the varying distributional effect of the cost burden of the pipelines. Clearly one set of persons (those in western US) would be negatively affected by slurry-pipeline use of scarce water resources, while a quite different set of people (those in midwestern and eastern US) would be positively affected by allegedly economical means of receiving much-needed western coal. In other words, those who would bear some of the greatest costs of this technology are a different group from those who would receive some of its greatest benefits. The calculated costs and benefits in the technology assessment, however, do not include the adverse effects of distributional inequities. The authors of the analysis merely assert: "Constitutional power is adequate to do that ["to make water available for use in a coal-slurry"], whether the source of the power is the inability of a state to thwart Federal policy . . . or the power of Congress"; the Federal government does not have to respect private rights regarding "allocation of navigable waters".[56]

Failure to include costs associated with disparate distributional effects in the study of coal-slurry technology is not surprising. As was pointed out in section three, economists do not define the aggregation assumption so as to include distributional effects. A number of the consequences of this conceptual omission were discussed in section four. As applied in this particular study, these consequences suggest that the assessment of coal-slurry pipelines is not as value-free as might be thought. For one thing, the omission of distributional effects associated with the technology might sanction employment of a utilitarian, rather than egalitarian, framework for calculating and evaluating costs and benefits.[57] Sanctioning distributive inequities, in turn, means that the normative bias of the assessment is toward possible violations of minority rights in order to serve the alleged good either of industry or of the majority.[58] (See Chapters Two and Eight of this volume for a longer discussion of the extent to which the theory underlying RCBA is utilitarian.)

Because of the pervasive use of the aggregation assumption in RCBA, it is not surprising that distributional effects are almost always ignored in technology assessments. Evaluations of pesticides, for example, typically fail to note the inequitable distribution of health costs when the chemicals are used. Women, children, those with allergies, sedentary persons, and farm workers, for example, all bear a higher cost than do other members of the population, as a result of employment of these toxic substances. So far as I know, however, no government or industrial technology assessment of pesticides treats these distributive effects.[59]

A similar situation occurs in the case of the allegedly most complete study of commercial nuclear technology (for generating electricity), the Rasmussen Report, known as WASH-1400. Here the costs of nuclear-generated electricity are computed and evaluated, but no reference is made to distributional effects. Persons living near a reactor, enrichment plant, waste facility, uranium mining camp, or reprocessing plant, for example, all bear a higher health cost from radioactively induced incidences of carcinogenic, mutagenic, and teratogenic death or injury. Infants, children, and members of future generations are also members of the class of persons likely to bear higher costs as a result of this technology. Yet, as in the case of the pesticide and coal-slurry assessments, distributional inequities resulting from use of nuclear technology are excluded from consideration in the cost-benefit framework.[60] If some critics of economic methodology are correct, however, then calculation of costs and benefits also ought to include consideration of the equity of distribution of the hazards, wealth, and opportunities resulting from employment of a given technology.[61]

Needless to say, the failure explicitly to include costs of distributional inequities in technology assessments can have wide-reaching, often unsuspected, implications for both public policy and foreign affairs. The authors of the recent Harvard Business School study of energy technologies point up one interesting consequence of US failure to calculate the costs of distributional inequities arising from US use of one-third of all oil consumed daily in the world. They note that inadvertence to this distributional problem places considerable pressure on the international oil market. Yet, such pressure is not in the best interests of the US, since it wants Japan and Europe to give up the breeder reactor. Other countries clearly will be unable to do so, however, so long as the US constricts the oil market, and so long as the foreign-policy consequences of their actions are not calculated as part of the cost of employing petroleum-based technologies.[62]

5.2. *Consequences of Using Subjective Feelings as Criteria*

Besides failure to include costs of distributional inequities, other aspects of the aggregation assumption also play a key role in how costs and benefits are calculated in technology assessments. One key notion built into the concept is that the criterion of being "better off" is how one feels subjectively. The assumption here is that the individual is the best judge of what policies will maximize benefits to him and minimize costs. Such a presupposition, however, does not always lead to actual optimization of individual well-being, as was noted in section three. To illustrate this point, consider the recent assessment of private automobile technology completed by the US Office of Technology Assessment. In the study, the authors note the costs of air pollution, and point out that, given the continuation of current trends (which they implicitly sanction), half of the US population will be exposed to extremely hazardous levels of air pollution in the year 2000 as a result of employment of private automobile technology.[63] They also point out the rising cost of petroleum, highway congestion, owning and operating a car, and accidents.[64] Next the costs and benefits are calculated, presumably as a sum of CV's, in order to determine what policy will maximize well-being. The goal here is to assess the feasibility of continued employment of private-transport technology, as opposed, for example, to increased reliance on mass transit. When this assessment is done, a surprising result occurs. The authors, in the name of the US citizens, employ the notion (built into CV) that each *individual's* feelings are to be taken as the criterion of what actually maximizes his well-being. They claim implicitly that, on this basis, continued use of private auto transport is more desirable than increased employment of mass transit because of the high cost persons place on more use of public transportation. Although no statistical studies were done, they maintain: "Americans have come to regard personal mobility as an inalienable right, and the automobile is viewed as the principal means to achieve this end".[65]

Because of the very information cited in the technology assessment, however, there is strong reason to believe that this subjective estimate of value, allegedly placed by individuals on personal, private-auto mobility, might not be accurate. It might not outweigh the *costs*, in terms of pollution, resource depletion, congestion, accidents, etc. Nevertheless, so long as each individual's *feelings* are taken as the criterion for the value placed on certain costs and benefits, then the quality of preferences is ignored, and authentic well-being cannot be distinguished from the satisfaction of wants.[66] These and other consequences, of employing the concept of CV to determine costs and benefits, were discussed earlier, in section three. To the extent that assessors of technology are not aware of these conceptual difficulties with calculating societal well-being in terms of individual feelings, however, to a similar degree will technology assessments fail to reflect the policy choices needed to maximize *authentic* well-being. This point seems obviously true in the case of automobile technology just cited.

Lest some democratically-minded person object, at this point, that the costs and benefits of technology must be determined on the basis of individual preferences, or paternalism will result, several facts should be noted. *First*, it is not my intent to argue that 'big brother' should set the price on the costs and benefits associated with private employment of autos. My point is that any technology assessment should note the methodological difficulties inherent in assuming (via the aggregation assumption) that preference is automatically a correct measure of well-being, and not just uncritically employ this notion built into standard RCBA methodology. *Second*, it is clear that there are some instances in which, problems of paternalism aside, individual preferences ought not naively be taken as indicators of authentic well-being. If a methodologically sensitive person were doing a technology assessment of microwave ovens or CT scanners,[67] for example, it could be argued that he would not be likely to employ individuals' subjective feelings about the costs of being subjected to unnecessary doses of ionizing radiation, eapecially if the individuals were ignorant of the hazards. Rather, the technology assessor would probably use the standard *BEIR* report as a basis for getting dose-response estimates of radiation-induced injury, and then would use these figures as a basis for computing the health costs associated with the injuries in question. Particularly in the case of technology assessments, it appears dangerous to use uncritically (without caveats) the CV notion that an individual's feelings are the best criterion for valuing costs affecting him, since many technology-related costs, e.g., pollution and its effects, are often not calculable in detail by the layman. Such caveats, however, are almost wholly absent in contemporary technology assessments.

5.3. *Consequences of Assuming That the Individual Can Best Judge His Welfare*

Another undesirable consequence, following from acceptance of the CV notion that the individual is best able to affix prices to the costs and benefits affecting him, also has great importance for the economic analyses in technology assessments. This consequence, discussed earlier in section three, is that wealthy and poor individuals do not enjoy the same freedom to price their own costs and benefits as they would like. As a result, when diverse individual CV's are summed, according to the aggregation assumption (for purposes of obtaining a Pareto Optimum), one can question whether they represent a homogenous parameter. Because many assessments fail to take account of the methodological difficulties arising as a consequence of employing the aggregation assumption, they are of limited value. Economists typically utilize subjective valuations of the worth of one's life, based on the occupation (e.g., coal mining) he is willing to accept, and then ignore the fact that these valuations are very likely a function of the economic constraints under which the individual must labor.

The subjective value placed on the life of a coal miner is an especially interesting case in point for two reasons. First, coal miners tend to put a lower dollar value on their lives than do members of other occupational groups.[68] And secondly, the prices affixed to the coal miners' lives are of critical importance, since these values are used in calculating relative costs and benefits of alternative technologies for generating electricity.[69] The discrepancies in valuation could easily affect the outcome of the comparative assessment. For example, if a high-technology mode of generating electricity (e.g., nuclear power) tended to employ more skilled workers, and if these more skilled employees lived under fewer economic constraints than those less skilled, then the skilled workers would probably place a higher price on their lives. As a consequence, costs of occupation-related deaths and injuries for this technology might be calculated to be far greater, for equal numbers of employees, than would similar costs for a low-technology means of generating electricity (e.g., coal). This suggests, not only that problems of equity (e.g., why should one's wealth determine the value of his life?) arise in comparative assessments of alternative technologies, but also that subjective notions of valuation (based on the aggregation assumption) could lead to false conclusions as to which technology in fact maximizes benefits and minimizes costs.

The same sorts of difficulties appear to arise in assessment of any technology or environmental impact in which large numbers of unskilled or economically constrained workers are employed. Occupational costs of pesticide toxicity, for example, might tend to be underestimated if CV's were typically determined by migrant agricultural workers' placing a value on their lives. Hence it would be well for assessors to take note of these methodological problems, in part because economic theorists themselves have often recognized them.[70] Also, those who apply economic methodology in environmental analyses and technology assessments might want to examine alternative approaches (other than the standard aggregation assumption) for expressing gains and losses. Mishan, for example, has proposed using social indicators and utility, not money, to express increments and decrements in well-being; these measures would not fall victim to the problem that, the higher the income group, the lower the marginal utility of a dollar gain or loss.[71]

5.4. *Consequences of Accepting the Market System*

As was probably evident from the discussion in section three, however, the most serious problem besetting the aggregation assumption appears to arise, not from subjective valuation of costs by individuals, but from employment of the market system in calculating costs and benefits. Built into the concept of compensating variation is the notion (3) that one's feeling of being 'better off' are measured by a sum of money determined by the individual and calculated according to market prices. On this interpretation, the aggregation assumption becomes enmeshed in all the problems associated with the thesis that market price = value. Particularly in the case of technology assessments, such a thesis can lead to factually erroneous conclusions and to bias in favor of the *status quo*. Let us examine some technology assessments to see why this is the case.

As has been frequently noted, although the US "has applied various assessment mechanisms", it has "relied primarily on the 'market' system for guiding and shaping the nature of new technological applications".[72] Market-based criteria for calculating costs and benefits are thus obvious in a number of contemporary technology assessments. In a recent technology assessment of railroads, for example, the authors note that the costs ascribed to various risks of the technology (derailments, for example) are judged as acceptable or not "through traditional marketplace operations".[73] This means that, although one of the five major "concerns" of the OTA study of railroad technology was "data collection",[74] no social costs, or externalities, of implementation of the technology were included in the cost-benefit calculations, since the market system excludes consideration of social costs.

The failure to calculate externalities (e.g., approximately 4,000 major evacuations per year because of accidents involving hazardous materials)[75] in computing the costs of railroad technology is a significant methodological problem for a number of reasons. For one thing, between 1.04 and 2.5 million carloads of hazardous materials are shipped annually, and roughly 65% of all tank cars loaded with liquefied petroleum gas, sulfuric acid, anhydrous ammonia, and liquid caustic soda are annually involved in the release of hazardous material.[76] Second, US railroads have approximately double the number of serious accidents, per train mile, as Canadian railroads; the US has developed no accident or risk data either to estimate social costs of these accidents or how to avoid them, while the Canadians have done both.[77] This means that US railroads appear to have a high accident rate, that the social costs of these accidents are not calculated as a basis for assessment of railroad technology, and that (on a US–Canadian comparison) it is possible that computation and analysis of social costs could positively affect railroad policy and rescue accidents. In the light of these facts, it is surprising that the most comprehensive and recent US government assessment of railroad technology neither calculates social costs, nor notes the probable effects of failing to do so, nor recommends that they be computed in the future, nor indicates that its authors recognize the methological difficulties with computing costs of the rail technology only on the basis of market considerations. Despite these methodological shortcomings, however, the report concludes that no new laws or regulations are needed to solve the problem of the high accident rate (which has doubled, per train mile, in the last ten years) of US railroads.[78]

What is questionable about this technology assessment is not only that its calculations appear to follow the aggregation assumption in failing to include social costs, but also that its authors seem not to see that this presents a great methodological difficulty with potentially serious effects on both public policy and human well-being. Most importantly, if the complete costs (including externalities) of employing railroad technology, according to existing US laws and regulations, are not *known*, how could the authors conclude that no new laws or regulations were needed to solve the accident problem? In ignoring externalities, and then in drawing a conclusion regarding minimizing costs, the assessors have fallen victim to some of the problems (discussed in section three) resulting from employing the methodological assumption that market price = real cost. As a consequence of this failing, it is likely that the actual costs of the rail technology will never be known, and hence that the railroads may not be operating cost-effectively. Moreover, the methodological assumption implicitly contributes to a bias toward the *status quo* (no new laws or regulations) and to a normative bias in favor of the railroad industry, as opposed to the public, who bear its social cost.

Much the same situation appears to have occurred in a recent OTA assessment of private-automobile technology. Certain social costs of the technology were mentioned (air pollution, noise, community disruption, death and injury, for example),[79] but calculation of the value of these externalities was neither made nor included in the overall cost-benefit assessment of the technology. In spite of the failure to include consideration of externalities, the authors of the report concluded both that extensive use of mass transit was not cost-effective, and that considerations of Americans' desire for personal mobility, via the auto, outweighed the costs incurred by such a massive system of private transport.[80] This suggests that, because the cost-benefit analysis presupposes the methodological validity of employing only market costs, (1) the true costs of the auto technology have not been calculated; (2) these costs cannot be evaluated against those of an alternative transportation technology; and (3) the assessment is biased in favor of the *status quo*, i.e., continuation of present trends in the use of private autos. The authors of this assessment would have done better, either to have included imperfect estimates of the value of these externalities within their considerations, or to have omitted them, but included careful statements as to the limits of their conclusions that were imposed by these omissions. Since they did neither, they appear to have fallen victim to the methodological and ethical problems (discussed in section three) associated with assuming that value = market price.

As in the evaluations of railroads and automobiles, the same serious consequences (of omitting consideration of external costs) occur in other technology assessments. Authors of one famous study of the costs and benefits of employing pesticides, for example, drew their conclusions on the basis of calculating only three (market) parameters: (1) the value of the average corn and soybean crop per pound of herbicide and insecticide; (2) the average cost per pound of herbicide and insecticide; and (3) the increased market price of corn and soybeans to consumers, as a result of reduced yield, if no chemicals had been used.[81] On the sole basis of benefit (1) and costs (2) and (3), the authors used the aggregation assumption to calculate the aggregate CV for the decrease in consumers' welfare, which they said would amount to $3.5 billion annually.[82] Obviously, however, carcinogenic, mutagenic, and teratogenic effects of pesticide use, as well as occupational costs and environmental damages, were ignored in the cost-benefit calculations.

Similar omissions of external costs occurred in a recent OTA assessment of coal-slurry pipelines. The report included an analysis, based on market parameters, for "the costs that pipelines will have to pay for water usage, such as pumping, transportation, and purchase price", and for the price of "the water's value in alternative uses".[83] Omitted in the calculations were the real costs of *any* use of the scarce western water supply, the price of resource depletion (especially for future water users), the benefits of water conservation, and the costs of water pollution through slurry use. Nevertheless, the authors of the technology assessment concluded that slurry transportation was a more cost-effective means of coal transport than was use of railways.[84]

One could go on, *ad infinitum*, citing technology assessments which adopt the prevailing methodological assumption that it is acceptable to omit external costs in the calculations of welfare economics. The previous illustrations ought to be sufficient, however, to exemplify the dangers associated with this practice and with application of the standard interpretation of the aggregation assumption. What is interesting in this regard is that, in technology assessments where externalities are included in the calculations, there appears to be less bias in favor of the *status quo*, fewer doubtful economic conclusions regarding costs and benefits, and less prejudice toward acceptance of the costs of the technology in question. Two excellent technology assessments illustrate this point well. In a recent OTA study of the technology of direct use of coal, the authors analyzed the social costs and benefits involved in Western coal development,[85] including aesthetic impacts.[86] After concluding their economic, technological, and social analysis, they drew a conclusion favorable to direct use of coal. Their conclusion, however, was couched in the context of warnings about the admitted uncertainties involved in "external costs, institutional and social constraints, and other nonmarket factors associated with coal use.[87] The authors cautioned that they were "not so clear" about the "validity of the . . . analysis" because of the methodological problems they encountered in estimating nonmarket costs such as externalities.[88]

Methodological awareness (of the problems associated with using only market criteria for assessing technology) also is evident in a recent OTA study of solar technology. In fact, the nonmarket considerations are evaluated so effectively that the authors explain how recognition of these parameters, and not simply market price, will be the major determinant of the costs and benefits of using solar technology in the future.[89] Some of the external benefits considered in this assessment include the labor intensiveness of various energy technologies, energy self-sufficiency, environmental impacts, and conservation of fossil fuels;[90] some of the externalities, on the other hand, are ill effects of onsite technology, regulatory barriers, and inadequate financial incentives.[91] As a consequence of consideration of all these social benefits and costs, one gets the impression that the economic methodology of this assessment does not force one into a bias that is pro-technology or pro status quo, as does the methodology in most other technology assessments.

Besides the tendency to omit external costs, another methodological consequence (mentioned in section three), of defining the aggregation assumption in terms of market values, is the failure to include consideration of the distorting effects of monopolies, subsidies, and market imperfections on the cost calculations. In one famous analysis (WASH-1224) of the relative costs and benefits of using coal, as compared to nuclear, technology for generation of electricity, for example, the effects of more than $100 billion in government subsidies of nuclear technology were ignored in the calculations. As a consequence of this omission, the authors of the report concluded that nuclear-generated electricity was more cost-effective than coal-generated power. If mere market prices had not been used as the basis of calculations, however, and if the nonmarket costs of government subsidies had been included in the prices of both means of generating electricity, then the opposite conclusion would have followed from the cost-benefit analysis. The new conclusion would be that coal-generated electricity is more cost-effective than nuclear-generated power.[92] As is obvious from this example, basing cost-benefit calculations on market prices alone can completely invalidate the conclusion drawn.

As well as threatening the validity of conclusions regarding the economics of various technologies, the aggregation assumption may also contribute towards a normative bias in policy formulation. Obviously, for example, if one uses market price as the measure of the cost of an item, and ignores government subsidies or regulations which lower that price, then he is not likely to have valid conclusions regarding the actual costs of that technology, as compared to less-subsidized or less-regulated technologies. Likewise subsidy or regulation, aimed at keeping the cost of using a technology high, is likely to result in lesser employment of the technology. On the other hand, subsidy or regulation, aimed at keeping the cost of using a technology low, is likely to result in greater employment of the technology. For all these reasons normative biases regarding future policy and practice are likely to result from ignoring the effects of subsidy or regulation.[93] Obviously, for example, more nuclear technology is likely to be used in the future if, relative to other technologies, the government subsidizes it more and these subsidies are ignored in cost-benefit computations.

The disastrous consequences of cost-benefit calculations which ignore implicit subsidies and favorable (to the consumer) price regulation are illustrated well by petroleum-based technologies. Stobaugh and Yergin, authors of the famous Harvard Business School assessment of various energy technologies, explain why this is so. They point out that, in early 1979, the average market price of US-produced oil, because it was kept down by government controls (keeping prices far below replacement costs), was $9 a barrel, while the world market price for oil delivered to the US was $15 per barrel. If all market distortions, externalities, subsidies, and effects of monopoly are taken into account, say Stobaugh and Yergin, the *real cost* of oil in early 1979 was approximately $35 per barrel.[94] Misleading econometric assumptions, such as the authenticity of market price, say Stobaugh and Yergin, have "helped to create some of the impasses and stalemates in US energy policy", because that policy has been based on methodologically deficient cost-benefit calculations.[95]

Some of the "impasses and stalemates" in US energy policy resulting from faulty economic-accounting methodology include: excessive promotion of the use of petroleum (between 1973 and 1979, US oil imports doubled);[96] a weakening of the dollar and increased likelihood of economic collapse;[97] and a failure to provide incentives for conservation and for use of solar technology.[98] Hence, for reasons of both *a priori* economic methodology (described in section three), and because of practical, social-political consequences of applications of this methodology, it makes sense for the aggregation assumption not to be based on summing market-determined CV's. This suggests that assessors would do well, either to analyze and to admit the methodological problems surrounding their use of the aggregation assumption in particular calculations, or to examine some of the alternatives to its employment. Regarding the inability of market prices to reflect accurately the costs associated with use of natural resources, for example, several economists have proposed a solution. They suggest that costs assigned to employment of nonrenewable resources be based on a social decision regarding a specific "depletion quota". Such a quota, they claim, could make conservation and recycling profitable in market terms.[99]

Once alternatives to market-based use of the aggregation assumption were considered, it would seem to be less likely that technology assessors fell victim to the epistemological and normative biases exemplified in this section. Epistemologically speaking, they would be more likely, both to avoid invalid conclusions based on incomplete cost data, and to note explicitly the cost-benefit *assumptions* on which their conclusions were contingent. They also probably would be better able to avoid begging the questions they were attempting to analyze, if their computations did not rest on prior acceptance of the aggregation assumption. Normatively speaking, once assessors of technology were more aware of the conceptual limits built into many of their accepted assumptions, they also might be more prone to avoid the (pro-technology, anti-conservation, pro-status quo) prejudice evident in some parts of the evaluations of the auto, rail, coal, nuclear, and pesticide technologies discussed here.

What the applications of all these conceptual points suggest is a point made earlier by John Kenneth Galbraith. As popular well-being improves, crude, market-based concepts become "progressively more inadequate as a basis for social judgment and as a guide to public policy".[100] If he is correct, then many of my criticisms of employment of the aggregation assumption, as a measure of welfare, are not necessarily evidence of a naive level of conceptualization in economics, but rather an illustration of the long way we have come in terms of social progress.

6. FUTURE DIRECTIONS AND THE AGGREGATION ASSUMPTION

If these reflections suggest one central proposition, it is that using risk-cost-benefit analysis, generally, and the aggregation assumption, specifically, unavoidably involves one in making value judgments. These judgments occur implicitly, by virtue of factors such as a hedonistic concept of compensating variation, as well as explicitly, as a result of the policy consequences following from methodological principles regarding market-based quantification of internal costs. That these normative judgments have been made in economics and in its applications to environmental analyses and technology assessments is significant, since "most economists have limited themselves to what they believed was a purely objective position". As Friedman and Mises put it: economics is value-free.[101]

If it is the case, however, that practitioners of RCBA cannot escape normative conclusions, then one of the potential values of this investigation is that it might help economists avoid, *not* the tendency to make such judgments, but rather the tendency to be *unaware* that they are doing so. Given an awareness of the normative presuppositions implicit in the aggregation assumption, several tasks face the researcher who wishes to mitigate the effects of these presuppositions. *First*, he should investigate how to evaluate distributional effects, either within or outside the scope of summing CV's in order to obtain a potential Pareto Improvement. *Second*, he should attempt to devise schemes for measuring welfare in which the maximum of all individual welfares does not obviously equal the maximum social welfare. As was suggested in section 5.3, using social indicators, rather than money, to express gradations in well-being might be one step in the right direction. *Third*, he should attempt to perfect frameworks for correcting market-price calculations used in RCBA. For example, employment of a 'depletion quota' might be one way to help cost natural resources more accurately.

In addition to these three *theoretical* tasks facing the practioner of RCBA, researchers face a *practical* task. This is to specify the constraints, on a particular RCBA conclusion, which have been imposed by virtue of traditional employment of the aggregation assumption. Such specifications might take the form of noting that, for example, in the OTA study cited earlier, *individual preferences* for personal mobility (via the automobile) do not necessarily represent the *social desirability* of this means of personal mobility. Such a note might constitute a warning against possible misuse of assessment conclusions regarding the automobile. In this way, at the level of applications, researchers have much to contribute regarding the methodological limits which the aggregation assumption imposes on their results.

7. SUMMARY AND CONCLUSIONS

If the preceding discussion is correct, then there are strong grounds for rejecting uncritical use of the aggregation assumption, the presupposition that summed CV's are a measure of welfare. More careful use of this assumption requires risk assessors and practitioners of RCBA to be aware of the normative biases implicit in the assumption and, if possible, to discover ways to mitigate these biases.

Specifically, researchers need to be aware that calculations made on the basis of this assumption: (1) ignore distributive effects; (2) implicitly define welfare in terms of egoistic hedonism; (3) specify what people prefer, not necessarily what is in their best interests; (4) aggregate CV's determined on the basis of different criteria; and (5) affirm market-based notions of risks, costs, and benefits. As a consequence of these five implicit presuppositions, use of the aggregation assumption is therefore likely to contribute to at least five questionable consequences: (1) acceptance of public policy sanctioning violations of minority rights in apportionment of technology-related risks, costs, and benefits; (2) support of assessment conclusions not maximizing *societal* welfare; (3) promotion of technological choices which meet citizens' demands rather than their needs; (4) assessment bias in favor of the preferences of those in higher income brackets; and (5) acceptance of public policy based on incomplete and incorrect representations of actual technological risks, costs, and benefits.

By careful attention to these presuppositions implicit in, and the consequences following from, the aggregation assumption, assessors may come closer to sorting out the inferences and values from the facts of risk analysis. This should lead, both to more analytic methodology and to better public policy. This sorting process should be made considerably easier by following the reforms outlined in Chapters Eight and Nine of this volume. In particular, the proposal to use ethically weighted versions of RCBA (see Chapter Eight) should help to take account of distributive effects of impacts; to define welfare in terms other than those of egoistic hedonism; to distinguish preferences from values; to specify the criteria for aggregating CV's; and to provide alternative notions of risks, costs, and benefits, in addition to those based on market considerations. If these goals can be accomplished, then we are likely to have both clearer and less question-begging RCBA methodology and better public policy.

NOTES

[1] Langdon Winner, *Autonomous Technology: Technics-out-of-Control as a Theme in Political Thought*, MIT Press, Cambridge, Massachusetts, 1977, p. 5. Hereafter cited as *AT*.

[2] J. K. Galbraith, *The New Industrial State*, Houghton Mifflin, Boston, 1967, p. 408; hereafter cited as *NIS*; and E. J. Mishan, *Welfare Economics*, Random House, New York, 1969, p. 5; hereafter cited as: *WE*.

[3] E. J. Mishan, *Cost-Benefit Analysis*, Praeger, New York; 1976, p. 391; hereafter cited as: *Cost-Benefit*. See M. W. Jones-Lee, *The Value of Life: An Economic Analysis*, University of Chicago Press, Chicago, 1976, pp. 6–14; hereafter cited as: *Value*. See also Mishan, *WE*, pp. 227–230

[4] Mishan, *Cost-Benefit*, p. 391; see also pp. 390–402.

[5] Mishan, *WE*, pp. 22–30.

[6] Jones-Lee, *Value*, p. 5.

[7] Mishan, *WE*, p. 113; see also pp. 107–113.

[8] Mishan, *Cost-Benefit*, p. 390; see also notes 42–45 in this chapter.

[9] Mishan, *Cost-Benefit*, p. 309, specifically affirms this aspect of compensating variation.

[10] See note 7 in this chapter.

[11] Jones-Lee, *Value*, p. 3 and Robert Coburn, 'Technology Assessment, Human Good, and Freedom', in K. E. Goodpaster and K. M. Sayre (eds.), *Ethics and the Problems of the 21st Century*, University of Notre Dame Press, Notre Dame, 1979, p. 109; hereafter cited as Coburn, 'TA', in Goodpaster and Sayre, *Ethics*.

[12] Oskar Morgenstern, *On the Accuracy of Economic Observations*, Princeton University Press, Princeton, N.J., 1963, pp. 100–101; hereafter cited as: *Accuracy*.

[13] Morgenstern, *Accuracy*, p. 101.

[14] Mishan, *Cost-Benefit*, p. 392.

[15] Mishan, *Cost-Benefit*, p. 393, makes a similar point about the poor.
[16] A. D. Biderman, 'Social Indicators and Goals', in R. A. Bauer (ed.) *Social Indicators*, MIT Press, Cambridge, 1966, pp. 131–132; hereafter cited as: Biderman, 'SI', in Bauer, *SI*. Jones-Lee, *Value*, p. 103, points out, however, that these distributional problems with CV sums ought not to be taken as evidence of "the moral poverty of cost-benefit analysis", since "a responsible decision-maker would normally be expected to consider distributional effects *together* with the results of a cost-benefit analysis".
[17] Cited in V. C. Walsh, 'Axiomatic Choice Theory and Values', in Sidney Hook (ed.), *Human Values and Economic Policy*, New York University Press, New York, 1967, p. 197 (hereafter cited as: Hook, *HV and EP*). One of the great classics in the last 25 years of debate on welfare economics is Dr. Arrow's *Social Choice and Individual Values.*
[18] See Milton Friedman, 'Value Judgements in Economics', in Hook, *HV and EP*, pp. 85–88 and E. C. Pasour, 'Benevolence and the Market', *Modern Age* 24 (2), (Spring 1980), 168–170; hereafter cited as: 'Market'. For criticisms of this claim, see Michael Freeden, 'Introduction', in J. A. Hobson, *Confessions of an Economic Heretic*, Harvester Press, Sussex, England, 1976, p. vi; hereafter cited as *Confessions*. See also Kenneth Boulding, *Economics as a Science*, McGraw-Hill, New York, 1970, p. 119; hereafter cited as: *EAAS.*
[19] S. S. Alexander, 'Human Values and Economists' Values', in Hook, *HV and EP*, p. 108.
[20] Coburn, 'TA' in Sayre and Goodpaster, *Ethics*, pp. 109–110, makes this same point.
[21] Gail Kennedy, 'Social Choice and Policy Formation', in Hook *HV and EP*, p. 142, makes a similar observation, as does John Ladd, 'The Use of Mechanical Models for the Solution of Ethical Problems', in Hook, *HV and EP*, pp. 167–168; hereafter cited as Ladd, 'Models'.
[22] This inadequacy is mentioned by M. A. Lutz and K. Lux, *The Challenge of Humanistic Economics*, Benjamin/Cummings, London, 1979, p. 4; hereafter cited as *Challenge.* They say it is characteristic, not only of the concepts of CV and Pareto Optimality in cost-benefit analysis, but also typical of 'conventional, mainstream economics' in general. John Ladd, 'Models', in Hook, *HV and EP*, p. 168, also holds the same position as Lutz and Lux.
[23] R. B. Brandt, 'Personal Values and the Justification of Institutions', in Hook, *HV and EP*, p. 37, and John Ladd, 'Models', in Hook, *HV and EP*, pp. 159, 166, also make this point.
[24] A similar observation is made by Gail Kennedy, 'Social Choice and Policy Formation', in Hook, *HV and EP*, p. 148.
[25] A related point is made by Coburn, 'TA', in Goodpaster and Sayre, *Ethics*, p. 111.
[26] R. B. Brandt, 'Personal Values and the Justification of Institutions', in Hook, *HV and EP*, p. 31, also makes this same observation; see p. 27.
[27] B. A. Emmett, *et al.*, 'The Distribution of Environmental Quality: Some Canadian Evidence', in D. F. Burkhardt and W. H. Ittelson (eds.), *Environmental Assessment of Socioeconomic Systems*, Plenum, New York, 1978, pp. 367–371, 374; hereafter cited as: Emmett, 'Distribution', in Burkhardt and Ittelson, *EA*.
[28] P. S. Albin, 'Economic Values and the Value of Human Life', in Hook, *HV and EP*, p. 97; Jones-Lee, *Value*, pp. 20–55.
[29] Mishan, *Cost-Benefit*, p. 318.
[30] Jones-Lee, *Value*, p. 39; see also p. 72. Mishan, *Cost-Benefit*, pp. 319–320.

[31] Mishan, *Cost-Benefit*, p. 318; see also 319–320. Jones-Lee, *Value*, p. 72.
[32] See M. C. Tool, *The Discretionary Economy: A Normative Theory of Political Economy*, Goodyear, Santa Monica, 1979, esp. pp. 208, 320–324, 334; hereafter cited as: *DE*. See also John Rawls, *A Theory of Justice*, Harvard University Press, Cambridge, 1973, pp. 14–15, 100–114, 342–350 (hereafter cited as *Justice*); and K. S. Shrader-Frechette, *Nuclear Power and Public Policy*, D. Reidel, Boston, 1980, pp. 31–35, 122–123.
[33] Pasour, 'Market', pp. 168–178.
[34] For discussion of the assumption of aggregation, see Sergio Koreisha and Robert Stobaugh, in 'Appendix: Limits to Models', in Robert Stobaugh and Daniel Yergin (eds.), *Energy Future: Report of the Energy Project at the Harvard Business School*, Random House, New York, 1979, pp. 238–265.
[35] See note 10. As Mishan, *Cost-Benefit*, pp. 391–392, puts it, there is a Pareto Improvement (measured as the algebraic sum of all CV's) "if men are better off *qua* consumers and producers .... Moreover the value of goods and 'bads' to men, either as consumers or producers (factor-owners) is not worked out from scratch. Market prices can be deemed to provide these values in the first instance, following which they can be corrected for 'market failure'.
[36] Adolf Lowe, 'The Normative Roots of Economic Values', in Hook, *HV and EP*, pp. 180; 171–173.
[37] Hobson, *Confessions*, pp. 39–40, and B. M. Anderson, *Social Value: A Study in Economic Theory Critical and Constructive*, A. M. Kelley, New York, 1966, pp. 26, 31, 162; hereafter cited as: *Social Value.*
[38] See K. E. Boulding, 'The Basis of Value Judgements in Economics', in Hook, *HV and EP*, pp. 67–69; hereafter cited as: 'Basis'. See also Anderson, *Social Value*, p. 24.
[39] See K. E. Boulding, 'Basis', in Hook, *HV and EP*, pp. 67–68. Morgenstern, *Accuracy*, p. 19, ties the "errors of economic statistics", such as price, in part to the fact of the prevalence of monopolies. In an economy characterized by monopoly, he says, statistics regarding price are not trustworthy because of "secret rebates granted to different customers". Moreover, he claims, "sales prices constitute some of the most closely guarded secrets in many businesses." For both these reasons it is likely not only that price ≠ value, but also that actual price ≠ official market price.
[40] See H. R. Bowen, Chairman, National Commission on Technology, Automation, and Economic Progress, *Applying Technology to Unmet Needs*, US Government Printing Office, Washington D.C., 1966, pp. v–138; hereafter cited as: *Applying Technology*. See also K. E. Boulding, 'Basis', in Hook *HV and EP*, pp. 67–68, and Mishan, *Cost-Benefit*, pp. 393–394. Externalities (also known as 'spillovers', 'diseconomies', or 'disamenities') are social benefits or costs (e.g., the cost of factory pollution to homeowners nearby) which are not taken account of either in the cost of the goods produced (e.g., by the factory) or by the factory owner. They are external to cost-benefit calculation, and hence do not enter the calculation of the market price. For this reason, says Mishan, (*The Costs of Economic Growth*, Praeger, New York, 1967, p. 53; hereafter cited as: *CEG*), "one can no longer take it for granted that the market price of a good is an index of its marginal price to society". Another way of making this same point (Mishan, *CEG*, p. 57) is to say that diseconomies cause social marginal costs of some goods to exceed their corresponding private marginal costs; this means that the social *value* of some goods is significantly less than the (private) market *price.*

[41] See K. E. Boulding, 'Basis', in Hook, *HV and EP*, pp. 67–68. See also E. F. Schumacher, *Small Is Beautiful: Economics as if People Mattered*, Harper, New York, 1973, pp. 38–49; hereafter cited as *Small.*

[42] There are no monetary-term values for natural resources because the 'cost' of using natural resources is measured in terms of low entropy and is subject to the limitations imposed by natural laws (e.g., the finite nature of non-renewable resources). For this reason, viz., the theoretical and physical limit to accessible resources, the price mechanism is unable to offset any shortages of land, energy, or materials. To assume otherwise is to commit the fallacy of "entropy bootlegging". (Nicholas Georgescu-Roegen, *Energy and Economic Myths: Institutional and Analytical Economic Essays*, Pergamon Press, New York, 1976, pp. xv, 10, 14–15; hereafter cited as: *EEM*. See also Shumacher, *Small*, pp. 41–49; Emmett, 'Distribution', in Burkhardt and Ittelson, *EA*, p. 363; and Bauer, *SOC*, p. 54.

[43] Mishan, *CEG*, p. 49; See also Tool, *DE*, pp. 280–285.

[44] Pasour, 'Market', pp. 168–178.

[45] Morgenstern, *Accuracy*, p. 26.

[46] This particular notion of 'specious accuracy' has been developed by Morgenstern, *Accuracy*, p. 62.

[47] A. D. Biderman, 'SI' in Bauer, *SI*, p. 111.

[48] Biderman, 'SI', in Bauer, *SI*, p. 97.

[49] This insight is discussed by R. A. Bauer, 'Detection and Anticipation of Impact: the Nature of the Task', in Bauer, *SI*, p. 46.

[50] M. C. Tool, *The Discretionary Economy*, Goodyear, Santa Monica, 1979, p. 334; hereafter cited as *DE*, p. 334, and Sidney Hook, 'Basic Values and Economic Policy', in Hook, *HV and EP*, p. 247, make these same points.

[51] See Pasour, 'Market', p. 168, and Milton Friedman, 'Value Judgments in Economics', in Hook, *HV and EP*, pp. 85–88.

[52] See Tool, *DE*, p. xvi. Because of these value premises and error components, it is increasingly clear that, to some degree, economic science "creates the world which it is investigating." (Boulding, *EAAS*, pp. 120–121.) One eminent expert in methodology believes that "economics has a long way to go before it will be ready" to admit the existence of these value premisses and error components. (Morgenstern, *Accuracy*, pp. 60–61; see also pp. vii and 7.)

[53] Bauer, *SOC*, p. 206.

[54] Congress of the US, Office of Technology Assessment, *A Technology Assessment of Coal Slurry Pipelines*, US Government Printing Office, Washington D.C., 1978, p. v; hereafter cited as: Congress, OTA, *Coal Slurry.*

[55] Congress, OTA, *Coal Slurry*, p. 15.

[56] Congress, OTA, *Coal Slurry*, pp. 131–132.

[57] For discussion of the utilitarian versus egalitarian frameworks for calculating and assessing costs and benefits of technology, see K. S. Shrader-Frechette, *Nuclear Power*, pp. 28, 31–34, 45–46, 61–62, 94–95, 99, 105–106, 149–150. See also J. S. Mill, *Utilitarianism, Liberty and Representative Government*, Dutton, New York, 1910, pp. 6–24; Jeremy Bentham, *The Utilitarians: An Introduction to the Principles and Morals of Legislation*, Doubleday, Garden City, New Yori, 1961, pp. 17–22; J. J. C. Smart, 'An Outline of a System of Utilitarian Ethics', in *Utilitarianism: For and Against*, (ed. J. J. C. Smart and B. Williams), Cambridge University Press, Cambridge, 1973,

pp. 3–74; Rawls, *Justice*, pp. 14–15; Charles Fried, *An Anatomy of Values*, Harvard University Press, Cambridge, 1978, 42–43; Charles Fried, *Right and Wrong*, Harvard University Press, Cambridge, 1978, pp. 116–117, 126–127; Alan Donagan, *The Theory of Morality*, University Press, Chicago, 1977, pp. 221–239; and Alasdair MacIntyre, 'Utilitarianism and Cost-Benefit Analysis: An Essay on the Relevance of Moral Philosophy to Bureaucratic Theory', in *Values in the Electric Power Industry*, (ed. K. M. Sayre), University of Notre Dame Press, Notre Dame, 1977, pp. 217–237.

58 See notes 57 and K. S. Shrader-Frechette, *Nuclear Power*, pp. 31–35.

59 For discussion of this point see K. S. Shrader-Frechette, *Environmental Ethics*, Boxwood Press, Pacific Grove, California, 1980. Chapter Eleven.

60 For an analysis of distributional inequities in cost-benefit studies of nuclear technology, see Shrader-Frechette, *Nuclear Power*, pp. 31–35, 61–62, 94–95, 149–150, 152–153.

61 Coburn, 'TA', in Goodpaster and Sayre, *Ethics*, pp. 115–116.

62 Daniel Yergin, 'Conservation: The Key Energy Source' (hereafter cited as: 'Conservation'), in Stobaugh and Yergin, *EF*, p. 137.

63 Congress, OTA, *Technology Assessment of Changes in the Future Use and Characteristics of the Automobile Transportation System*, 2 vols., US Government Printing Office, Washington D.C., 1979, Vol. 1, p. 16; hereafter cited as: *Auto I* or *Auto II.*

64 See Congress, OTA, *Auto I*, p. 31, for example.

65 Congress, OTA, *Auto II*, p. 25.

66 Lutz and Lux, *Challenge*, p. 4, note: "Conventional, mainstream economics has not been able to adequately deal with values because it has not seen economics in terms of needs . . . what *has* been its focus? The answer is wants".

67 See Congress, OTA, *Policy Implications of the Computed Tomography (CT) Scanner*, US Government Printing Office, Washington D.C., 1978; hereafter cited as *Scanner.*

68 See Jones-Lee, *Value*, p. 39, for a table on values ascribed to human life.

69 See, for example, US Atomic Energy Commission, *Comparative Risk-Cost-Benefit Study of Alternative Sources of Electrical Energy*, Report No. WASH-1224, US Government Printing Office, Washington D.C., 1974; hereafter cited as: AEC, WASH-1224.

70 Jones-Lee, *Value*, pp. 150 ff., for example, notes that the value of life, in cost-benefit analyses, is a function of the wealth of the individual whose life is being "priced".

71 Mishan, *Cost-Benefit*, pp. 403–406.

72 C. H. Mayo, 'The Management of Technology Assessment', in *Technology Assessment*, (ed. R. G. Kasper), Praeger, New York, p. 78; hereafter cited as *TA.*

73 Congress, OTA, *An Evaluation of Railroad Safety*, US Government Printing Office, Washington D.C., 1978, p. 37; hereafter cited as: *RR.*

74 Congress, OTA, *RR*, p. 156.

75 Congress, OTA, *RR*, pp. 14, 141.

76 Congress, OTA, *RR*, pp. 14, 141.

77 Congress, OTA, *Railroad Safety — US–Canadian Comparison*, US Government Printing Office, Washington D.C., 1979, pp. vii–viii; hereafter cited as: *RR US-C.*

78 Congress, OTA, *RR*, p. xi.

79 Congress, OTA, *Auto II*, pp. 75, 295, esp. p. 251; Congress, OTA, *Auto I*, pp. 15–16. According to Bowen, *Applying Technology*, p. V–138, the most frequently used cost for air pollution, in terms of damage to property (but *excluding* health costs, absence from work, etc, which "constitute the most significant economic loss of all"), is $65 per

capita per year. This is an annual cost of more than $12 billion in the US, and it still excludes the most important monetary losses.

[80] Congress, OTA, *Auto II*, p. 228.

[81] Congress, OTA, *Pest Management Strategies*, vol. 2, US Government Printing Office, Washington D.C., 1979, pp. 68–71, 79–81; hereafter cited as: *Pest M.S.*

[82] Congress, OTA, *Pest M.S.*, pp. 79–81.

[83] Congress, OTA, *Coal Slurry*, p. 84.

[84] See note 94.

[85] Congress, OTA, *The Direct Use of Coal*, US Government Printing Office, Washington D.C., 1979, pp. 316–324; hereafter cited as: *Direct Use.*

[86] Congress, OTA, *Direct Use*, pp. 323–324.

[87] Congress, OTA, *Direct Use*, p. 372.

[88] Congress, OTA, *Direct Use*, p. 372.

[89] Congress, OTA, *Application of Solar Technology to Today's Energy Needs*, Vol. 1, US Government Printing Office, Washington D.C., 1975, p. 3; hereafter cited as: *Solar I.*

[90] Congress, OTA, *Solar I*, pp. 12, 22–23.

[91] Congress, OTA, *Solar I*, p. 3.

[92] For a complete analysis of this example, including actual cost parameters involved, see K. S. Shrader-Frechette, *Nuclear Power*, pp. 108–134.

[93] D. C. North and R. L. Miller, *The Economics of Public Issues*, Harper and Row, New York, 1971, pp. 138–140; hereafter cited as: *Economic.*

[94] Stobaugh and Yergin, 'The End of Easy Oil' (hereafter cited as: 'End'), in Stobaugh and Yergin, *EF*, pp. 9, 11.

[95] Stobaugh and Yergin, 'End', in Stobaugh and Yergin, *EF*, p. 11.

[96] Stobaugh and Yergin, 'End', in Stobaugh and Yergin, *EF*, p. 4.

[97] Stobaugh and Yergin, 'End', in Stobaugh and Yergin, *EF*, p. 6.

[98] See note 134; Stobaugh and Yergin, 'Conclusion: Toward a Balanced Energy Program', in Stobaugh and Yergin, *EF*, p. 227; M. A. Maidique, 'Solar America', in Stobaugh and Yergin, *EF*, p. 211.

[99] Lutz and Lux, *Challenge*, pp. 305–307; see also pp. 297–304.

[100] J. K. Galbraith, *NIS*, p. 407.

[101] Lutz and Lux, *Challenge*, p. 3.

CHAPTER SIX

RCBA AND THE ASSUMPTION OF PARTIAL QUANTIFICATION

1. INTRODUCTION

Until this century, there was usually great harmony between the predominant *ethics* of the culture and the underlying assumptions of its writers on *economics*. Adam Smith mirrored the atmosphere of 18th-century egoism, for example, while Bentham epitomized the climate of 19th-century utilitarian reform. In this century, however, there have been few parallels between ethics and economics.[1] Perhaps this is because economics has in part been assimilated to 'positive science' in the Comptean sense, while most of the questions of ethics arise explicitly only in a nonpositivist framework.

The disjunction between ethics and economics has created especially grave difficulties in the area of technology assessment (TA) and environmental-impact analysis (EIA), as the three preceding chapters have illustrated. Assessors typically base their analyses primarily on engineering-cost estimates[2] which they erroneously describe as "value-neutral", "objective", or "atheoretical".[3] In reality, however, nearly all TA'a and EIA's have been dominated by key econometric methods, notably risk-cost-benefit analysis (RCBA),[4] whose ethical assumptions have been ignored, in large part because they were not recognized.[5]

Reaction to the widespread US requirement of basing policy on risk-cost-benefit analysis (RCBA) has been mixed. Critics have rejected this "trial by quantification",[6] while its proponents have warned that it provides virtually the only clear way of dealing with the problem of reducing risks to society.[7] Much of the controversy surrounding this method focuses on quantification, the major problem besetting RCBA methodology.[8] According to Friedman, quantification is at the heart of what he calls "the perennial criticism of 'orthodox' economic theory",[9] viz., that it assumes that all parameters are homogenous.

2. THE PROBLEM OF QUANTIFICATION

Basically, the controversy over quantification concerns the way that risks, costs, and benefits are represented. Proponents of complete quantification maintain that, whenever RCBA is used, all risks, costs, and benefits ought to be represented either by cardinal or ordinal measures. Their opponents claim that such measures ought not be used for all parameters. Regardless of how they believe that various risks, costs, and benefits ought to be represented, members of both camps maintain that the benefits of a particular project ought to exceed the risks and costs.[10]

Although economists attempt to quantify many parameters in assessing the risks, costs, and benefits of various projects, they generally oppose complete quantification and adhere to a methodological view which I call 'the assumption of partial quantification'.[11] This assumption is that no quantitative values ought to be placed on qualitative, nonmarket, or subjective risks, costs, and benefits, such as the aesthetic impacts of technology. Welfare economists and RCBA practitioners accept this principle because they claim to "recognize, at least implicitly, the intractability of quantifying [all risks, costs, and] benefits"; because nonmarket parameters cannot be "measured with honesty", says Mishan, most economists do not extend their quantification to include them.[12] Instead, he sanctions the dominant practice of providing "a physical description" of the nonmarket factors and of listing them separately from the quantitative variables.[13]

US courts have followed this majority view of economists and have consistently upheld the sufficiency of RCBA results which were part quantitative and part qualitative.[14] For example, "the courts unanimously agree that the 1969 NEPA does not require environmental impacts to be converted to monetary values .... The preferable view is to require quantification of each factor to the extent possible under existing methodologies".[15]

The obvious problem, however, is the degree to which quantification is possible. Humans have always been fascinated by numbers, and economists are no exception. Plato, for example, apparently believed that, regardless of their formation, all concepts were arithmomorphic, viz., all resembled number, as his friend Xenocrates taught. Although he was troubled by definition in one dialogue after another, he apparently believed that all concepts were capable of being defined. In the *Philebus*, he set out the task of the philosopher who would be great: "He who never looks for numbers in anything will not himself be looked for in the number of famous men".[16] Galileo's advice to astronomers and physicists echoes one of the same Platonic themes: "science is measurement". The first attempts of mathematical economists, such as Walras and Jevons, however, to introduce mathematical models for explaining human behavior met with overwhelming criticism concerning "the legitimacy of forcing human nature into the rigid frame of a mathematical structure".[17] Despite numerous improvements in the mathematical techniques used, these same criticisms continue to be directed at (the minority of) RCBA practitioners who do not accept the assumption of partial quantification and who, instead, attempt to place numerical values on nonmarket, as well as market, factors in their analyses. Thus the attempt to measure or quantify subjective or nonmarket parameters, "a century-old problem" in the behavioral sciences,[18] continues to the present.

In this essay, I wish to challenge the assumption of partial quantification and to argue for acceptance of the view that all parameters ought to be quantified whenever RCBA is used. To accomplish this, I will investigate the arguments against the assumption of partial quantification and show where they go wrong in supporting this 'majority position'. Next, I will provide some positive arguments for complete quantification. Finally, I will look at some cases in which quantification has been incomplete and at some in which it has been complete, in order to see what practical insights can be obtained regarding the assumption of partial quantification. I will argue that, although no quantification of nonmarket parameters ever approaches being ontologically adequate, although 'price' never equals 'value', and although fallacies of aggregation can occur as a result of complete quantification, it is *practically necessary* to quantify all relevant assessment parameters whenever one compares market risks, costs and benefits as a basis for public-policy decisions. Even though I basically support analytical, rather than intuitive, modes of risk assessment and technology-related evaluation,[19] I will argue neither in favor of analytical methods generally nor in favor of monetary quantification specifically. Analytical assessment techniques, especially those utilizing common monetary parameters, are susceptible to many complicated methodological problems which I cannot evaluate here. My argument, instead, is a limited, conditional, and strategic one. *If* market parameters are quantified and compared in RCBA, *then* it is reasonable to attempt to quantify everything, rather than to provide merely a physical description of nonmarket items and to leave them out of consideration in aggregating costs and benefits. While avoiding (in this essay) the general question of the *theoretical* desirability of risk-cost-benefit analysis and quantification, I would like to address the *practical* desirability of relinquishing the assumption of partial quantification, once one has decided in favor of using RCBA.

This modest conditional thesis is more significant then might appear, since federal courts in the US have recently rejected proposed environmental rules and technological regulations, e.g., on benzene exposure, on the grounds that no cost-benefit data were presented to justify them.[20] For this reason, and because of widespread RCBA use mandated by the courts, by NEPA, and by US regulatory agencies, the question of whether nonmarket parameters ought to be quantified in risk-cost-benefit analysis is one which potentially touches nearly every technology assessment or regulatory-agency rulemaking decision in the US.

3. ARGUMENTS AGAINST THE ASSUMPTION OF PARTIAL QUANTIFICATION

In support of their view that nonmarket or qualitative risks, costs, and benefits ought not to be expressed in quantitative terms in RCBA, economists and philosophers rely on a number of rationales which I call, respectively, (1) the argument from objectivity, (2) the argument from misuse, (3) the argument from alternatives, (4) the argument from simplicity, (5) the argument from politics, (6) the argument from horse and rabbit stew, (7) the argument from dehumanization, and (8) the argument from arbitrariness. Each of these rationales is deficient, however, either because it rests on a number of faulty assumptions or because it focuses on a point not at issue. Let us consider these arguments, one by one.

3.1. *The Argument from Objectivity*

Perhaps the most common reason for not representing qualititive or nonmarket factors in quantitative terms is "the argument from objectivity". Used by many neoclassical economists, this view is that, since only the market system provides the most objective means for quantifying certain risks, costs, and benefits, and since whatever has no market price has no clearly *objective* value, therefore what has no objective value ought not to be represented as objective by means of expressing it in quantitative form.[21] F. A. Hayek explains this faith in market pricing by claiming that the market alone is able to utilize

> the knowledge of particular circumstances widely dispersed among thousands or millions of individuals . . . the market and the competitive determination of prices have provided a procedure by which it is possible to convey to the individual managers of productive units as much information in condensed form as they need in order to fit their plans into the order of the rest of the system . . . having all the individual managers of businesses convey . . . the knowledge of particular facts which they possess [through any means other than market pricing] is clearly impossible . . . the market and price mechanism provide in this sense a sort of discovery procedure which both makes the

utilization of more facts possible than any other known system, and which provides the incentive for constant discovery of new facts which improve adaptation to the ever-changing circumstances of the world in which we live;

in other words, for Hayek and related scholars, market prices provide a "spontaneous ordering mechanism" for the goods of the world.[22]

Formulated in this manner, the argument from objectivity rests on a number of key assumptions, all of which are highly questionable. One such assumption is that market price provides the most objective procedure for determining value. As was already argued in the previous chapter (section 3.3), this assumption is false. Using price as a measure of value gives price an incorrect philosophical or ontological status.[23]

Economically speaking, there are also a number of reasons why market prices are not objective measures of values. For one thing, market prices are subject to the distorting effects of monopoly,[24] and they do not include social costs, or externalities, within them.[25] Also, they represent the effects of speculative instabilities in the market, rather than accurate costs/prices,[26] and they fail to provide or include monetary-term values for natural resources and for "free goods" (e.g., air) or "public goods" (e.g., national defense), even though these items obviously have great value.[27] All this means that, for both philosophical and economic reasons, the most objective measures of risks, costs and benefits are not necessarily their market prices. It can be argued, as several generations of welfare economists have done,[28] that the most objective measures of certain costs, e.g., of generating electricity, can only be obtained by taking account of key nonmarket parameters, such as social costs, e.g., the pollution caused by generation of electricity.

In fact, it is standard procedure, "in evaluating any [RCBA] project", for the economists to " 'correct' a number of market prices" and to "attribute prices to unpriced gains and losses that it [the project] is expected to generate".[29] These 'corrections' are achieved by a number of techniques, such as 'willingness to pay', all of which fall under the procedure known as "shadow pricing".[30]

Given all these deficiencies in the market, and the use of nonmarket techniques to correct them, it is not clear, either that the market provides the most objective means of pricing risks, costs, and benefits, or that nonmarket forms of pricing must be rejected as subjective. This, in turn, means that it is doubtful that there is a hard and fast line between objective and subjective values when they are identified, respectively, with market and nonmarket pricings.

In response to this attack on his assumption that market prices provide the most objective means of valuing risks, costs, and benefits, the proponent of the argument from objectivity may make a slightly different point. He may claim that, to introduce any other pricing system renders the results 'subjective', since nonmarket pricing is "inconsistent with accepted procedures" for aggregating and comparing RCBA parameters.[31] In response to this claim, that the objectivity of RCBA is best assured by using only one pricing mechanism for quantities that are to be aggregated, there are several obvious objections. One is that the decision to add nonmarket and market risks, costs, and benefits appears no more methodologically suspect than does the decision to add different market risks, costs, and benefits. For example, if one added only market parameters in comparing the risks, costs, and benefits of nuclear, as opposed to coal, generation of electricity, and if the labor costs associated with coal were arbitrarily, unpredictably, and unfairly kept low because of successful union busting, could one truly say that the uncorrected labor costs for coal were truly comparable to those for nuclear, even though both were market parameters?[32] I don't think these two parameters are comparable and that they should be uncritically aggregated and compared. The reason, however, is not that one is 'nonmarket', but that any parameters, even market ones, are subject to distortions which somehow need to be accounted for in RCBA.

An even more compelling reason for rejecting the argument from objectivity, and its appeals to consistency, is that accepting it would lead one to accept, uncritically, not only distorted market prices, but also a distinct bias in the RCBA. This bias arises because of the positivistic underpinnings of the argument.[33] According to these positivistic underpinnings, "only facts are 'objective', and values are merely 'subjective' "; as a consequence, "there can be no rational process for deciding questions of value".[34] But because proponents of the argument from objectivity often believe that only facts (e.g., market prices) are objective, they ignore all risks, costs, and benefits having some evaluational component. Hence, in an attempt to produce firm, quantitative results, proponents of this argument produce conclusions with a pro-industry or pro-commercially-viable bias.[35] This is because nonmarket parameters, such as adverse spillovers, are not quantified and hence not treated as factual. More will be said later about how failure to quantify nonmarket parameters in RCBA results in this distinct bias. The main insight, however, should be noted. If nonmarket means of quantification are suspect because they are allegedly neither factual nor objective, then the consequence of neglecting to quantify all but market parameters is that one becomes trapped in a 'positivistic paralysis'; he believes that ignoring given factors is more desirable than dealing with them as best one knows how. The problem with this thesis can easily be seen if one considers whether RCBA aggregates are more distorted by their complete omission of nonmarket parameters or by their including admittedly imperfect values for them. Provided that one knows whether the omitted values are costs or benefits, adverse or positive effects, it is obvious that some information about them is better than none at all and that some attempt to include them in the quantitative analysis is better than their skewing the analysis because they are completely ignored. More about this point will be discussed when the arguments in favor of complete quantification are treated.

3.2. *The Argument from Misuse*

Many economists, of course, realize that market prices do not always provide the most objective means of valuing risks, costs, and benefits, and their objections to quantifying nonmarket parameters do not rest on any assumptions that market prices are objective, while nonmarket ones are subjective. Rather, theorists such as Boulding, Morgenstern, and Rowe, who subscribe to (what I call) the argument from misuse, maintain that partial quantification helps to avoid further bias and arbitrariness. Kenneth Boulding, for example, warns of the danger of using quantitatively measurable economic indices, especially in terms of money.[36] He believes that people are likely, when they see such measures, to forget all the assumptions and *caveats* built into their employment, and instead to use them uncritically in deliberating about complex problems of public policymaking. Likewise, Rowe worries that the impact of the value judgments implicit in certain quantifications might be "masked by improper data comparisons due to oversimplification of the problem".[37] Morgenstern is concerned about what he calls "specious accuracy" in economics. This is the tendency to fix a precise, quantitative value on something which is either lacking in definition or classification or which, when quantified, is "irrelevant for whatever immediate purpose may be considered".[38] This lack of definition or lack of usefulness of the quantified factor is forgotten simply because one is overtaken by the impressiveness of the quantities – which nevertheless have only a "specious accuracy".

The alleged reason why one is likely to misuse and misunderstand the quantified nonmarket parameters is that many people fall victim to what Self calls "the myth of numerology". This is the belief that rationality requires counting and measurement. Once one subscribes to this myth, says Self, the numbers *distract* him from the harder task of examining relationships between value judgements and the relevant information and between value judgments themselves. For him complete quantification is merely a "slipshod way of dodging the issues"[39] The issues are dodged because, although the quantities bear an imperfect relationship to the nonmarket concepts they are designed to represent, people treat the surrogate quantities as though they were identical to the original concepts. As Bauer puts it, people forget the inferences they have made in assigning the numbers and they create "a spurious aura of hardness about the data".[40]

Certainly the argument from misuse is correct in noting that people are likely to forget the limits on the questionable procedures of quantifying nonmarket factors. Its proponents fail to appreciate, however, that not quantifying nonmarket parameters can often cause more serious consequences, in terms of the accuracy of the RCBA conclusions, than can misuse or misrepresentation of those parameters, once they are quantified.[41] Amory Lovins and other authors have shown how excluding certain costs, e.g., nonmarket ones, can change the whole outcome of a RCBA. As Lovins puts it, "the sort of change about which all the economists in the world, laid end to end, might never agree – can change tens of billions of dollars' benefit into a new cost, or vice versa".[42] If Lovins is right, and I think he is, then the argument from misuse fails for the same reason as does the argument from objectivity. Both rest on the erroneous assumption that it is better to exclude certain items from the RCBA aggregation than to include them, imperfectly quantified, and run the risk of their being misused. In the case of the argument from misuse, this assumption is particularly misguided because, if it is possible to *misuse* an imperfectly quantified factor, simply because it is quantified, then it is also possible to misuse an imperfect aggregate, simply because one is likely to forget that it fails to include values for nonmarket parameters. Hence, if the argument from misuse is admitted to count against quantifying nonmarket factors, then the same argument weighs even more heavily against failure to quantify all parameters, because the limits on the incomplete RCBA aggregate also are likely to be forgotten, but with more devastating consequences than misusing nonmarket parameters. In the case of partial quantification, the whole RCBA results, their very conclusions, are likely both to be misused as well as to be incorrect because of their omissions.

3.3. *The Argument from Alternatives*

One of the most powerful arguments supporting the majority of practitioners of RCBA in their belief that only market parameters ought to be quantified is (what I call) 'the argument from alternatives'. Proponents of this argument, such as Hare and Self, maintain that, instead of quantifying nonmarket parameters, they have other procedures open to them, such as the 'trial-design model'. In this model, there are no complete, all-inclusive cost-benefit calculations. Instead, "the designer just produces more or less detailed particular designs for the client to look at, all of which he certifies as at least feasible and attaches perhaps a rough costing to them". This procedure is like buying shoes, except that the options are not tried on.[43] Another way of characterizing alternatives to complete quantification is to appeal to 'planning', a "trial-and-error process" of trying to cope with nonmarket factors by choosing the policy option which promotes beneficial relationships of people and land.[44] For proponents of the 'argument from alternatives', nonmarket RCBA factors must be dealt with in terms of the planner's social judgment, rather than in terms of distinct quantities.[45] Planners simply make a number of judgments about the relative merits of different ways of valuing nonmarket parameters, and then they incorporate these judgments into the RCBA conclusions. In this way, they claim, they can achieve "imaginative constructions" for social policy and can truly synthesize all the aspects of a typical RCBA problem.

Powerful though it sounds, the argument from alternatives fails to provide a clear basis for solving problems addressed by RCBA. As one critic put it succinctly, all such 'alternatives' are "substantively void and a recipe for obscurantist intuitionism".[46] Because the planner or the practitioner of the trial-design model relies on what he admits are "much less rigorous intellectual theories ... and a rag-bag of applied sciences and empirical knowledge",[47] it is obvious that he is capable of being biased towards his own social vision in how he evaluates nonmarket factors in RCBA. This, in return, makes it likely that the very bias which the planner seeks to avoid, in condemning complete quantification, is likely to arise in more subjectivist, arbitrary, and unclear forms than it arises when economists use established procedures for quantifying nonmarket parameters. The very vagueness and open-endedness of planning and trial-design, which seem dependent upon how well their RCBA conclusions 'work' or 'fit', is a recipe for arbitrariness. Moreover, it is not clear even how the arbitrary conclusions of a chosen 'trial-design' RCBA might be evaluated. As a proponent of the argument from alternatives put it: "Comprehensive planning produces a closely interlinked mesh of results which cannot be unscrambled for pricing purposes".[48] If the results of alternatives to complete RCBA quantification cannot be "unscrambled", then presumably they must be accepted at face value. But this face-value acceptance precludes analysis. Instead, it must be built either upon an acceptance of an argument from authority or upon some intuition. In either case, the acceptance cannot be the basis of public analysis and evaluation. At least in the case of quantifying nonmarket parameters, the public is able to assess rationally whether it agrees with the evaluation. They know precisely with what they disagree in the RCBA, and that disagreement can be made the subject of fruitful exchange among economists, policymakers, and the public. In the case of alternatives to complete RCBA quantification, neither the basis of disagreement nor the means of resolving it is clear. For this reason, proponents of the argument from alternatives are likely to espouse a more authoritarian, less democratic, more obscure means of valuing risks, costs, and benefits than are the proponents of complete RCBA quantification.

3.4. *The Argument from Simplicity*

For proponents of the argument from simplicity, however, the very precision and clarity afforded by quantifying nonmarket parameters is precisely what they have against complete quantification. For them, this lack of obscurity in representing quantitative factors in RCBA is bought at too great a price: the assumption that all aspects of a *qualitative* risk, cost, or benefit can be expressed by a simple *quantity*. As one proponent of the argument from simplicity puts it, it is "most unlikely" that such a simplification, viz., representing a complex concept by a number, is epistemologically acceptable.[49]

In support of their claim that complete quantification leads to unacceptable oversimplifications, proponents of the argument from simplicity are able to point to the obvious difficulties associated with various means of quantifying nonmarket parameters. One such procedure is to represent the 'cost' of a human life as the discounted earnings which the person is likely to receive for the remainder of his working days. Rice, Lave, Seskin, and others have used this criterion for measuring life,[50] and they have been criticized widely. A number of authors have showed how accepting this methodological assumption for costing life leads to counterintuitive, and very likely false, consequences.[51] Some of these consequences, for example, are that the lives of old people and of housewives have no quantitative value. Thus, conclude proponents of the argument from simplicity, the attempt to quantify nonmarket parameters is doomed because it is based on the naive and "ludicrous premise" that a single number can enable one to "judge alternative courses of action".[52] Proponents of RCBA, say Myrdal and others, should be more sensitive to "the kinds of data to be collected" and should "press harder for empirical data" and a "censorious" research approach, rather than rely on numbers alone.[53]

In response to the argument from simplicity, proponents of complete quantification are likely to agree that all forms of measuring or quantifying qualitative parameters are indeed simplistic and at least partially incorrect. For them, however, advocates of this argument miss the crucial point. This is that the oversimplification and difficulty involved in quantification are irrelevant. The real question is "Can the judgment [regarding quantification] be made"?[54] So long as the quantifying judgment can be made, and they believe that it can, then practitioners of RCBA can continually improve the quantifies they use by attempting to account for more and more aspects of the concepts they represent numerically. I know of no proponent of complete quantification who believes that his numbers are not gross oversimplifications of the concepts they represent. Hence, to charge that such quantities oversimplify the costing/judging situation does not count as an objection against complete quantification. In fact, it would be quite suspicious if welfare economists or practitioners of RCBA had discovered how to measure the value of human life and other intangibles, when philosophers have stewed over such issues for centuries.

Rather, the purpose of admittedly oversimplified quantification is to provide a basis for more sophisticated quantification and thereby to exhibit relevant information useful for rational debate about the best means of quantitatively capturing key aspects of qualitative phenomena. Faulty attempts at quantification are necessary, moreover, if one is not to beg the whole question of whether RCBA conclusions can be arrived at in precise, quantitative manner. Not to quantify, on the basis of the argument from simplicity, is merely to prejudge the issue of whether numbers indeed can help to make RCBA policy choices clearer, more efficient, and more accessible to public debate.

One of the faulty assumptions undergirding the argument from simplicity is very likely the belief that oversimplification is to be avoided in policy analysis to the same degree, and for the same reasons, as it is to be avoided in theoretical science. However, the nature of the problems facing the theoretical scientist and the practitioner of RCBA are quite different. Because he is usually not dealing with a problem demanding immediate practical action, a theoretician has the luxury of declining to use any quantities which do not capture the complexity of the phenomenon he wishes to study. Moreover using imperfect quantities is unlikely to help him in situations where he is seeking exact predictive power over certain phenomena, since inexact or oversimplified numbers or concepts often lead to incorrect predictions which can be falsified.

The situation of the policymaker using RCBA is disanalogous to that of the scientist in at least two respects. *First*, the nature and the urgency of the problems he is addressing (e.g., whether RCBA indicates that coal - or nuclear - generated electricity is more desirable) mean that the policymaker must act, on as clear a basis as possible, prior to having complete knowledge of many complex parameters. A decisionmaker does not have the luxury of saying that his procedures are too simplistic to deal with the phenomena about which he must formulate policy. This means he must use admittedly simplistic formulations, but, at best, formulations whose implicit assumptions are clear for all to see, criticize, and revise. Thus the relevant criticism of a policymaker wont to quantify qualitative parameters is *not* that he is oversimplifying the situation, because he already knows that he is doing so. Rather, the relevant criticism is that such oversimplifications are less clear, less accessible to the public, and less open to subsequent modification/improvement than are other simplifications. In my opinion, it would be hard for an opponent of complete quantification to show that his non-quantitative or intuitive alternatives to numerical measures were clear, more accessible to the public, and more open to modification than was use of particular quantities to represent qualitative phenomena.

The policymaker using RCBA is disanalogous to the theoretical scientist, wary of oversimplification, in a second respect. Not only is the scientist more concerned with *epistemological* problems of accurate measurement/representation, while the RCBA policymaker is more concerned with *practical* problems of making the best decision in a situation of imperfect knowledge where there are no wholly desirable options, but also the criteria for successful scientific decisions are quite different from those for successful practical decisions. The scientist is often judged by whether he is right or wrong in what he alleges or predicts. In science, the laurels of fame go to the great discoverer who produces verifiable results. For the policymaker, however, the case is quite different. Often his RCBA results cannot be verified because what counts as a more desirable policy option is frequently a function of one's social values. (For example, a policymaker's current decision to follow the 'soft', rather than the 'hard' energy path might be vindicated in the future if his descendants believe that decentralization outweighs all other evaluative considerations, even the extraordinary costs involved in the soft path. The future might not vindicate the soft-path choice if, on the other hand, subsequent generations place great value on centralized energy sources.

The differences in practical, versus scientific, decisions, however, are not merely differences in the degree to which science is, and policy is not, verifiable. The differences also arise as a consequence of what counts as 'good policy'. 'Good science' often amounts to following methods which result in what are thought to be substantively correct conclusions, either because they accord with other accepted scientific laws, principles, and interpretations or because they can be verified in some sense. 'Good policy', on the other hand, frequently cannot be assessed on substantive grounds, but must be judged in terms of procedural criteria. These include, for example, whether the policy was arrived at through rational debate concerning clear alternatives and whether the contributions of any particular interest group were excluded *a priori*.

If indeed the criteria for good policy are different from those for good science in just these ways, e.g., regarding the degree to which its conclusions may be verified and the extent to which substantive, rather than procedural, norms are used for evaluating it, then the argument from simplicity surely does not count against complete quantification in RCBA as it might if used against some form of quantification in the theoretical sciences. In fact, proponents of the argument from simplicity appear to fail to distinguish the differences between policy conclusions and those of science and, for this reason, fail to see the heuristic power of precisely the RCBA quantifications which they view as overly simplified. More about this will be discussed later, in connection with the arguments in favor of complete quantification.

3.5. *The Argument from Politics*

In addition to the argument from simplicity, other objections to complete quantification arise because RCBA is erroneously viewed as more like a theoretical science than a practical, policy science. One of the most common of these is (what I call) 'the argument from politics'. Used by theorists such as Mishan, McCarey, and Lovins, the gist of this objection is that political decisions, about the value of subjective or nonmarket parameters, should be left to the political process and not be brought into the economics of the RCBA. In the words of Mishan, to quantify nonmarket factors is to wrong society; it is "to place the evaluation of the [RCBA] project in the hands of the economists" rather than in the hands of the people and the political process.[55] There are at least three reasons, according to proponents of the argument from politics, why complete quantification should be avoided:

1. in moving political decisions to the economic sphere, it gives power to the *experts* which should reside in the people;[56]
2. in allowing economists to quantify qualitative parameters, one gives too much power to persons who are able to make largely *subjective* judgments;[57]
3. in allowing practitioners of RCBA to quantify nonmarket parameters, one "effects a *deception* and sacrifices meaning on the altar of quantification, in order to save face".[58] (The "deception" is the suggestion that all RCBA parameters were arrived at consistently, by means of the market, when they were not, because nonmarket measures were determined subjectively.)

If indeed complete quantification had as a necessary consequence the displacement of decisionmaking power from the people to economists, and if that displacement of political power resulted in subjectivism and deception, then proponents of the argument from politics would have a powerful reason not to quantify nonmarket parameters. It is not clear, however, that such disastrous consequences follow from the quantification of qualitative factors in RCBA. This is because the real objection made by proponents of the argument from politics appears to be, not to *quantification*, per se, but to who performs it. Their claims about displacement of political power from the people to economists would be addressed were it the case that economists provided a number of alternative approaches to quantifying particular qualitative parameters, spelling out the assumptions and consequences of each, but then left the choice among these alternative quantifications to the people.

Such a procedure of providing alternative approaches would also address claim (2) of the argument from politics, because it could no longer be alleged that quantification of nonmarket parameters allowed economists to impose their subjective preferences on the political process. (The final reason (3) frequently used by proponents of the argument from politics has already been answered, in connection with the discussion of the argument from objectivity in section 3.1. of this chapter.) Thus it appears that while proponents of the argument from politics have hit upon a good insight, viz., that policy ought to be decided by the people and not by economists, their application of this insight to the problem of complete RCBA quantification is misdirected. This is because the illicit transfer of political power arises not from quantification *per se*, but from the failure to provide alternative quantification schemes and from the failure to allow the public to decide among these schemes.

Moreover, if one is worried about the political assumptions of RCBA practitioners dominating the policy evaluation, as proponents of the argument from politics are, then it seems that one way to avoid such biased assumptions might be to use explicit numerical criteria for each aspect of the RCBA. As one theorist put it, "every practical judgment in policy affairs is based on a structure of concepts that is largely implicit and poorly understood".[59] If, by using numbers, one can "get at" the qualitative factors relevant to such policy judgments, then those factors seem more likely to be clearly understood. And if they are more clearly understood, then to that degree is it less likely that purely political or arbitrary judgments can be made concerning them. This is because people can easily debate whether or not the weight assigned to a given factor, relative to other factors in the RCBA, has been represented accurately within a particular cardinal or ordinal measurement scheme.

It is far more difficult for people to debate the pros and cons of an implicit intuitive criterion for measuring the worth of qualitative factors than to debate the pros and cons of alternative measurement schemes. One reason is that persons are often unable to formulate the particular intuitive criterion which they are employing. If they cannot state and explain the criterion, then is is difficult for others both to understand and to assess it. Another reason why it is difficult to debate the merits of an intuitive criterion for assessing nonmarket parameters is that, even when such a criterion can be formulated clearly, participants in the debate are likely to approach it in terms of radically different assumptions. Because of the differences in these assumptions, there would be no common denominator, either for assessing the merits of different evaluations of a particular nonmarket parameter or for arriving at some solution to this evaluative problem.

For example, suppose that two persons were debating a particular intuitive criterion for valuing the nonmarket benefit of allowing sacred Indian burial grounds to remain intact and secure from any sort of development, even under eminent-domain provisions. Suppose, further, that both persons agreed that the value of this land was incommensurable and hence incapable of being represented either within cardinal or ordinal measurement schemes. However, suppose that one person held libertarian assumptions about property rights and the role of the state, and that the other person held Marxist assumptions on both these points. Because of the differences in their assumptions, it is unlikely that these persons could arrive at some rational compromise or reasoned resolution of their problem of valuing the Indian land. Indeed, it is difficult to see how either of them could even make logical contact with the points of the other. At a minimum, resolution of their valuing problem seems to demand at least an ordinal 'measurement' of the benefit they attribute to the land. Otherwise, there could be no possibility of understanding whether there were conditions under which each of them was willing to compromise so as to agree with the other on the value. In other words, without at least an *ordinal* representation of how a person valued all risks, costs, and benefits of a particular project, including the benefit of protecting sacred Indian land, someone could not understand the alleged intuitions of the other, much less attempt to come to some common valuation with him. To the extent that an allegedly political compromise on valuing a particular item is rational and rule-governed, and understood by all parties to the dispute, then to that degree is at least some sort of ordinal measurement of various parameters presupposed. This is because any complex RCBA is bound to have factors which are more or less important, relative to each other. And if they are more or less important, they can be represented ordinally.

Even if the proponents of the argument from politics are right, however, and such evaluative compromises do not require or presuppose at least an ordinal measurement scheme, purely political means of valuing nonmarket parameters seem highly undesirable. If weighing the merits of the subjective parameters in a RCBA simply is left to the vagaries of the political process, without those subjective parameters being formulated in terms of a number of alternative measurement schemes (cardinal or ordinal), then the political process of evaluating those parameters appears likely to be more arbitrary and more time-consuming than otherwise might be the case. In part this is because cardinal or ordinal schemes instantly reveal the weight assigned to one factor over another, whereas intuitive schemes are rarely so explicit, particularly in cases in which one must intuitively represent hundreds of different factors. Perhaps the non-ordinal, non-cardinal relationships among them could be spelled out, but only after a lengthy and complicated, and therefore impractical, explanation.

Moreover, because they are clear and explicit, measurement schemes for all RCBA parameters offer a great deal of 'citizen protection'. The protection afforded reasonable social decisionmaking, by the use of cardinal or ordinal measures of all RCBA parameters, is somewhat similar to the protection afforded civil liberties by the use of a national *Constitution*. Both provide a known framework, admittedly imperfect, against which one can be protected from the inconsistencies of 'politics' – mere popular opinion, administrative whim, or uninformed majority rule. Such protection is essential, as will be seen later (section 4.4), because circumstances seldom permit satisfaction of the conditions necessary for morally acceptable, participatory democracy.[60]

3.6. *The Argument from Horse and Rabbit Stew*

Admittedly, however, complete quantification is not seen by a majority of welfare economists or practitioners of RCBA as a protective means of systematizing social valuation of qualitative parameters. According to Mishan, Lovins, Georgescu-Roegen, Shackle, Skolimowski, Myrdal, and others, one ought not to attempt to quantify nonmarket factors because there is no common denominator between what is quantitative and what is qualitative. For Mishan, mixing two unlike things, tangible market items and intangible nonmarket goods, gets one into the problem of "horse and rabbit stew". Because the two types of items are dissimilar, says Mishan, either the scientific "rabbit" (the quantifiable entities) flavors the whole stew, and the unscientific "horseflesh" (the "unquantifiable" entities) is lost or vice versa. There simply is no rational way to combine the two, because there is no comparability between them, says Mishan. Technically put, the alleged problem is that one cannot obtain the scalar measurability of diversely composed collections.[61]

This objection touches the core of the problem with quantification and is part of "the classical argument against utilitarianism and especially against the Benthamite felicific calculus".[62] How does one aggregate and compare various risks, costs, and benefits if, as Mishan says, many goods "elude attempts to translate them into money values"?[63]

For example, how might one translate the value of his home into monetary terms? This is exactly the objection Lovins raises when he comments on the work of the Roskill Commission on the third London airport. He points out how the group attempted to measure the various risks, costs, and benefits associated with alternative airport sites. According to Lovins, 38 percent of the Londoners interviewed said that no amount of money could compensate them for the loss of their homes in order to build the airport.[64] Hence, for Lovins, to place a cost on the loss of these homes simply cannot be done. Whenever certain intangible or qualitative risks, costs, and benefits are quantified, say the proponents of the argument from horse and rabbit stew, then the resulting aggregations and comparisons are simply illicit, because one cannot make diverse things commensurable. As one economist put it: "psychical magnitudes are not precise".[65] Hence, for advocates of this argument, to quantify nonmarket parameters, aggregate them, and then compare them to market items in a RCBA is to do what no good cook would do: attempt to combine things which will not mix and which will yield a flawed and biased "cuisine".

The first, and most obvious, response to the proponents of the argument from horse and rabbit stew is that one need assign neither monetary nor cardinal values to all RCBA parameters in order to conduct a risk-cost-benefit analysis. One might wish to represent all the factors in terms of utility, for example, or to use an ordinal measurement scheme. Even if one did choose to employ a monetary framework, however, it is not clear that it would face the crucial problems alleged by the proponents of this argument. Admittedly, one cannot accurately measure intangibles. Nearly everyone, both proponents and opponents of this argument, would admit that there is always a qualitative residual hidden behind the numerical pattern used to represent nonmarket items.[66] Admittedly, market and nonmarket items are not strictly comparable in any quantitative way.

The problem with the argument from horse and rabbit stew, however, is that it misses the crucial point at issue in RCBA quantification, and it misses it in much the same way as does the argument from simplicity. This is that the methodological problems associated with measuring or comparing alleged "incomparables" are irrelevant. The real question is: "Can the judgment [regarding quantification] be made"?[67] The judgment is made every day, both by laymen and scholars. In an ordinary sense, it is made implicitly whenever people choose to pay certain insurance premiums, to accept particular court settlements, or to spend their money in one way rather than another. Likewise, scholars repeatedly make judgments about particular ways of quantifying allegedly qualitative items, and they do so by means of psychophysical scaling,[68] inferring social costs,[69] and evaluating the welfare of various persons.[70] As Harsanyi points out, "there is no doubt about the fact that people do" this; the only issue is what logical basis there is for such judgments.[71]

When one is faced with a practical problem concerning allocation of limited resources, as he is in every RCBA, his refusal to compare alleged 'incomparables' simply leads to paralysis when be should be taking steps toward action. As was pointed out in response to the argument from simplicity (see section 3.4), proponents of complete quantification do not naively believe that they can measure intangibles perfectly and compare them with tangibles. They do believe, however, that it is necessary to make a decision regarding various societal problems of allocation, and they know that they will never have adequate facts or an adequate methodological basis for doing so. Hence, in order not to beg the question of whether complete quantification in RCBA is the best way, among a number of flawed ways, to make this decision, they attempt to quantify all parameters. Not to quantify nonmarket factors, on the basis of the argument from horse and rabbit stew, is merely to prejudge the issue of whether numbers indeed can help make RCBA policy choices clearer, more efficient, and more accessible to public debate. If one does not attempt to quantify all parameters, he will never know the limits of this numerical method, as compared to other methods of evaluation.

As was also pointed out in response to the argument from simplicity, one errs if he believes that, because alleged incommensurables cannot be aggregated or compared in theoretical science, therefore they cannot be or ought not be compared in RCBA. Both because of the differences in the types of problems and applications addressed by the theoretical scientist, versus the practitioner of RCBA, and because of the differences in criteria for assessing good scientific results, versus good RCBA results, the argument from horse and rabbit stew does not necessarily count against complete RCBA quantification, as it might count against certain moves in theoretical science.

Moreover, although quantities and qualities are incommensurable in a strict scientific sense, there is no doubt that people do employ them in decisionmaking as if they were commensurable. This is because the practical consequences of acting as if they were incommensurable are much more problematic than those of acting as if they were commensurable. For example, if one were to say that the value of a human life were *incommensurable* with the value of any tangible or market goods, then it is not clear that policymakers could provide a clear, rational basis, either for allowing any public risk or for requiring certain levels of pollution-control expenditures in order to save human lives. Instead, they would likely have to rely on *ad hoc* judgments as to why a given level of pollution was/was not acceptable, at a given cost for abatement, regardless of the content and status of other judgments about such acceptabilities in related areas.

Also, if a person really did believe that market and nonmarket goods were incommensurable, to the extent that they could *never* be rendered comparable, then such a belief would preclude not only complete quantification is RCBA but also *any way* of assessing the risks, costs, and benefits of alternative decision scenarios. Obviously *any* comparison of alternatives involving 'incommensurables' leaves a residual unaccounted for by the theory of comparison. The problem is not unique to quantification. If one used some nonquantitative, philosophical evaluation scheme for assessing alternatives, and if that scheme were based upon the criterion of maximizing equity, for example, then the proponent of an alternative philosophical scheme could always charge that it failed to account for some non-equity 'residual' found in the incommensurable factors. Likewise, if one used some scheme based upon the criterion of aesthetics for deciding upon alternative risks, costs, and benefits, then the proponent of an alternative scheme could always charge that it failed to account for some non-aesthetic 'residual' found in the incommensurable factors. This is because *any evaluative scheme* for RCBA must employ certain simplifying assumptions, if it is to provide a decision procedure, rather than a mere discussion, relating to available policy alternatives. Any simplifying assumptions, in return, are bound to have a residual not accounted for by the theory. This is why those who focus on the methodological probolems associated with complete quantification miss the point. The point is whether the evaluation can be carried out at all, and which evaluation scheme provides the most consistently acceptable framework for decision, i.e., which provides the best logical basis for evaluations. When RCBA parameters are completely quantified, of course there are numerous nonquantitative criteria (e.g., equity, the sacredness of life) at work in providing the basis for particular, alternative ordinal or cardinal measurements. Quantification is not a *substitute* for use of these other criteria, but a means for clarifying the various policy alternatives/decision procedures they generate. Thus clarified, alternative RCBA decision procedures are more susceptible to democratic control.

Moreover, in claiming that nonmarket parameters ought not to be quantified because they are incommensurable, proponents of the argument from horse and rabbit stew make highly debatable assumptions about measurability and about the precision and commensurability of economic parameters generally. With respect to measurability, they assume that the assignment of numerals to objects other than by the procedures involved in 'fundamental' or 'derived' measurement is not measurement at all. (Fundamental measurement occurs when the axioms of additivity can be shown to be isomorphic with the manipulations performed upon objects, e.g., in measurement of length and weight. Derived measurement takes place by means of numerical laws relating fundamental magnitudes, e.g., in measuring density by the ratio of mass to volume.)[72] Obviously, however, a strong case could be made for the claim that measurement also includes the assignment of numerals to objects by means of an explicit procedure, e.g., preference-ordering according to some ordinal theory. If so, then it is not clear that nonmarket parameters are not measurable, at least in this ordinalist sense.

Also, in alleging that nonmarket parameters are not measurable, in some sense, proponents of the argument from horse and rabbit stew appear to forget that even market parameters are not commensurable, and hence not measurable, in their sense. Even though market prices, for example, provide some explicit values for goods, they mean something different for each person. Moreover, whenever these prices are averaged, the average is (strictly speaking) true only of a mathematical entity composed of particulars, for each of which the average may be false. Because of the way in which averages of market prices implicitly ascribe a sort of permanence to composite entities, even though these entities have no determinate, measurable character in the sense assigned to them, it is not clear that one can consistently allow many different measures of market items but disallow all measures of nonmarket items. Hence, if the argument from horse and rabbit stew is correct, then it likely proves much more than its proponents wish, which is reason to doubt it.

3.7. *The Argument from Dehumanization*

Another argument against complete quantification is similar to the argument from horse and rabbit stew in that it rests on the assumption that nonmarket items are somehow 'beyond quantification'. Proponents of this approach (the argument from dehumanization), such as Hampshire, MacLean, and Wolff, maintain that quantification of nonmarket parameters dehumanizes our experience. Speaking of Nozick's use of utility theory, Wolff, for example, says such rationalization through quantification is "creepy". According to Wolff, use of mathematical techniques to deal with qualitative phenomena rests on the questionable assumption "that a hitherto uninvaded sphere of human activity should be ... ready for the extension into it of these [quantitative] models and methods".[73] The human sphere of nonmarket phenomena (e.g., the value of life, health, death, love, well-being), says Wolff, ought not be subjected to "the dehumanization of quasi-economic rationalization".[74]

Another philosopher, Stuart Hampshire, echoes the same sentiments when he complains that the "large-scale computations in modern politics and social planning bring with them a coarseness and grossness of moral feeling, a blunting of sensibility, and a suppression of individual discrimination and gentleness, which is a price that they will not pay for the benefits of clear calculation".[75] Other writers speak of the extension of quantification into qualitative areas as the "new philistinism".[76]

University of Maryland philosopher, Douglas MacLean, claims that the biggest opposition to such extensions of procedures of quantification is not political but philosophical. The problem, says Maclean, is *not* that we can never admit that the costs of saving lives in certain situations are too high. Rather, it is that our symbolic needs are thwarted by replacing moral and political thought with computational methods.[77] His point is that we have to look at specific RCBA cases in just such a way that systematic use of quantified risk assessment discourages us from doing. We have to figure out how to incorporate the kinds of values, like sacredness, that are not economic values".[78]

The brunt of the charge made by proponents of the argument from dehumanization appears to be that advocates of complete quantification attempt illicitly to reduce human values to economic ones. There are a number of reasons, however, why this charge is misdirected. For one thing, proponents of complete quantification do not believe that "every person has his price". As Self puts it, they do not believe that every person can or ought to put a price on his spouse, honor, or principles. They do believe, however, that "almost all individuals have multiple goals or values which often conflict" and that most goals "require the use of scarce resources". They also know that any "use of scarce resources to pursue a goal can be described as its cost; and if some of these resources could be switched to one or more other goals . . . [then] an opportunity cost is involved – namely the various degrees of goal achievement or satisfaction that these resources used in a different way could produce".[79] This being so, proponents of complete quantification are not attempting somehow to reduce human values to economic ones, but to get a handle on how to spend societal monies. Such expenditures could be made on the basis of human, political, religious, economic, ecological, or other values. The point is not to reduce all the myriad types of values to economic ones, but simply to rank them by means of a common denominator which lends itself to clarity, understanding and amendment by the populace. In examining what individuals might be willing to pay for safety in certain areas, for example, the proponent of complete quantification is not attempting to value life, but to provide a rational framework for public debate as to how to spend public monies. In other words, he is operating on the assumption that choice is helped by a common medium of accounting and that, by using this medium, he is saying nothing about values, but merely helping to elucidate the choices of society, e.g., as to whether it wishes to build center medians on all highways with two-directional traffic.[80]

Like it or not, almost any decision about how to spend public funds is an implicit decision about how much health or safety society can afford.[81] If a legislature decides not to spend the money to build highway medians, for example, then it is implicitly placing a lower monetary ranking on the affordability of saving lives by the medians than on the affordability of saving lives by means of other government expenditures. Likewise, if it decides not to build the median because it is too expensive, even though its absence is responsible for 100 deaths annually in the community, then it is not reducing these lives to a monetary value but implicitly affirming whether it can afford to save them.

Attacks on the alleged reduction of humanistic to economic values in RCBA are misdirected for at least four reasons. *First*, no 'valuation of *life*' is taking place (but merely an affirmation of a scheme for deciding how to spend finite resources. For example, if one ascribes a quantitatively larger benefit to saving lives in situation 'A' rather than in situation 'B', no 'valuation of life' need be taking place. Rather, one might judge that those in situation 'A' are in greater need than those in 'B', or that they have a greater right to be saved, or any one of a number of other reasons). Quantitative measures enable one to ascribe a value to certain *actions* in certain situations, a process not reducible simply to valuing *lives*.

Second, such monetary accounting procedures are implicit and accepted in many other actions that we perform. For example, though life is priceless, few people refrain from buying life insurance on the grounds that such a purchase is dehumanizing. Moreover, they view this purchase, not as an illicit 'valuation of life', but as a practical decision as to what action they can afford.

Third, the argument from dehumanization is philosophically unsound because it fails to take account of the fact that even *market prices* of items do not represent their real value, and that economists usually do not believe that they do. The problem is not that values of market items are easily obtainable, while values of nonmarket items cannot be obtained (and then only arbitrarily) except through a 'dehumanizing' quantification. For the proponent of complete quantification, real valuation takes place neither in the case of market, nor in that of nonmarket goods. Rather, the 'monetization' of both represents merely an accounting scheme, and one that is admittedly flawed.

If quantification of nonmarket parameters is condemned as dehumanizing, then because of the flawed accounting scheme, quantification of market parameters must be condemned for the same reason. If it is dehumanizing to put a price on one's safety, then it is also dehumanizing to put a price on one's time, as obviously is done in many market transactions, whether one is charging a consulting fee or buying a dishwasher. This is in part because all market goods have a human component (e.g., the time saved by my dishwasher, which may or may not be adequately represented by its purchase price). It is also because many goods lie within, or outside, the market for reasons solely of historical accident. It is largely an historical accident, for example, that my blood has a market price, even though its value to a dying person is not captured by the price paid for it. It is also an historical accident that various aesthetic qualities, such as the varieties of trees planted on acreage, often do not have a market price, even though their value likewise is not captured by the cost of purchasing, planting, and nourishing the trees.

As the dishwasher/blood/tree examples suggest, proponents of the argument from dehumanization fail to recognize both that *any* quantification (market or nonmarket) is flawed, and that quantification is simply a modeling technique, not a criterion for ontological valuation. In attacking only the monetization of qualitative goods, these proponents thus err both in consistency (since their arguments provide grounds for attacking market quantification as well) and in fact (since the quantification is not a valuation).

Perhaps the root of charges of dehumanization, misdirected as they are, is the widespread feeling that much in life is beyond money, and that money is the root of all evil. One could agree with these almost-theological stances, however, and yet employ a number of quantification schemes for making decisions about public spending. Moreover, if one is 'tainted' with evil because of his use of such schemes then, as was previously suggested, virtually everyone, and not just proponents of complete quantification, are so tainted.

Perhaps another reason why proponents of the argument from dehumanization err is that they fail to distinguish the realms of individual, from collective, choice, even though complete quantification is more applicable in the latter realm. Admittedly, *individuals* do not need an accounting scheme or modeling technique to decide how to spend their money. Their choices are easier, because their individual decisions regarding costs and pricings are integrated in terms of their consciousness and experience. An individual has a value system which enables him to make decisions which often require no quantification. Collective choices, exercised through government, however, are more difficult. They must follow a different logic because there is no unified value system, short of quantification, which provides clear decision criteria. Moreover, implicit quantification of nonmarket parameters is evident in at least two typical societal situations. First, every time public funds are spent for 'A', a societal cost is incurred, and there are fewer funds for 'B'. At that point an *implicit* price is placed on the resources which a community can spare for any item. Second, if there is to be a sophisticated society, able to compensate its members for risk and loss and to spend its funds so as to enhance safety, then one must calculate monetary compensations for admittedly human goods. I suspect that many proponents of the argument from dehumanization would sanction government regulation and monetary compensation, but would not sanction the conditions (e.g., complete quantification) necessary to realize them. Compensation and regulation require an extremely advanced stage of social, political, and economic integration. But this integration, in turn, requires an advanced system of social planning and calculating. Proponents of the argument from dehumanization cannot have it both ways. If they want the benefits of compensation and regulation, then they must accept the burden of quantifying numerous costs and benefits essential to that compensation and regulation.

3.8. *The Argument from Arbitrariness*

Another line of attack used by opponents of complete quantification is the argument from arbitrariness. Here they claim that nonmarket parameters simply cannot be quantified because they are arbitrary or subjective. The precise sense in which proponents of this argument use the term 'arbitrary' generally falls into three categories. They claim that the quantification is arbitrary, either because (1) it is ontologically impossible to accomplish; or because (2) it is determined by societal norms, which themselves are arbitrary; or because (3) it is arbitrary from the economists' perspective.

Proponents of (1) maintain that "the world continuum" has no joints where, as Plato thought, a good carver could separate one thing from another. Hence, to attempt to quantify nonmarket parameters, they say, is to use numbers to cut "artificial slits" in the world continuum which is a whole.[82] According to Self and Georgescu-Roegen, both proponents of the argument from arbitrariness, many people nevertheless attempt to cut such artificial slits. They do so, says Self, because they subscribe to the "myth of numerology". According to this myth, one can weigh alternative values only by converting them to numbers. The myth is false, says Self, because there is no *independent* basis for assigning numbers to nonmarket factors; those assigned merely reflect the *qualitative* preferences already made. He claims that we know this 'reflection' to be so because, if one doesn't like the resulting sum of risks, costs, and benefits, he can simply change the quantitative weights assigned to each before aggregating them. Moreover, says Self, we do not make "point counts" when attempting to make a *personal* decision about our individual risks, costs, and benefits. For him this constitutes further proof that there is no independent (of our own preferences) ontological basis for quantifying nonmarket parameters.

What of this argument that complete quantification is ontologically arbitrary? Insofar as its proponents claim that there is no ontological basis for quantification, they are correct. Insofar as they maintain that, without such as ontological basis, complete quantification ought not to be attempted, they are incorrect. As is already evident, proponents of complete quantification make no claims about economists' abilities to grasp ontological reality through quantity. Indeed, as consideration of the argument from dehumanization revealed, economists do not attempt to ascribe ontological *values* to things by quantifying them. They attempt merely to employ various quantitative methods so as to make the problem of resource allocation more tractable. They attempt to provide mathematical models to aid in the decision-making process. They do not believe that such models ontologically describe the world, but merely that they facilitate thinking about highly complex policy decisions.

If assigning quantities to qualitative factors *did* require some independent, ontological criterion for assignment, then undesirable consequences would follow. For example, then courts could never award financial settlements for loss of lives, persons could never defend purchasing a particular type of health-insurance policy, and it would be impossible to justify certain expenditures to enhance the aesthetic quality of one's surroundings. If Self is correct, then none of these actions, all of which presuppose placing a quantitative value on a qualitative parameter (life, health, beauty), could be called legitimate. This is because there is no independent ontological criterion for the quantitative assignments. There are, however, good reasons for defending particular quantitative assignments, whether regarding court settlements after accidental deaths or regarding certain RCBA parameters. These good reasons include whether the assignment is consistent with other assignments made in similar cases (e.g., other instances of accidental death) and whether it is consistent with the preferences of reasonable people. Hence the proponent of the argument from arbitrariness faces a dilemma. Either he must condemn all quantifications (even socially beneficial ones, such as those used for insurance purposes). Or he must admit that one needs no ontological criterion in order to justify the desirability of quantifying nonmarket parameters. To choose the former course is obviously inconsistent with reasonable social policy.

Moreover, the fact that *personal* decisions do not always require 'point counts' does not show that complete quantification is arbitrary and undesirable for *societal* decisionmaking. Individual decisions are unified by one's own values and preference-orderings. Societal decisions require some analytic framework (like quantification), precisely because myriad values and preference-orderings of numerous persons are at stake. These have to be reconciled and unified somehow, as they already are for each individual. Hence the absence of quantification in making many personal decisions does not count against the desirability of quantification in making societal decisions. The two decisionmaking cases are fundamentally disanalogous.

Likewise, the fact that persons can change the numerical representations of qualitative parameters, if they do not like the resulting aggregations, is not a compelling argument against quantification. The purpose of complete quantification is not to determine immutable numerical assignments, but (a) to determine the *relative* weights of various parameters; (b) to set forth these assignments in clear, straightforward fashion, so as (c) to allow debate about these assignments and so as (d) to provide a framework for amending these assignments on the basis of democratic social policy. Clearly some amendments of particular quantifications can be justified and some cannot. If one wanted to amend the number assigned to the value of one death from cancer, solely in order to show that a particular technology was not justified in RCBA terms, then his argument would be circular and question-begging. However, if one wanted to amend the number assigned to the cost of one death from cancer, on the grounds that a higher cost would be required to offset the average medical, employment, and burial costs of the average cancer victim, then such an argument would be more plausible. The mere fact that it is possible to 'juggle' nonmarket quantifications to justify one's preconceived conclusions does not prove that quantification, *per se*, is always suspect. It proves merely that it, like every other basis for policymaking, can be misused. For all these reasons, appeal to ontological arbitrariness does not appear to threaten use of complete quantification in RCBA.

Another variant of the argument from arbitrariness, however, is that quantification of nonmarket parameters is suspect because it is based on 'arbitrary' societal norms for assignment of numbers. As both Self and Lovins point out, whether a given item is counted as a cost or as a benefit is a matter of societal values. For some persons, generation of electricity is a *cost* to be measured in terms of thermal pollution and resource depletion, while for others it is a *benefit* to be measured in terms of creating jobs and providing energy. Hence, they claim, there is no independent, nonarbitrary way to quantify parameters based on societal norms.[83]

If complete quantification ought not to be attempted, however, because it is based on arbitrary societal norms as to how to count costs, risks, and benefits, then this same rationale undercuts any form of societal decision-making. If the problem is that there are no nonarbitrary societal norms for evaluating risks, costs, and benefits, then any method of policymaking, not just RCBA, is suspect. This is because even a nonquantitative method could be criticized on the grounds that it was based on arbitrary societal norms. Hence, those who attack complete quantification, on grounds that what is societally arbitrary cannot be quantified, face a dilemma similar to that confronting those who worry about the ontological arbitrariness of quantification. Either they must condemn all types of societal decisions regarding risks, costs, and benefits (even those based on nonquantitative methods), because they reflect or rely only on arbitrary societal norms. Or they must admit that one needs no nonarbitrary societal norm in order to justify the desirability of quantifying all RCBA parameters. Since societal decisions regarding risks, costs, and benefits must be made, it is obvious that, even if there is no independent, nonarbitrary societal norm, there must be some procedure for making those decisions.

Other proponents of the argument from arbitrariness proscribe complete quantification on the grounds that it is arbitrary from an economic point of view.[84] From their perspective, to introduce any quantification other than that based on market parameters is arbitrary, since nonmarket pricing is "inconsistent with accepted procedures" for aggregating and comparing RCBA parameters.[85] This variant of the argument from arbitrariness, however, amounts to one version of the argument from objectivity, already criticized in section 3.1 of this essay. Most of the difficulties faced by the earlier argument also confront this variant of the argument from arbitrariness. Hence it also provides little support for the view that one ought not to attempt complete quantification.

The big problem with all these variants of the argument from arbitrariness, in addition to those already noted, is that they charge RCBA with arbitrariness in a way that is misdirected. In claiming that RCBA quantifications of qualitative parameters are arbitrary, proponents of this argument fail to distinguish arbitrariness of method from arbitrariness of calculations. Clearly, RCBA is indeed arbitrary in the latter sense, but not in the former. This is because the *method* of valuing qualitative parameters can be clearly specified according to any one of a number of alternatives (ability to pay, social efficiency, preference-orderings, for example). The *precision* with which any factor can be calculated by means of one of these methods, however, cannot be clearly specified and is arbitrary in that sense. This means that RCBA is indeed a nonarbitrary method in that consistent techniques yield imprecise results. These imprecise results are not wholly arbitrary, however, since the value judgments underlying them (e.g., defining cost in terms of willingness to pay), however controversial, are nevertheless clear and consistent. The arbitrariness of RCBA consists only in that these value judgments cannot be precisely applied for technical reasons.

Moreover, if one is worried about the alleged arbitrariness of RCBA calculational results, then the solution might not be to foreswear complete quantification. It might be, instead, to obtain multiple studies, each done by a different interest group, prior to public selection of one of the studies as most representative or most desirable. In other words, perhaps arbitrariness is best dealt with by debating the merits of *alternative* RCBA quantifications rather than by avoiding quantification of parameters altogether. In this way, the public might be able to dissect the assumptions, uncertainties, and value judgments underlying various options for quantification.[86]

4. ARGUMENTS IN FAVOR OF COMPLETE QUANTIFICATION

If the arbitrariness of RCBA quantification is best addressed by debating the merits of *alternative* quantification schemes, then perhaps many of the objections to complete quantification are best interpreted, not as obstacles to numerical representation of qualitative parameters, but as pleas for more reasonable *uses* of those representations (e.g., as one of many possible numerical alternatives, each open to debate). This suggests that, if quantification were used properly, it might be possible to make a strong case for it. Such a case could be supported by a number of rationales, which I call, respectively, (1) the argument from uniformity; (2) the argument from utility; (3) the argument from democracy; (4) the argument from intuitions; (5) the argument from Gresham's Law; (6) the argument from clarity; (7) the argument from heuristic power; (8) the argument from circularity; and (9) the argument from realism. Let us consiser each of these arguments.

4.1 *The Argument from Uniformity*

The argument from uniformity, one of the oldest used to support complete quantification, rests on the claim that the valuation process central to welfare economics always implies reducing a heterogeneous aggregate to a common, uniform measure, such as money. Otherwise, proponents of the argument maintain, it is impossible to compare or aggregate costs and benefits, a procedure necessary for accurate assessment.[87] Expressed in this way, the argument can be traced back to Aristotle. In the *Nicomachean Ethics* (1133a–b), Aristotle argued that all things which were exchanged must be somehow comparable and therefore measurable by one thing. He concluded that there could be no exchange if this equality and commensurability were not present. This unit of equality and commensurability, Aristotle believed, was *demand*, "which holds all things together" (1133a).

Following Aristotle's insight, numerous proponents of complete quantification have argued that qualitative factors *must* be quantified, if they are to be commensurate with market parameters.[88] Moreover they have claimed that numerical assignments based on preference-orderings are a desirable way to reduce all items to the unit of demand.

Basically, the argument from uniformity is that, unless some common denominator, some uniform method for comparing diverse RCBA factors is employed, it is impossible to compare or to evaluate these parameters relative to each other. Even when alternative risks, costs, and benefits are compared qualitatively, there must be something, some quality or criterion, by means of which all these risks, costs, and benefits can be compared. Given that alternatives may be ordered by this criterion, indeed that they *must* be ordered if a decision is to be reached, it is reasonable to assign numbers to these orderings. But once one has assigned numbers, then one has accepted complete quantification.

4.2 *The Argument from Utility*

Commensurability and quantification are not just *epistemological* requirements for societal decisionmaking, however; they are also required for very pragmatic reasons. According to proponents of the argument from utility, complete quantification is absolutely necessary, if government is to intervene in the economy and take account of social costs relative to policymaking. Nash puts the point succinctly: Firms seeking to maximize utility or profit will ignore the costs (of their actions) to society. But this means that state intervention is required. Government must develop a system of taxes and subsidies to bring about equality between private and social costs. But bringing about this equality, in order to have optimality in the neoclassical sense, requires a quantitative measure of nonmarket, social costs.[89] Some quantitative measure of social costs is required, whether for designing taxes to secure marginal social-cost pricing, for setting standards to regulate output, for determining the optimum size of projects, and so on.[90]

Of course, if one doubts the Pigovian policy rule – that externalities necessitate government actions – then one cannot argue for complete quantification on the grounds that these actions require it. Burton, for example, maintains that this policy rule is incorrect.[91] He, Cheung, and Rowley all argue that net improvements in welfare are not accomplished by taking some income from certain (A) persons and giving it to others (B). If A inflicts harm on B, they claim, the question is not "How should we *restrain* A?". They say that to avoid the harm to B would be to inflict harm on A. For them, the relevant question is whether A should be allowed to harm B or whether B should be allowed to harm A. According to Rowley, "The task is to avoid the more serious harm".[92]

If indeed this is the relevant point, then it is not clear that externalities necessitate government actions. However, there are reasons to doubt that the issue is "the more serious harm". For one thing, the desirability of an action may be determined on the basis of who has the *right* to perform it, apart from whether performance or nonperformance results in the more serious harm. *Second*, to take as more desirable the action which avoids the more serious harm, and thereby to follow a minimax strategy, is not arguably superior to following a maximin strategy. *Third*, to take as more desirable the action which avoids the more serious harm is to focus on the desirable or undesirable consequences of the action and to ignore the moral quality of the act itself (e.g., whether one was justified in performing the action). Hence it does seem reasonable to affirm that externalities necessitate corrective government actions.

If proponents of the argument from utility are right, such corrective actions are best based on quantified RCBA, and not on some theory about neoclassical externalities. To treat nonmarket items as nonquantifiable externalities would render policymakers less able to deal with two sorts of problems: the fact that pollutants often exert their effects through an accumulated stock, rather than a flow, of the residual; and the fact that the accumulated stock of the residuals is indestructible. Because of these two problems, the individual himself cannot be left to decide, via neoclassical theories of externalities, what his preferences are regarding these pollutants. In other words, because pollutants behave as a public bad, because the jurisdiction of this bad is often extensive and irreversible, and because the individual has no means of personally monitoring his intake, he alone cannot be left to decide social policy on the basis of his preferences.[93] Government intervention, with requisite quantification, is necessary for such policymaking. This is because clearly defined, extensive pollution controls are often necessary.

In this regard, consider the example of planning for radioactive pollution control according to the $1000 per man-rem criterion of the NRC. Although I would be the first to argue that the monetary cost, assigned to the action of allowing a one man-rem radiation exposure, is seriously inadequate, nevertheless the procedure of fixing such an amount, as a basis for pollution-control expenditures, appears to be necessary. For any society in which a zero-risk pollution standard is not recognized, this seems to be the only practical way to provide government and industry with a clear criterion for required action. Hence the real question appears to be, not whether some quantitative value ought to be used, but how its assignment can be made more consistent, rational, equitable, and democratic through informed decisionmaking involving the public.[94]

4.3 *The Argument from Democracy*

In other words, the real question is not the theoretical one of whether non-market parameters can legitimately be quantified. Rather, it is the practical question of what alternatives there are to quantification. According to proponents of the argument from democracy, all alternatives to quantification run the risk of allowing mere politics, and not reason, to take over the policy process. For them, despite all the obvious problems with attempting to quantify qualitative or subjective costs and benefits, the major argument in its favor is that the alternatives to quantification are even less desirable. As one student of economic methodology expressed it: it is "often politically convenient to avoid quantitative statements, since it is then easier to assert that the policy was successful".[95] In the absence of publicly justified quantifications of costs and benefits, it is more likely that purely political, undemocratic, misleading, or secretive methods of policymaking will be employed.[96] For instance, consider one famous analysis of nuclear technology, WASH-1400. Here, authors writing under US Nuclear Regulatory Commission contracts argued that atomic energy is safe and cost-effective. since the per-year, per-reactor probability of a core melt is only 1 in 17 thousand. Because explicit numerical parameters were both used for this probability and for the costs and benefits of nuclear-generated electricity, it has been much easier for concerned citizens to investigate the issue according to established principles of scientific methodology. Without the presence of quantitative values affixed to specific parameters, controversies over nuclear technology would be reduced to vague generalities or to debate over the credentials of various spokespersons.

The great benefit of quantification is that it allows the moral issues involved to become overt and explicitly represented. This explicit numerical representation, in turn, helps to create open, rational debate over the values and preferences underlying the quantification. Without explicit quantification of all parameters, both fuzzy politics and *ad hoc* decisionmaking are likely to result. Moreover, without clear, numerical criteria in terms of which policies can be debated, and decisions can be publicly reviewed, the regulatory process is likely to be both interminable and poorly understood.[97] If there are no clearly accepted principles for how to cost or to represent qualitative parameters numerically, then it is likely to take a long time to 'get a regulation through' the government.[98] Therefore, for reasons of clarity, efficiency, and promoting rational debate among the public, quantification of all parameters appears desirable.

In response to this argument from democracy, opponents are likely to maintain that quantified RCBA does not promote clarity, efficiency, and democratic control of policy procedures and results. As one opponent put it, complete quantification is highly susceptible to abuse because the preparation of quantified RCBA requires specialized skills possessed only by experts. This being so, use of quantified RCBA means that experts are able to 'snow' people.[99]

Admittedly, such opponents are correct in believing that particular quantitative methods may not be understood by the entire populace. The relevant question, however, is whether an explicit measurement scheme is *more* accessible and understandable than an implicit, nonquantitative, perhaps unformulated, means of aggregating and weighting particular RCBA parameters. In evaluating a policy decision based on a RCBA which is not wholly quantified, citizens may not be told how the resultant weighting was accomplished, or they may be told different things by different policymakers. More likely, the policy decision may not represent a rational step in any consistent and understandable long-range plan, but may be simply the incoherent "patchwork" product of numerous compromises and deals. In other words, in the face of nonquantitative forms of policymaking, the citizen may be overwhelmed by the difficulties and intricacies of the political process and by the failure to use a precise or consistent model. However, in the face of quantitative RCBA, the problems facing the citizen reduce primarily to a lack of technical expertise. The latter difficulty is much more easily overcome than the former, since the conditions for alleviating it are clear. The quantitative methods may pose technical obstacles for the layman, but once the methods are understood, the results to the RCBA are open for all to see. The problem with using some unspecified, nonquantitative means of weighting RCBA parameters is that the difficulties besetting them are rarely merely methodological. One cannot simply master some quantitative technique and then be certain that he will not be 'snowed' by experts. Hence it may well be that quantification serves the interests of democratic control more than does the failure to quantify completely.

If it is the case that policymaking cannot get away from value judgments, then perhaps the greatest danger comes, not from making alleged value judgments through explicit quantifications, but from being unaware of what those values are when they are left vague and unquantified. One reason why this is so is that not assigning quantitative parameters "leaves the decisionmaker free to impose his own values or a range of values in aggregating risks, costs and benefits".[100]

Moreover, if quantitative parameters are not openly assigned and made the subject of public review, the economist or risk assessor can confidently anticipate a request that he somehow "organize the raw data". In the presence of a bewilderingly complex array of data, some quantitative and some qualitative, the scientist/assessor is sure to be asked by the policymaker to "provide some method by which the large variety of consequences expected from each policy be weighted in some way so as to enable the politician to compare the overall merits of the alternative policies". This request obviously places arbitrary powers in the hands of the expert or economist.[101] If nonmarket parameters were explicitly represented by alternative numerical schemes, however, then the potential for behind-the-scenes, undemocratic valuation of the parameters would be much less. Moreover, if the assessors would disclose the implications of various quantification schemes for alternative policies, then they would enable the people or their representatives to select desirable and consistent policies – policies which accorded with the wishes of the majority of the electorate.

4.4 *The Argument from Intuition*

Underlying the problem that failure to quantify all parameters often results in undemocratic ways of valuing them is a bigger difficulty. This is that, even if qualitative factors were somehow weighted politically, without the use of cardinal or ordinal measures, the results would likely be *rationally unsatisfying* and *morally uncompelling*.

Intuitive policy evaluations, accomplished through some political/democratic means, would likely be *morally uncompelling* because they would not meet the conditions necessary for participatory democracy to be an instance of procedural moral acceptability. We all probably believe that the acceptability of policy has something to do with how it is (or could be) arrived at. We believe that there are conditions, under which decisionmakers participate, which confer moral acceptability upon any agreed-upon policy that results. When we examine these conditions (e.g., noncoercion, rationality, acceptance of terms, disinterestedness, joint agreement, universality, community self-interestedness, equal and full information, nonriskiness, possibility, counting all votes, and voice), however, we discover that it is doubtful that they can be fully satisfied by real people in real decisionmaking situations.[102] But if these conditions are necessary to confer procedural moral acceptability on intuitive policy evaluations, and if these conditions cannot in practice be met, then it is likely that these evaluations are not morally compelling.

Political means of valuing nonmarket parameters likewise would probably be *rationally unsatisfying* because they could be based on power plays, emotion, failure to heed consequences of one's proposed actions, vested interests, and lack of information. Intuitive valuations, accomplished through some political means, could easily represent the enshrinement of the worst opinions. Even though cardinal or ordinal schemes for dealing with RCBA parameters leave much to be desired, they are superior to intuitive methods of policy assessment at least in that they require an organized, systematic approach. Measurement schemes at least force one to do analysis, however imperfect, while intuitive frameworks have no clear procedure, and certainly not an analytic one, for rationally approaching a problem. As Lave put it, "In the end the question is, 'Do you want to do analysis about this, or do you want to stay with your hunches?' If you want to do analysis . . . you've got to deal with it somehow".[103] People can rationally argue about the merits of different policy analyses, provided that they employ the same methodology. People are far less able to argue about the merits of different intuitions about the value of policy parameters, because the intuitions at issue are usually incommensurable. That is, different intuitions about valuing various risks, costs, and benefits normally rest on incompatible assumptions about values, e.g., Rawlsian egalitarianism versus Nozickian libertarianism.

People can only rationally argue about the merits of theories/views which are somehow commensurable. The most obvious way to render two theories commensurable is to allow proponents of each of them to assign quantitative values or rank orderings to common parameters, so as to have a basis for understanding the comparative weightings of the components of the two theories.

However, to argue for quantified RCBA (and hence for cardinal or ordinal measurement schemes as a way of representing risks, costs, and benefits), so that different theories about valuing them become commensurable, may appear to be open to a substantial objection. This objection is that persons who reject intuitive or judgmental forms of policymaking, in favor of quantified RCBA, thereby sanction only one theory of rationality, and a narrow, rigid one at that.

Proponents of quantified RCBA obviously do not wish to reduce all forms of rationality to one. Rather, their argument is that, of the many forms of rationality, quantified RCBA, including ordinal representation schemes, is the best means of *societal* decisionmaking in a *democracy*. Admittedly, there is no substitute for intuition or human judgment. Everyone uses it daily. Intuition is most successful, however, in individual situations of judgment, or cases in which all the decisionmakers share the same relevant values. In other words, intuition works best when the decisionmakers operate from some unified perspective. When persons do not share the same perspective or values, however, as is almost always the case among citizens of a region or country, decisionmaking is more difficult. Some formal, to some extent artificial, means of unifying decisionmakers and rendering their values commensurable, must be introduced. Quantified RCBA is the best of these means, because it provides for dividing a project into component risks, costs, and benefits, and for evaluating alternative weighting schemes for each of these risks, costs, and benefits. Moreover, *any* formal scheme that specified the relevant decision parameters, and provides a means for alternative weightings of those parameters, based on commensurable theories, would be equally acceptable. The point is that intuition provides neither for specification of relevant parameters nor for consideration of alternative weightings and rankings (ordinal measurement).

Admittedly, intuition might provide a means for policymaking and for evaluating the various risks, costs, and benefits of a particular project, if there were some *procedural* way of guaranteeing that all points of view on the project might be heard, considered, and evaluated. With suitable safeguards of procedural justice, applied to political negotiation, policymaking about technological projects could rely completely on human judgment and intuition, informed by occasional use of quantitative results. Citizens, presumably, could debate the merits of alternative, nonquantitative criteria for evaluating such projects.

Persuasive as it sounds, there are major problems with an appeal to intuition *cum* political negotiation informed by procedural justice, as an alternative to quantified RCBA. One problem (already mentioned) is that such forms of policymaking would likely be morally uncompelling. Another obvious problem is that, with so many technological projects to be evaluated, and with democracy involved in much other business, it is practically impossible to relegate policymaking to the political process. The whole point of using some analytic methodology is to avoid the inefficiency and practical impossibility of relying on forms of policymaking which are wholly political. Hence, to argue for a political mechanism fueled by intuition is to ask for a return to inefficiency and to choking the wheels of democracy. Given that all citizens cannot make all decisions, and even that all federal legislators cannot make all decisions about technology policy, we seem required to use a decision framework which is part analytic-quantitative and part political. It must be understandable to, and accessible by, most citizens and open to modification by them. Intuition does not meet this requirement, as was argued earlier, because it provides no unified perspective from which to view many alternative theories of value. And intuition, plus political negotiation safeguarded by procedural justice, together, do not meet this requirement because they are both morally uncompelling and unworkable from a practical point of view. Hence proponents of quantified RCBA are not sanctioning only one theory of rationality, but the most workable means of rationally making decisions in a democracy, given that other forms of decisionmaking are morally uncompelling, are effective only for cases involving individuals, or for situations with no practical constraints.

Just as opponents of complete quantification want to rely on intuition and not cardinal or ordinal representation to guide their policymaking, so also, in an earlier era, traditionalists fought the progressive extension of mathematics into new fields. "The irrationals, the surds, the imaginaries, and the negatives are numbers that still bear names reminiscent of protest – protest against outlandish practice and against the writing of unauthorized absurdities. But orthodoxy bent to chaos by rationalizing the use of irrationals and by imagining a broader domain in which imaginaries and negatives could serve a proper elements".[104] Likewise, it could be argued that the concept of measurement in RCBA ought to be enlarged so as to include the assignment of numerals to qualitative items according to some (any) consistent rule. If indeed practice dictates this enlargement, and if intuitive assessments of policy parameters are the only alternatives to it, then the proponents of the argument from intuition have a powerful case.

4.5. *The Argument from Gresham's Law*

One reason why it is desirable to enlarge the concept of measurement used in RCBA is that purely qualitative factors are often devalued if they are not expressed in quantitative form. In the face of some RCBA risks, costs, and benefits which are quantified and some which are not, the tendency of most persons is to ignore those in the latter group, often because they don't know how to take account of them. Without cardinal or ordinal representations of nonmarket costs and benefits, a modern-day, double-barreled Gresham's Law operates: "monetary information tends to drive out of circulation quantitative information of greater significance, and quantitative information of any kind tends to retard the circulation of qualitative information".[105] If this is true, then failure to quantify nonmarket parameters serves the interests of those who wish either to consider only market costs and benefits, or to limit the scope, and therefore the validity, of risk-cost-benefit analysis.

Our responses to RCBA's which are part quantitative, part qualitative, illustrate Gresham's Law because of our psychological tendency to believe that whatever factors we are best able to measure and order are most important. Even though we all maintain that qualitative parameters are important, the fact is that they appear not to be *counted* as such because people do not know how to deal with them. As will be clear from the examples discussed later (section 5), failure to represent all RCBA parameters by some cardinal or ordinal measure means that the qualitative factors are simply ignored by most people.

When qualitative parameters are not counted as important, and when the factors which they represent are adverse (e.g., ignored spillover effects), failure to quantify all RCBA parameters results in a bias toward narrowly commercial projects.[106] In other words, excluding consideration of non-measurable or uncalculatable effects results in "a serious understatement of the costs" of an enterprise.[107] This means, for example, that when natural resources are (in effect) given a zero price/cost because they are treated as allegedly qualitative, nonmarket goods, a number of undesirable consequences follow: (1) There is a discrepancy between private and social costs and a suboptimal allocation of environmental, production, labor, and capital factors. (2) The opportunity costs are not fully appreciated, e.g., when water is used for dumping pollutants. (3) The prices of goods with pollution nuisances do not reflect their environmental nuisance. And (4) because prices are skewed, goods are not allocated efficiently.[108]

If indeed it is true that, following Gresham's Law, people tend to discount qualitative information, then the real question is not whether complete quantification is arbitrary. Rather the real question is whether it is more arbitrary to quantify imperfectly (but have the qualitative factors counted) or not to quantify nonmarket items at all (and have them ignored). Proponents of the argument from Gresham's Law believe that it is better to have qualitative parameters counted and therefore better to quantify them, rather then to have them ignored. If they are correct in this belief, then it is misguided to avoid complete quantification on the grounds that one ought not to give "an arbitrary value" to a qualitative item.[109] It is misguided because a greater arbitrariness, even error, arises from not counting the qualitative factors at all.

Of course, it could be claimed, by opponents of the argument from Gresham's Law, that people ought to be trained not to ignore qualitative factors when they are presented with quantitative data. While theoretically correct, this response appears to overestimate people's ability not to be dominated by quantitative thinking. Moreover, so long as they are dominated (rightly or wrongly) by such quantitative thinking, then strategies for counting risks, costs, and benefits will have to take account of the consequences of this domination. For practical reasons, complete quantification may be necessary.

4.6. *The Argument from Clarity*

Besides people's tendencies to overestimate the importance of quantitative data, there is another strong, *practical* argument in favor of complete quantification. This 'argument from clarity' comes down to the trite observation that "the devil you know is better than the devil you don't know". The main reason for quantifying all parameters is that the methodological assumptions about their values are more obvious, easier to 'get at', and therefore easier to criticize than are the presuppositions behind qualitative, subjective, or intuitive means of formulating costs and benefits.

Opponents of the argument from clarity disagree that quantification helps one to uncover the assumptions beneath his valuations. As Self puts it, the use of numbers does not help one to think more carefully. He claims that if a person thinks narrowly, quantification will not help him. But, he says, if a person thinks broadly, quantification is not needed.[110] This response to the argument from clarity does not work, however, and for several reasons.

First, Self erroneously assumes that a person finds himself in only one of only two possible positions: either he thinks broadly about RCBA or he thinks narrowly. In reality, it is probably closer to the truth to say that people think neither narrowly nor broadly, but that there are gradations both in their narrowness and broadness and in the ways in which this narrowness and broadness is exhibited. For example, a person might be extremely narrow in the ways he views factors A, B, C, D, and E, middle-of-the-road in the ways he views F, G, H, I, J, K, and L, but moderately broad in ways he sees factors M, N, O, P, Q, and R. If this plurality of multivariate positions indeed describes where people stand on being 'broad' or 'narrow', then it is clear that quantification might help them to expand and to clarify their ways of thinking about things. Because Self explains thinking in simplistic, binary terms, it is not clear that he has proved anything regarding the desirability of quantification.

In a second response to proponents of the argument from clarity, Self discounts the value of complete quantification because he says that the *numbers* themselves do not clarify. Rather, he maintains that the clarity value of complete quantification lies in the systematic listing and use of supporting *reasons* for a particular quantification, and not in the numbers themselves.[111]

What this claim ignores, however, is that the status of supporting reasons, in the case of alternative numerical representations, and in the case of alternative nonquantitative valuing schemes, is not analogous. It is obviously easier, 'cleaner', and more precise to argue about which numerical representation for a particular parameter is best, than it is to argue about which qualitative valuing scheme is most desirable. This is because, although people will disagree about both quantitative and qualitative valuings, the debate over particular cardinal or ordinal assignments is narrower, and thus more manageable, than debate over qualitative assessment. Debate over quantifying an item is over how to rank a *particular parameter*, relative to others. Debate over qualitative evaluations, however, is over how to value *all the components* in a decision scheme. Quantification provides more clarity precisely because it divides the decision problem into numerous components (risks, costs, benefits), the quantification of each of which can be debated. Without such division into components, the potential for misunderstanding, unresolved disagreement, and talking at cross purposes seems much greater.

4.7. *The Argument from Heuristic Power*

From the point of view of the practicing economist, it also seems likely that complete quantification is desirable. As Edgeworth put it, the most valuable economic theory is that which is mathematical.[112] Use of mathematics is to be prized, not necessarily because everything is amenable to mathematical treatment – it may not be – but because economists can only check the correctness of their views by attempting to develop mathematical theories having predictive power. According to proponents of the argument from heuristic power, quantification of all parameters will enable one to develop precise predictive theories governing those parameters. Granted, those predictions may not work. Nevertheless, economists learn something about why these predictions do or do not work, to better or worse degrees, by observing the effects of their assigning particular numbers to certain factors.

As science develops, it no longer merely *investigates* the world. Instead its practitioners *create* the world which it is investigating. Especially in the social sciences, like economics, almost all we can know is what we create ourselves. Prediction in these social sciences "can be achieved only by setting up consciously created systems which will make the predictions come true".[113] Thus, whether the predictions are achievable or not, the only way to learn this is to attempt to make correct predictions and to see what happens if one assumes that certain parameters have a given numerical value.

4.8. *The Argument from Circularity*

Not to attempt to learn what one can from the 'thought experiment' of complete quantification is simply to beg the whole question of whether complete quantification is useful. Thus, for proponents of the argument from circularity, complete quantification is desirable because it enables one to avoid the unscientific, metaphysical, and question-begging presupposition that quantification is impossible. Granted, quantification may be impossible, say proponents of this argument. But those who despair will not succeed, if it is possible.[114] Just as electricity was previously thought immeasurable, but was later found to be so, perhaps utility, welfare, or societal preference will be discovered to be measurable. Whether or not it is, we can only find out by trying, not by begging the question of measurability.

If one is so wedded to the question-begging approach of not attempting to measure nonmarket parameters, and if one believes that he is fundamentally constrained from using mathematics to gain insights about problems of resource allocation, then for him there is no way to quantify all RCBA parameters so that this constraint is removed. This means that, for this sort of opponent of complete quantification, the world must be viewed in a particular way. But if this is the case, then his anti-quantification 'paradigm' has become a metaphysics in the worst sense of the term. The problem for the person who wants to use RCBA in making public policy decisions is *not* how to generate models of preference-orderings under the constraint that scalar comparisons cannot be made. Rather the non-question-begging problem facing him is *whether*, and if so, precisely *why* such scalar comparisons are desirable, what kind of system or 'world' might permit such comparisons, and what the results of making these comparisons might be.[115]

4.9. *The Argument from Realism*

In other words, proponents of complete quantification are not necessarily optimistic about the consequences of assigning numerical representations to all RCBA parameters. They do believe, however, that it is at least reasonable to ask how one might numerically represent the many factors competing for his time or his money. Moreover, say proponents of the argument from realism, it may be far more realistic to attempt to obtain such measures than merely to assign subjective valuations to them. After all, subjective (nonquantitative) valuations, or valuations not measured at least ordinally, are quite artificial, at least in one sense. One can assign them without estimating how much something is worth in terms of *other claims* upon his time or his money.[116] Quantification, or at least an ordinal ranking, is necessary in order to obtain a general rule for evaluating programs relative to each other.[117] Without such an ordinal ranking, realism is not served. Real-world policy must be made with finite numbers of dollars and finite numbers of policy alternatives. In the light of these finite figures, we can talk about assigning a quantitative representation to allegedly qualitative figures.

The usefulness of such numerical assignments can be seen quite easily, even if one is uncertain about their precision. Take the case of a proposal to build a newer, larger airport in a major city.[118] Suppose that the RCBA for the airport is completed, but that (typically) it ignores certain qualitative costs (e.g., airport noise) because they are 'unpriceable' or 'incommensurable'.

Suppose further that, having ignored these nonmarket costs, the study shows that the benefits of the airport project exceed the risks and costs by $10 million a year. However, if one considers that approximately 500 thousand families in the immediate vicinity of the airport suffer the 'qualitative' costs of aircraft noise, traffic congestion, and increased loss of life, then it is not clear that the qualitative costs can be ignored. In fact, if the 500 thousand families each suffered $20 annually because of these nonmarket costs, then the benefits of the airport could not be said to exceed the costs and risks. This means that, even if there is no *nonarbitrary* means of quantifying the airport's qualitative costs, arbitrary assignment of the $20 per family figure gives the public something to debate and a framework for ordering their preferences. It is much easier to ask: "Is it worth $20 to us to be rid of the airport noise, congestion, and traffic for one year?" than is is to ask whether the pros outweigh the cons regarding the new airport. All the difficulties of arbitrariness, objectivity, and incommensurability aside, it is much easier to think about the airport problem if we do so in this quantitative way, and use the numerical assignments hypothetically, than if we do not have the benefit of the '$20 consideration'.

5. THE CONSEQUENCES OF PARTIAL QUANTIFICATION IN ACTUAL RCBA'S

What the argument from realism suggests is that, although there may be epistemological reasons against complete quantification, there are a number of practical considerations in its favor. This whole issue of practicality brings us to an interesting question. "What have been the practical consequences of following the methodological assumption favoring only partial quantification?" If proponents of complete quantification in RCBA are right, and especially if proponents of the argument from democracy and the argument from Gresham's Law are right, then RCBA studies employing only partial quantification ought to exhibit several telling characteristics. One likely characteristic is that nonquantified costs and benefits will be ignored in the analysis. Consideration of representative technological studies illustrates both that many assessors have in fact failed to consider whatever parameters were not quantified, and that their omissions have led to a number of undesirable results. Some of the consequences include: (1) a failure to determine the actual RCBA desirability of various technological programs and policies; (2) a tendency to neglect consideration of social and political solutions to pressing environmental and technical problems; (3) an apparent willingness to draw specific conclusions unwarranted by the data; (4) a reluctance to investigate the social costs of technology; and (5) a tendency to exhibit pro-industry, pro-technology, pro-status-quo bias in the analyses. Let us see how these consequences are exhibited.

Consider first a recent OTA assessment of whether to develop the flows of the Uinta and Whiterocks Rivers in central Utah. The purpose of the study was to determine, on RCBA grounds, whether such development, for recreational uses, irrigation, and municipal and industrial water supplies, was desirable. Surprisingly the calculations included as the *only costs* of the project, those for construction of a dam, reservoir, and canal. The nonmarket hazards of pollution and congestion, likely to result from industrialization, were not considered. Moreover the authors of the report assumed that undeveloped land had no market value for recreational benefits, and they ignored the qualitative, negative social, political, and enviromental effects of development. Instead they concluded that the aggregate benefits of development obviously far outweighed the costs.[119]

Likewise the authors of recent OTA studies on pest management, for example, ignored all the 'qualitative' social, medical, and environmental hazards of using chemicals. Instead they included only three easily quantifiable parameters (cost of pesticides, , and value of crops with and without employment of herbicides and insecticides). As a consequence, they concluded that use of chemical pest control was not only cost-effective, but more cost-effective than biological means of crop management. (Costs of biological control were not calculated.)[120]

A similar, erroneous methodology was followed in the OTA assessment of oil tanker technology. The study contains *no calculations* for the costs of oil spills (including damage to recreational beaches, to the commercial fishing industry, and to aesthetic quality.)[121] In fact, the economic consequences of accidents were ignored completely; the only costs calculated, those regarding safety improvement, were for various types of ship design, navigational aids, and alternative control systems for the vessels.[122] Since the report ignored the social and economic parameters that were difficult to calculate, it is not surprising that *social* (e.g., restriction of travel in some enclosed waters) and *economic* (e.g., mandatory increases in liability coverage as a deterrent to operation of some ships) solutions to the oil spill problems were also ignored. Instead only technological solution (e.g., improved navigation systems), whose costs were easily quantifiable, were considered in the analysis. This suggests that application of the accepted methodological principle, of not quantifying nonmarket parameters, may lead to ignoring crucial RCBA considerations. Were these omissions taken into account, it is not clear that the new RCBA results would support the conclusions of the original assessment. It also suggests that, precisely because their costs and benefits are difficult to quantify, social and political solutions to the relevant problem may be ignored.

Certain nontechnological solutions were assuredly omitted from consideration, for example, in a recent RCBA study of railroad safety done by the OTA. The authors of the report specifically denied that one type of social-political solution, increased regulation of the railroads, was cost-effective.[123] The denial is puzzling, however, because regulation intuitively appears to be at least a possible solution to the number-one cause of increased accidents: deferred maintenance, especially on tracks. Regulation, in the form of mandatory use of the "Hazardous Information Emergency Response" form (used by Canadian railroads), also seems to be at least a possible solution to the problem that 65 percent of all US railway cars loaded with hazardous materials are involved annually in accidental releases of these substances.[124]

Because the assessors omitted RCBA studies of programs whose costs and benefits are difficult to quantify – regulatory, inspection, and track-maintenance programs – there was no hard data to prove that regulation was/was not cost-effective. The authors even admitted that railroad policy, especially as regards safety, was based on market-determined risks and benefits and not on inclusion of other assessment parameters, such as (nonmarket) social cost.[125] They also explicitly admitted that the reason, why the cost-effectiveness of certain programs was not calculated, was that they followed the methodological principle under consideration and therefore eschewed parameters and goals which were not 'measurable'.[126] Such items were not included, they said, because of "data gathering difficulties".[127] This leads one to suspect that, in the absence of market-based RCBA 'proof' of means to reduce risks or to lower social costs, the industry or technology is assumed innocent until proven guilty. Precisely because qualitative social costs are ignored, while quantitative benefits are emphasized, the policy of not quantifying nonmarket parameters tends to produce a pro-technology bias among policymakers.

Similar pro-technology bias is clearly evident in a recent OTA study of coal-slurry technology. Only easily quantifiable market costs were considered (e.g., pumping water for use in the pipelines),[128] while more qualitative costs were *not* priced (e.g., any use of water where it is a scarce natural resource, as in the western US).[129] As a consequence of this partial quantification, the authors of the assessment were readily able to draw a conclusion in favor of the technology. They claimed: "slurry pipelines can, according to this analysis, transport coal more economically than can other modes [of transportation]".[130]

The same sort of unwarranted conclusion was drawn in another study, an OTA assessment of the technology for transport of liquefied natural gas (LNG). The authors concluded that existing US Coast Guard standards were cost-effective, i.e., adequate to avoid catastrophe as a result of LNG ship failure.[131] This conclusion was reached in spite of (and perhaps on account of?) the fact that RCBA calculations were not done for the social and enviromental costs of a major LNG spill. They were not done, presumably, because researchers claimed they would be too subjective; instead they said that decisions should be based on "nonquantitative approaches".[132] As a result, the assessors implicitly judged that the benefits outweighed the costs. They sanctioned the status quo, a continuation of the use of the technology as currently employed,[133] even though LNG transport accidents in the past have killed hundreds of persons and caused billions of dollars in damage.[134] In the light of such potential hazards, one wonders how "nonquantitative approaches" are sufficient to justify a 'business as usual' conclusion.

A particularly interesting case of pro-technology bias, apparently resulting from the failure to quantify all RCBA risks, costs, and benefits, is evidenced in a recent assessment of the computed tomography (CT) scanner. Although the assessors considered market factors such as purchase price of the machines, operating costs, and profits to the hospital as a result of their use, they excluded the nonmarket social, medical, and environmental parameters necessary for determining the true costs and benefits of a scan. They noted, however, that high hospital profits were made on CT scans,[135] that head scans tended to be used in 90 percent of cases unnecessarily,[136] and that a typical CT scan exposed the patient to 30 rads of radiation.[137] They failed, however, to compute the alleged subjective or qualitative costs of increased cancers and genetic injuries occasioned by such exposure.

Failure to calculate the nonmarket costs of CT scans is a significant omission. For example, in the 30-year period following one CT-scan exposure to 30 rads of ionizing radiation, 2 of every 100 persons exposed will contract cancer simply because of the CT scan. (This calculation is based on the standard *BEIR* report, *Biological Effects of Ionizing Radiation* of the NAS (National Acadeny of Sciences), used by the US government to yield dose-response statistics.) If there is a latency period (for cancer) of 30 to 40 years, and if .0002 cancers per year are induced by exposure to one rad of radiation, then 2 of the 100 persons exposed to 30 rads will contract the disease from this exposure alone.[138]

Despite their omitting discussion of such social costs (cancer) in the RCBA, the assessment team concluded that computed tomography scans are "a relatively safe and painless procedure".[139] Obviously, however, CT safety is a function of a particular radiation risk as measured against the specific benefit of more accurate radiological diagnosis. Since the assessment team mentioned criteria for situations in which the CT scan was likely to be negative (suggesting, therefore, cases in which its use was contraindicated because of radiation risk),[140] the risks, costs, and benefits of use of the scans in different situations easily could have been assessed in the RCBA. Because the hazards posed by cancer and genetic damage were not quantified, however, the nonmarket cost of radiation risk was ignored.[141] As a consequence, this technology assessment did not accurately characterize the risks, costs, and benefits of the scanners. This means that, although the benefits do not outweigh the risks (of CT scans) in some situations, the assessment failed to note this fact. If the assessment is followed, it may cause CT scans to be misused in exactly those situations.

Failure to quantify nonmarket costs has also generated a bias in favor of the status quo in various studies of solar energy, nuclear power, and coal-generated electricity. In the US government report, WASH-1224, for example, in which the risks, costs, and benefits of nuclear- versus coal-generated electricity were compared, the authors concluded that atomic energy was a more cost-effective means of producing power. However, they did not include the nonmarket costs of radioactive waste storage in their calculations. If this one allegedly 'nonmarket cost', waste storage (as given on the basis of actual US government expenditures), had been quantified and included in the computations, when all other parameters remained the same, then the opposite conclusion would follow. That is, coal, and not nuclear power, could be shown (by an even wider margin) to be the cheaper source of electricity.[142]

Likewise, solar power can be shown to be more, or less, cost-effective than conventional energy sources, depending on whether allegedly subjective parameters (e.g., health risks) are, or are not, quantified and included in the RCBA. This means that failing to quantify (and therefore ignoring) nonmarket costs can bias assessments in favor of conventional energy technologies. When nonmarket parameters are quantified and included in the computations for the cost-effectiveness of various energy technologies, solar power and conservation (for the first 10 million barrels per day of oil equivalent) are cheaper than all other conventional energy sources. But when nonmarket parameters (especially social costs) are excluded, on the other hand, then solar and conservation are said to be more expensive than conventional power sources.[143] According to the US Office of Technology Assessment (the OTA), though onsite solar devices *could supply* "over 40 percent of US energy demand by the mid 1980's".[144] whether they *will do so*, in fact, will depend upon factors such as the public's perceptions of their costs. Hence, future energy policy will be determined, in part, by whether all the costs of various energy technologies are quantified and considered. This, in itself, is a compelling argument for complete quantification.

Just as future policy regarding solar power may be skewed erroneously by failure to quantify nonmarket costs and, as a consequence, ignoring those costs, so also future policy regarding the automobile is likely to be controlled in part by failure to quantify nonmarket costs and, as a consequence, ignoring them. Consider, for example, the likely policy consequences of British, versus American, assessment procedures regarding quantification. Whereas British government studies of auto technology always quantify and include the nonmarket costs of accidents in their RCBA calculations, US-government RCBA analyses of the same technology typically refrain both from quantifying nonmarket costs and from taking them into account.[145] The British *Transport Policy* bases its calculations for the nonmarket costs of accident-caused pain and suffering, for instance, on the assumption that these costs are approximately 50% as great as total property damages for the accident and 600% higher than total medical expenses.[146] This suggests that, if the British assessments quantify and include such auto costs, while the US studies do not, then US conclusions may likely underestimate the costs of automobile technology. This suggests further that, if the British figures are good estimates of the real costs of accidents, and if US RCBA's followed them and employed complete quantification, then assessors of automobile technology might have to revise the essentially pro-auto policy embraced in the most recent US studies.[147]

One of the major reasons, for the pro-auto conclusions of the US reports, appears to be that their authors have employed the methodological principle of not quantifying, and therefore not including, allegedly subjective or nonmarket parameters in their aggregations of risks, costs, and benefits. In the most extensive and up-to-date US government assessment, for example, the authors include only the individual's yearly *market* cost (for depreciation, insurance, gas, etc.) of owning and operating a car.[148] The calculations exclude prices for items such as air pollution, noise, risk of death and injury, and resource depletion.[149] Neither these nonmarket cost factors, nor the nonmarket benefit of 'personal mobility' were assigned quantitative values in the RCBA. Yet the assessors discounted the significance of the nonmarket, qualitative costs, and emphasized auto benefits. They concluded that the benefits of private transport outweighed both the benefits of mass transit and the total costs of employing the automobile.[150] In yet another case, failure to quantify all nonmarket parameters appears to have caused the qualitative social costs to be ignored, with the result that the benefits appear to outweigh the costs of the technology.

6. CONCLUSION

If the insights afforded by these few technology assessments and environmental-impact analyses are typical, then they suggest that following the assumption of partial quantification leads to a number of undesirable consequences. These include the tendency to ignore nonmarket, qualitative costs, with the consequence that the benefits of the technology appear to outweigh the costs. This means that the RCBA is likely to be weighted in favor of a pro-industry, pro-technology solution to the policy questions it addresses. This, in turn, suggests that reasonable, disinterested policymakers might not be serving their own best interests if they oppose complete quantification. If the RCBA studies just discussed are representative – and further study will be needed to discover whether they are – then partial quantification is likely to lead to assessors' ignoring qualitative, nonmarket costs. It may well be that complete quantification is the best way to insure a reasonable, complete, disinterested discussion of all relevant RCBA parameters.

Earlier I pointed out that many economists and philosophers opposed quantification of nonmarket parameters, on the grounds that such pricing was an example of 'money fetishism' and 'economic Philistinism'. This essay suggests that failure to quantify allegedly qualitative costs, however, results in their being excluded from technology assessments and environmental-impact analyses. As a consequence, human risk, pain, suffering, and death are thereby ignored. If failure to quantify indeed causes us to ignore such factors, then nonquantification also results in a very practical sort of 'economic Philistinism', and perhaps one more serious than that arising from imperfect attempts to assign cardinal or ordinal measures.

NOTES

[1] A. L. Macfie, 'Welfare in Economic Theory', *The Philosophical Quarterly* 3 (10), (January 1953), 59.

[2] S. Koreisha and R. Stobaugh, 'Appendix', in *Energy Fugure* (ed. by R. Stobaugh and D. Yergin), Random House, New York, 1979, p. 234; hereafter cited as: Appendix, in *EF*.
[3] See, for example, E.C. Pasour, 'Benevolence and the Market', *Modern Age* 24 (2), (Spring 1980), 168. TA's and EIA's are widely held to be unbiased, outside the realm of policy or value judgments, nonpartisan, and objective (Congress, US Office of Technology Assessment, *Annual Report to the Congress for 1977*, US Government Printing Office, Washington, D.C., 1977, p. 4; hereafter cited as: Congress, OTA, *AR 1977*.
[4] As was mentioned earlier, in Chapters One and Two, all US regulatory agencies now base their assessments in part on RCBA, and the 1979 NEPA has been interpreted to require a risk-cost-benefit assessment. See Peter Self, *Econocrats and the Policy Process*, Macmillan, London, 1975, p. ix; hereafter cited as: Self, *PPCBA*. See also L. J. Carter, 'Dispute over Cancer Risk Quantification', *Science* 203 (4387), (March 30, 1979), 1324–1325; hereafter cited as: Carter, Dispute. See also C. Starr and C. Whipple, 'Risks of Risk Decisions', *Science* 208 (4448), (June 6, 1980), 1118; hereafter cited as Starr and Whipple, Risks. Finally, see Joel Yellin, 'Judicial Review and Nuclear Power', *George Washington Law Review* 45 (5), (August 1977), and J. R. Luke, 'Environmental Impact Assessment for Water Resource Projects', *George Washington Law Review* 45 (5), (August 1977), 1106–1107; hereafter cited as: Luke, *EIA*.
[5] See Koreisha and Stobaugh, Appendix, p. 11.
[6] Self, *PPCBA*, p. ix.
[7] Lester Lave, 'Public Perception of Risk', in Mitre Corporation, *Symposium/Workshop ... Risk Assessment and Governmental Decision Making*, The Mitre Corporation, McLean, Virginia, 1979, p. 577; hereafter cited as: Lave, Public, and Mitre Corporation.
[8] S. Gage, 'Risk Assessment in Governmental Decision Making', in Mitre Corporation (note 7), pp. 11–13. See Sheldon Samuels, 'Panel on Accident Risk Assessment', in Mitre Corporation (note 7), p. 391.
[9] Quoted by Alexander Rosenberg, *Microeconomic Laws: A Philosophical Analysis*, University of Pittsburgh Press, Pittsburgh, 1976, p. 155. See also N. Georgescu-Roegen, *Analytical Economics*, Harvard University Press, Cambridge, 1966, p. 186, who makes the same point; hereafter cited as: Georgescu-Roegen, *AE*.
[10] See E. J. Mishan, *Economics for Social Decisions*, Praeger, New York, 1972, pp. 11–14; hereafter cited as: Mishan, *ESD*.
[11] For confirmation of the fact that welfare economists generally do not quantify nonmarket costs and benefits, see R. M. Hare, 'Contrasting Methods of Environmental Planning', in K. E. Goodpaster and K. M. Sayre (eds.), *Ethics and the Problems of the 21st Century*, University of Notre Dame Press, Notre Dame, 1979, pp. 64–68 (hereafter cited as: Hare, Methods, and Goodpaster and Sayre, *Ethics*); E. J. Mishan, *Welfare Economics*, Random House, New York, 1969, p. 86; hereafter cited as: WE. See also M. W. Jones-Lee, *The Value of Life: An Economic Analysis*, University of Chicago Press, Chicago, 1976, pp. 21–28 (hereafter cited as: *Value*); and L. H. Mayo, 'The Management of Technology Assessment', in R. G. Kasper (ed.), *Technology Assessment: Understanding the Social Consequences of Technological Applications*, Praeger, New York, 1972, p. 78 (hereafter cited as: Kasper, *TA*). Employment of this principle of nonquantification raises an interesting epistemological issue. (1) Can welfare economists be said not to be using the notions of Pareto Optimum and 'compensating variation,' since they do not include all cost-benefit parameters in their calculations? (2) Or, on the other hand, may

they be said to employ modified versions of these two concepts, since they are not practically usable as defined in economic theory? Whether either (1) or (2), or neither, is the case will not substantially affect the discussion in this section. Although most economists would probably agree with (2), the point of examining the methodological principle here (regarding nonquantification of some parameters) is to assess its desirability and not to determine its status as Pareto-based or not.

[12] H. P. Green, 'Cost-Risk-Benefit Assessment and the Law', *George Washington Law Review* 45, (5), (August 1977), 904–905; hereafter cited as: Green, CRBA. See also Self, *PPCBA*, pp. 78–79. E. Mishan, *Cost-Benefit Analysis*, Praeger, New York, 1976, pp. 160–161; hereafter cited as: Mishan, *CBA*. See also Mishan, *ESD* p. 21, and E. Rotwein, 'Mathematical Economics', in S. R. Krupp (ed.), *The Structure of Economic Science*, Prentice-Hall, Englewood Cliffs, 1966, p. 102; hereafter cited as: Rotwein, *ME*, and Krupp, *SES*.

[13] Mishan, *CBA*, pp. 160–161. The economists' practice of not quantifying nonmarket parameters is also generally adhered to by government regulators. According to one observer, "nobody in Washington puts a dollar value on lives, pain, or injuries." (Fred Hapgood, 'Risk-Benefit Analysis', *The Atlantic* 243 (1), (January 1979), 36.

[14] T. C. Means, 'The Concorde Calculus', *George Washington Law Review* 45 (5), (August 1977), 1044.

[15] Luke, EIA (note 5), p. 1108.

[16] Plato, *Philebus*, tr. with notes by J. C. B. Gosling, Clarendon Press, Oxford, 1975, pp. 7, 9 (16c–17b, 18a–18c).

[17] Georgescu-Roegen, *AE*, (note 9), p. 171.

[18] S. S. Stevens, 'Measurement, Psychophysics, and Utility', in C. W. Churchman and P. Ratoosh (ed.), *Measurement: Definitions and Theories*, John Wiley, New York, 1959, p. 36; hereafter cited as: Stevens, Measurement, and Churchman and Ratoosh, *Measurement.*

[19] Although I will not argue this broader point here, my claim is that rational assessment procedures require an unambiguous means of comparing policy alternatives. Unless various options can be expressed in terms of a 'common denominator' (e.g., ordering, preference-ranking, rating, quantification), one not necessarily based on a 'numerical' or monetary system, then there is little assurance that technology assessment will be as rational or as objective as might be possible. For a discussion of analytical, versus intuitive, modes of risk assessment, see C. Starr and C. Whipple, Risks (note 4), pp. 1114–1119.

[20] L. J. Carter, 'Dispute over Cancer Risk Quantification', *Science* 203 (4387), (March 30, 1979), 1324–1325.

[21] This argument is given in L. H. Mayo, 'The Management of Technology Assessment', in Kasper, *TA* (note 11), p. 78.

[22] F. A. Hayek, 'The New Confusion About "Planning"', in E. F. Paul and P. A. Russo (eds.), *Public Policy*, Chatham House, Chatham, New Jersey, 1982, p. 307.

[23] J. A. Hobson, *Confessions of an Economic Heretic*, Harvester Press, Sussex, England, 1976, pp. 39–40; hereafter cited as: *Confessions*. See also B. M. Anderson, *Social Value: A Study in Economic Theory Critical and Constructive*, A. M. Kelly, New York, 1966, pp. 24–26, 31, 162; hereafter cited as: *Social Value*. See K. E. Boulding, 'The Basis of Value Judgments in Economics', in Sidney Hook (ed.), *Human Values and Economic Policy*, New York University Press, New York, 1967, pp. 85–88; hereafter cited as: Hook, *HV and EP*, and Boulding, 'Basis'.

[24] See K. E. Boulding, 'Basis', in Hook, *HV and EP*, pp. 67–68. Oskar Morgenstern, *On the Accuracy of Economic Observations*, Princeton University Press, Princeton, N. J. 1963, p. 19 (hereafter cited as: *Accuracy*), ties the "errors of economic statistics", such as price, in part to the fact of the prevalence of monopolies. In an economy characterized by monopoly, he says, statistics regarding price are not trustwothy because of "secret rebates granted to different customers". Moreover, he claims, "sales prices constitute some of the most closely guarded secrets in many businesses". For both these reasons it is likely not only that price ≠ value, but also that actual price ≠ official market price.

[25] R. C. Dorf, *Technology, Society, and Man*, Boyd and Fraser, San Francisco, 1974, pp. 223–240 (hereafter cited as: TSM), and H. R. Bowen, Chairman, National Commission on Technology, Automation, and Economic Progress, *Applying Technology to Unmet Needs*, US Government Printing Office, Washington, D.C., 1966, pp. v–138; hereafter cited as: *Applying Technology*. See also K. E. Boulding, 'Basis', in Hook, *HV and EP*, pp. 67–68, and E. J. Mishan, *CBA* (note 12), pp. 393–394. Externalities (also known as 'spillovers', 'diseconomies', or 'disamenities') are social benefits or costs (e.g., the cost of factory pollution to homeowners nearby) which are not taken account of either in the cost of the goods produced (e.g., by the factory) or by the factory owner. They are 'external' to cost-benefit calculation, and hence do not enter the calculation of the market price. For this reason, says Mishan (*The Costs of Economic Growth*, Praeger, New York, 1967, p. 53; hereafter cited as: *CEG*), "one can no longer take it for granted that the market price of a good is an index of its marginal price to society". Another way of making this same point (Mishan, *CEG*, p. 57) is to say that diseconomies cause social marginal costs of some goods to exceed their corresponding private marginal costs; this means that the social *value* of some goods is significantly less than the (private) market *price*.

[26] See K. E. Boulding, 'Basis', in Hook, *HV and EP*, pp. 67–68, and E. F. Schumacher, *Small Is Beautiful: Economics as if People Mattered*, Harper, New York, 1973, pp. 38–49; hereafter cited as: *Small*.

[27] There are no monetary-term values for natural resources because the 'cost' of using natural resources is measured in terms of low entropy and is subject to the limitations imposed by natural laws (e.g., the finite nature of nonrenewable resources). For this reason, viz., the theoretical and physical limit to accessible resources, the price mechanism is unable to offset any shortages of land, energy, or materials.

[28] See, for example, A. C. Pigou, *The Economics of Welfare*, Macmillan, London, 1932, and Mishan, *WE* (note 11).

[29] Mishan, *ESD* (note 12), p. 69.

[30] See Self, *PPCBA* (note 2), pp. 76–86; see also pp. 71–72.

[31] A. L. Sorkin, *Economic Aspects of Natural Hazards*, Lexington Books, Lexington, Massachusetts, 1982, p. 62; hereafter cited as: Sorkin *EANH*.

[32] D. W. Pearce, 'Introduction', in Pearce (ed.), *The Valuation of Social Cost*, George Allen and Unwin, London, 1978, p. 2; hereafter cited as Pearce, 'Introduction', and Pearce, *Valuation*.

[33] See Martin Hollis and Edward Nell, *Rational Economic Man*, Cambridge University Press, London, 1975, who claim that neoclassical economics rests essentially on a positivistic philosophy of science.

[34] R. M. Hare, Methods (note 11), p. 76.

[35] Mishan, *CBA* (note 12), p. 161, notes this fact.

[36] Quoted by B. M. Gross, 'Preface', in R. A. Bauer (ed.), *Social Indicators*, MIT Press,

Cambridge, 1966, p. xiii; hereafter cited as: Bauer, *SI*. See also Bauer, 'Detection and Anticipation of Impact: the Nature of the Task', in Bauer, *SI*, pp. 36–48; R. M. Hare, 'Contrasting Methods of Environmental Planning', in Goodpaster and Sayre, *Ethics*, pp. 64, 65, and D. E. Kash, Director of the Science and Public Policy Program, University of Oklahoma, in Congress of the US, *Technology Assessment Activities in the Industrial, Academic, and Governmental Communities*. Hearings Before the Technology Assessment Board of the Office of Technology Assessment, 94th Congress, Second Session, June 8–10, 14, 1976, US Government Printing Office, Washington, D.C., 1976, p. 198. Hereafter cited as: Congress, *TA in IAG*.

[37] W. D. Rowe, *An Anatomy of Risk*, John Wiley, New York, 1977, p. 431; hereafter cited as: Rowe, *Risk*.

[38] O. Morgenstern, *On the Accuracy of Economic Observations*, Princeton University Press, Princeton, 1963, pp. 35, 63; hereafter cited as: Morgenstern, *Accuracy*.

[39] Self, *PPCBA*, pp. 91–92; see also pp. 78–79.

[40] Bauer, Detection, in Bauer, *SI* (note 36), p. 46.

[41] See K. S. Shrader-Frechette, *Nuclear Power and Public Policy*, D. Reidel, Boston, 1980, pp. 55–59. Here the author shows that if nonmarket costs (for storing radioactive wastes) are added to the RCBA then, contrary to the accepted government conclusion, nuclear-generated electricity is more expensive, per kilowatt-hour, than that from coal.

[42] A. Lovins, 'Cost-Risk-Benefit Assessment in Energy Policy', *George Washington Law Review* 45 (5), (August 1977), 930; hereafter cited as: Lovins, CRBA.

[43] R. M. Hare, Methods (note 11), pp. 64, 68, 70.

[44] Self, *PPCBA*, pp. 165–171.

[45] Self, *PPCBA*, p. 166.

[46] Self, *PPCBA*, p. 171.

[47] Self, *PPCBA*, p. 169–170.

[48] Self, *PPCBA*, p. 166.

[49] Self, *PPCBA*, p. 89.

[50] L. Lave and E. Seskin, 'Air Pollution and Human Health', *Science* 169 (3947), (1970), 723–733 and D. Rice, *Estimating the Cost of Illness*. PHS Publication No. 947–6, US Government Printing Office, Washington, D.C., 1966.

[51] See E. Rappoport, 'Remarks on the Economic Theory of Life Value', in D. Okrent (ed.), *Risk-Benefit Methodology and Application*, UCLA School of Engineering and Applied Science, Los Angeles, 1975, pp. 609–613. See also Rowe, *Risk* (note 37), pp. 225–226.

[52] B. M. Gross, 'The State of the Nation', in Bauer, *SI* (note 36), p. 168.

[53] Gunnar Myrdal, *Against the Stream*, Random House, New York, 1973, p. 168; hereafter cited as: Myrdal, AS. See also note 52.

[54] Stevens, Measurement (note 18), p. 61.

[55] Mishan, *ESD* (note 12), p. 19. See also Mishan, CBA (note 12), p. 382.

[56] Lovins, CRBA (note 43), p. 938.

[57] T. C. Means, 'The Concorde Calculus', *George Washington Law Review* 45 (5), (August 1977), 1061–1062; hereafter cited as: Means, CC. See also W. R. McCarey, 'Pesticide Regulation', *George Washington Law Review* 45 (5), (August 1977), 1093–1094; hereafter cited as: McCarey, Pesticide.

[58] Mishan, *ESD* (note 12), pp. 19–20.

[59] A. D. Biderman, 'Social Indicators and Goals', in Bauer, *SI* (note 36), p. 101.

60 Norman S. Care, 'Participation and Policy', *Ethics* 88 (1), (July 1978), 316–337, has an excellent discussion of this point; hereafter cited as: Participation.
61 Mishan, *CBA* (note 12), p. 160. G. L. S. Shackle, *Epistemics and Economics*, Cambridge University Press, Cambridge, 1972, pp. 45–47; hereafter cited as: Shackle, *E. and E*. Georgescu-Roegen, *AE* (note 9), p. 196, makes a similar point.
62 Gunnar Myrdal, *The Political Element in the Development of Economic Theory*, Harvard University Press, Cambridge, 1955, p. 89; hereafter cited as: Myrdal, *PED.*
63 Mishan, *ESD*, p. 21. Similar points are raised by Self, *PPCBA*, pp. 73–80; Myrdal, AS (note 53), p. 149; Lovins, CRBA (note 42), pp. 925–926; Georgescu-Roegen, *AE* (note 9) pp. 17ff., 47ff.; E. Rotwein, 'Mathematical Economics', in S. R. Krupp (ed.), *The Structure of Economic Science*, Prentice-Hall, Englewood Cliffs, 1966, p. 102; hereafter cited as: Rotwein, ME, and Krupp, *Structure*. See also Green, CRBA (note 12), p. 905.
64 Lovins, CRBA (note 42), p. 927.
65 A. Radomysler, 'Welfare Economics and Economic Policy', in K. J. Arrow and T. Scitovsky (eds.), *Readings in Welfare Economics*, Homewood, Illinois, Irwin, 1969, p. 89; hereafter cited as: Radomysler, Welfare, and Arrow and Scitovsky, *Welfare.*
66 Georgescu-Roegen, *AE* (note 9), p. 52.
67 See note 54.
68 See, for example, S. S. Stevens, Measurement (note 18), pp. 36–42.
69 See, for example, Pearce, 'Introduction' (note 32) and A. L. Sorkin, *Economic Aspects of Natural Hazards*, Lexington Books, Lexington, Massachusetts, 1982, pp. 59–74; hereafter cited as: Sorkin, *EANH.*
70 See, for example, Richard Brandt, 'The Concept of Welfare', in S. R. Krupp, *Structure* (note 64), pp. 257–276.
71 John C. Harsanyi, 'Cardinal Welfare, Individualistic Ethics, and Interpersonal Comparisons of Utility', in K. J. Arrow and T. Scitovsky, *Readings in Welfare Economics*, R. D. Irwin, Homewood, Illinois, 1969, p. 55; hereafter cited as: Harsanyi, CW, and Arrow and Scitovsky, *Welfare.*
72 See Stevens, Measurement (note 18), pp. 21–22.
73 R. P. Wolff, 'The Derivation of the Minimal State', in Jeffrey Paul (ed.), *Reading Nozick*, Rowman and Littlefield, Totowa, N.J., 1981, pp. 99; hereafter cited as: Wolff, Derivation, in Paul, *RN.*
74 Wolff, Derivation, p. 101.
75 Stuart Hampshire, 'Morality and Pessimism', in Hampshire (ed.), *Public and Private Morality*, University Press, Cambridge, 1978, p. 5.
76 B. M. Gross, 'The State of the Nation', in Bauer, SI (note 36), p. 168. See also H. Skolimowski, 'Technology Assessment as a Critique of Civilization', in R. S. Cohen, *et al., PSA 1974*, D. Reidel, Boston, 1976, pp. 459–465, esp. p. 461.
77 D. MacLean, 'Quantified Risk Assessment and the Quality of Life', in Dorothy Zinberg (ed.), *Uncertain Power*, Pergamon Press, New York, 1983, Section V; hereafter cited as: MacLean, QRA.
78 MacLean, QRA, Section V.
79 Self, *PPCBA* (note 3), p. 69.
80 This same point is made by numerous economists. See, for example, Self, *PPCBA*, p. 68, and A. Kneese, S. Ben-David, W. Schulze, 'A Study of the Ethical Foundations of Benefit-Cost Analysis Techniques'. Working Paper, unpublished, August 1979, p. 23; hereafter cited as: EF.

[81] This same point is made by Pearce, 'Introduction' (note 32), p. 3.
[82] N. Georgescu-Roegen, AE (note 9), p. 33. See also G. L. S. Shackle, *Epistemics and Economics*, University Press, Cambridge, 1972, pp. 8–9.
[83] See Self, *PPCBA* (note 3), pp. 70–75 and A. Lovins, 'Cost-Risk-Benefit Assessment in Energy Policy', *George Washington Law Review* 45 (5), (1977), 925ff.; hereafter cited as CRBA.
[84] E. Mishan, *CBA* (note 12), p. 407.
[85] See note 31.
[86] H. P. Green, 'Legal and Political Dimensions of Risk-Benefit Methodology, in D. Okrent (ed.), *Risk-Benefit Methodology and Application*, UCLA School of Engineering and Applied Science, Los Angeles, 1975, UCLA-ENG 7598, pp. 287–289; hereafter cited as: *RBM.*
[87] Boulding, 'Basis' (note 23), p. 64. See also Anderson, *Social Value* (note 23), p. 13.
[88] L. Lave, 'Discussion', in Mitre Corporation (note 7), p. 181.
[89] C. A. Nash, 'The Theory of Social Cost Measurement', in Pearce, Valuation (note 32), p. 8; hereafter cited as: Nash, SCM.
[90] Pearce, 'Introduction' (note 32), p. 7.
[91] J. Burton, 'Epilogue', in S. Cheung, *The Myth of Social Cost*, Institute of Economic Affairs, Lancing, Sussex, 1978, p. 90; hereafter cited as: Burton, Epilogue, in Cheung, *Myth.*
[92] C. K. Rowley, 'Prologue', in Cheung, *Myth*, pp. 11–12. See also Cheung, *Myth*, p. 21.
[93] This same point is made by D. W. Pearce, *Valuation* (note 32), pp. 134–135.
[94] This example is taken from K. S. Shrader-Frechette, *Nuclear Power and Public Policy*, D. Reidel, Dordrecht, 1983, pp. 115–116; hereafter cited as: *Nuclear Power.*
[95] Morgenstern, *Accuracy* (note 24), p. 125.
[96] J. Primack and F. von Hippel, *Advice and Dissent: Scientists in the Political Arena*, Basic Books, New York, 1974, p. 33.
[97] C. Starr, *Current Issues in Energy*, Pergamon, New York, 1979, p. 11.
[98] L. Lave, 'Discussion', in Mitre Corporation (note 7), p. 177, makes this same point.
[99] This objection was raised by Dr. Stanley Carpenter (of Georgia Institute of Technology) in a private conversation. For a related objection and a response to it, see Alex C. Michalos, 'A Reconsideration of the Idea of a Science Court', *Research in Philosophy and Technology* 3 (1980), 26–27.
[100] A. Van Horn and R. Wilson, *The Status of Risk-Benefit Analysis*, Energy and Environmental Policy Center, Cambridge, Massachusetts, Harvard University, 1976, discussion paper, p. 4.
[101] Mishan, *CBA* (note 12), p. 383.
[102] This is an important point in my argument. Rather than go into an extended discussion and defense of it here, which would take me too far afield, I recommend the reader to an excellent article, Care, Participation (note 60). Care establishes the point in question.
[103] L. Lave, 'Panel: Public Perceptions of Risk', in Mitre Corporation (note 7), p. 577.
[104] S. Stevens, 'Measurement' (note 18), pp. 18–19.
[105] B. M. Gross, 'The State of the Nation: Social Systems Accounting', in Bauer, *SI* (note 36), p. 222. See also p. 260, where Gross discusses the 'selectivity-comprehensiveness paradox'. There is a tension, perhaps resulting from the prevalence of the application of Gresham's Law, between choosing to measure quantitatively only a few

parameters ('selectivity') and deciding to attempt to measure quantitatively a more comprehensive list of items ('comprehensiveness'). When one opts for the former, he gets an *exact*, but *irrelevant* indicator. When he chooses a more comprehensive list of values to quantify, he obtains a more relevant (i.e., applicable, realistic, or usable) indicator, but a much less exact one, since he necessarily encounters more difficulty in quantifying qualitative or subjective factors. Hence, even if Gresham's Law is avoided via quantification, one still faces considerable difficulty in the form of this paradox. Authors who make similar points include Mishan, CEG, p. xx; John Davoll, 'Systematic Distortion in Planning and Assessment', in D. F. Burkhardt and W. H. Ittelson (eds.), *Environmental Assessment of Socioeconomic Systems*, Plenum, New York, 1978, p. 12 (hereafter cited as: Burkhardt and Ittelson, *EA*); A. R. Tamplin and J. W. Gofman, *Population Control Through Nuclear Pollution*, Nelson-Hall, Chicago, 1970, p. 82; and R. A. Bauer, 'Detection and Anticipation of Impact: the Nature of the Task', in Bauer, *SI* (note 36), p. 35.

[106] E. Mishan, *ESD* (note 10), p. 109.

[107] A. Sorkin, EANH (note 31), p. 62.

[108] H. Siebert, *Economics of the Environment*, D. C. Heath, Lexington, Massachusetts, 1981, pp. 16–18.

[109] This is Mishan's line of reasoning. He claims (*CBA*, p. 267) that it is better to describe the effect than to give it an arbitrary value through quantification. He favors providing "a physical description of the spillovers and some idea of their significance", but not quantifying them, because they can't be "measured with honesty" (pp. 160–161).

[110] Self, *PPCBA* (note 3), p. 92.

[111] Self, *PPCBA* (note 3), p. 92.

[112] F. Y. Edgeworth, *Papers Relating to Political Economy*, Royal Economic Society, London, 1925, III, p. 182.

[113] K. Boulding, *Economics as a Science*, McGraw-Hill, New York, 1970, pp. 120–121.

[114] This same argument is used by W. S. Jevons, *The Theory of Political Economy*, Kelley and Millman, New York, 1957, pp. 7–10.

[115] A similar point is made by C. W. Churchman, 'On the Intercomparison of Utilities', in S. Krupp, *SES* (note 12), p. 256.

[116] A similar point is made by Self, *PPCBA* (note 3), p. 83.

[117] A similar point is made by R. C. Lind, 'The Analysis of Benefit-Risk Relationship', in Committee on Public Engineering Policy, *Perspectives on Benefit-Risk Decision Making*, National Academy of Engineering, Washington, D. C., 1972, p. 110.

[118] Mishan, *CBA* (note 12), p. 161, uses a similar example to illustrate a somewhat different point.

[119] Congress, *TA in IAG* (note 36), pp. 248–250.

[120] Congress, OTA, *Pest Management Strategies*, vol. 2, US Government Printing Office, Washington, D.C., 1979, pp. 48–51, 68–81.

[121] Congress, OTA, *Oil Transportation by Tankers: An Analysis of Marine Pollution and Safety Measures*, US Government Printing Office, Washington, D.C., 1975, pp. 26–37, 173; hereafter cited as: *Oil Tankers.*

[122] Congress, OTA, *Oil Tankers*, pp. 38–71.

[123] Congress, OTA, *An Evaluation of Railroad Safety*, US Government Printing Office, Washington, D.C., 1978, p. xi; hereafter cited as: *RR.*

[124] Information concerning these problems was taken from Congress, OTA, *RR*, pp. 14, 141–161, and Congress, OTA, *Railroad Safety – US–Canadian Comparison*, US Government Printing Office, Washington, D.C., 1979, pp. vii–xi; hereafter cited as: *RR-US-C.*

[125] Congress, OTA, *RR*, p. 37.

[126] Congress, OTA, *RR*, p. 160.

[127] Congress, OTA, *RR*, pp. x–xi, 125.

[128] Congress of the US, Office of Technology Assessment, *A Technology Assessment of Coal Slurry Pipelines*, US Government Printing Office, Washington, D.C., 1978, p. 84; hereafter cited as: Congress, OTA, *Coal Slurry.*

[129] Congress, OTA, *Coal Slurry*, pp. 84, 99. For discussion of the problem of 'pricing' natural resources, see M. A. Lutz and K. Lux, *The Challenge of Humanistic Economics*, Benjamin/Cummings, London, 1979, pp. 297–308, esp. 305–307.

[130] Congress, OTA, *Coal Slurry*, p. 15.

[131] Congress, OTA, *Transportation of Liquefied Natural Gas*, US Government Printing Office, Washington, D.C., 1977, p. 42; hereafter cited as: *LNG.*

[132] Congress, OTA, *LNG*, p. 62. See also pp. 63, 66.

[133] See note 131.

[134] Congress, OTA, *LNG*, p. 8.

[135] See Congress, OTA, *Policy Implications of the Computed Tomography (CT) Scanner*, US Government Printing Office, Washington, D.C., 1978, pp. iii, 9, 105; hereafter cited as: *Scanner.*

[136] See Congress, OTA, *Scanner*, pp. 8, 67, 71, 105; this conclusion is based on the facts that most head scans are done merely because of headaches and, in the absence of other abnormalities, are almost always negative. Even with other symptoms, up to 90% of all head scans are negative.

[137] Congress, OTA, *Scanner*, p. 38. This dose is 177 times greater than the average annual dose of radiation to which a person is exposed.

[138] Calculational data from government dose-response studies and from the BEIR report may be found in K. S. Shrader-Frechette, *Nuclear Power* (note 94), p. 26; see also p. 115 for calculations regarding genetic deaths of offspring when parents are exposed to radiation; genetic deaths are higher, by a factor of 10, than induced cancers when equal amounts of radiation exposure occur. Cost-estimate data for cancers and genetic deaths may be computed on the basis of the discussion throughout Jones-Lee, *Value* (note 11).

[139] Congress, OTA, *Scanner*, p. 105.

[140] Congress, OTA, *Scanner*, pp. 8, 71.

[141] See note 138.

[142] For a complete analysis of this example, see K. S. Shrader-Frechette, *Nuclear Power* (note 94), pp. 49–68.

[143] Stobaugh and Yergin, 'Conclusion', in Stobaugh and Yergin, EF, pp. 216–233, esp. p. 227; see also Congress, OTA, *Application of Solar Technology to Today's Energy Needs*, Vol. 1, US Government Printing Office, Washington, D.C., 1975, pp. 3, 12, 21; hereafter cited as: *Solar I.*

[144] Congress, OTA, *Solar I*, p. 3.

[145] For British statistics, see M. R. McDowell and D. F. Cooper, 'Control Methodology of the U.K. Road Traffic System', in Burkhardt and Ittelson, *EA*, pp. 279–298. For US

data, see Congress, OTA, *Technology Assessment of Changes in the Future Use and Characteristics of the Automobile Transportation System*, 2 vols. US Government Printing Office, Washington, D.C., 1979, vol. 1 pp. 16, 21, 25–31; hereafter cited as: *Auto I or Auto II.*

[146] See note 145.

[147] See Congress, OTA, *Auto I*, p. 25.

[148] Congress, OTA, *Auto I*, p. 31.

[149] Congress, *Auto II*, p. 251; see also pp. 75–295.

[150] Congress, OTA, *Auto I*, p. 25.

CHAPTER SEVEN

THE PROBLEM OF REGIONAL EQUITY

INTRODUCTION

Ever since at least 1788 when Hamilton, Madison, and Jay published the federalist articles, there has been lively debate over whether civic decision-making in the United States ought to be based on principles of local autonomy or national supremacy. To the political scientist and legal philosopher, the issue focuses on the assests and liabilities of decentralized versus centralized government. For the ethician or moral philosopher, one of the key difficulties is whether regional or local equity (equality of consideration among regions or locales) is either possible or desirable within a centralized system. To the welfare economist, this question involves whether the national government can and ought to distribute the various social, political, industrial, technological, and environmental costs and benefits of federal projects equally among all geographical areas.

Geographical balancing in the distribution of government expenditures, such as military procurement and pork-barrel public works projects, has always characterized the US congressional committee system.[1] In more recent years, states such as California and Massachusetts have formally adopted balancing strategies designed to control urban growth and to maximize 'regional equity' within their borders. By defining 'fairness' in terms of geographical access to development and to environmental protection, these states have minimized the adverse distributive effects of government actions.[2]

Surprisingly, analysis of geographical or regional equity (analysis of the costs and benefits variously affecting different locales) "is seldom done" in technology assessments and environmental-impact analyses.[3] Obviously, however, distributive impacts of technology- and environment-related projects fall differently on different communities. A substantial amount of sulfate pollution in the eastern states, such as Pennsylvania, for example, is the result of emissions from coal-fired plants located hundreds of miles westward in Ohio and West Virginia.[4] Likewise, for instance, much of Los Angeles' continued commercial and industrial development is dependent upon its importing scarce water from other areas of the Southwest, many of which have the same problem.[5]

As I will argue in this chapter, the problems of regional equity posed by cases of sulfate pollution and scarce water resources provide typical examples of the methodological and ethical difficulties faced by contemporary technology assessors and environmental-impact analysts. Methodologically speaking, assessors have yet to deal adequately with the secondary effects of technology, such as the distributive impacts "upon those who . . . simply happen to be in the way".[6] To remedy this deficiency, I will argue that assessment methodology ought to be expanded so as to provide for evaluation of the extent to which a given project enhances regional equality and autonomy.[7] Ethically speaking, the widespread failure to analyze and evaluate impacts on various geographical areas also leads to a number of difficulties. I will argue that this shortcoming not only compromises the allegedly nonpartisan character of technology assessment and environmental-impact analysis, but also suggests that policymakers are largely unaware of the ethical dimensions of questions involving regional distributions of costs and benefits. To help resolve these problems, I will suggest several means of improving technological and environmental analyses. If my suggestions are correct, they should lead to a more democratic procedure of public policymaking.

2. METHODOLOGICAL PROBLEMS WITH ANALYSIS OF DISTRIBUTIVE IMPACTS

One reason why geographical divergences in distributional impacts have not been treated adequately, if at all, in most technology assessments and environmental-impact analyses is that the methods used to measure social impacts remain problematic. There simply are no sophisicated means of distributed cost-benefit analysis, as opposed to well developed methods of aggregation.[8]

Aggregation is a simplifying assumption built into cost-benefit analysis. It stipulates that nonhomogenous data (e.g., costs of both *on*shore and *off*shore oil production) may be lumped together (e.g., to provide a measure of the costs of oil production) for purposes of theoretical convenience. Despite the fact that use of this econometric assumption enables one to fit the complexities of the real world into variables that can be handled by a simple mode, it leads to inaccuracies. In the oil-production illustration just cited, for instance, uncritical use of aggregated data might lead one to conclude, for example, that production of natural gas was cheaper than production of domestic oil for generating electricity. In reality, however, it could be that natural gas provided a cheaper power source than offshore-produced oil, but a more expensive one than onshore oil.[9]

Although use of the assumption of aggregation can lead to obviously false conclusions, its inaccuracies are less susceptible to detection when one employs aggregated data which are thought to be homogenous in all relevant respects or which are so highly technical that differences within them may not be known or understood. This often occurs, for example, when complex data are used by researchers who did not develop them. When aggregated data are combined with other statistics, the limits of their validity are even less likely to be recognized.[10]

Despite this fact, most economists continue to use the principle of aggregation. They define 'public welfare', for example, as an "aggregate of preferences".[11] Account is taken, neither of individual deviations from this aggregate, nor of the undesirable consequences of following a method based on what might be called the "tyranny of the majority". As a result, currently available econometric data and models are not wholly applicable to pressing problems of socioeconomic change. As one assessor put it:

> Aggregated national economic and census statistics say nothing about pockets of poverty, depressed communities, sick industries, or deprived social groups. These are averaged out, and so long as the averages appear favorable, there is no indication of, or data on, regional or local problems.[12]

Because risk-cost-benefit analysis, with its attendant use of aggregation, has been narrowly conceived to eschew the evaluation of distributional effects, it is understandable that assessors of technological and environmental impacts are usually able only to pay lip service to the fact that distributional consequences ought to be analyzed.[13] The US Office of Technology Assessment (OTA), however, stipulates that distributive impacts be evaluated. The third criterion (of eleven) used in authorizing specific OTA projects reads: "How significant are the costs and benefits to society of the various policy options involved, and how will they be distributed among various impacted groups?"[14]

Despite this criterion and the alleged assessment expertise of the OTA, most TAs and EIAs (including those of other government agencies, such as the Environmental Protection Agency) have either ignored or inadequately treated questions of the equity of geographical distribution. In the recent OTA study of coal-slurry pipelines, for example, the assessors made several brief, qualitative references to the fact that coal-producing areas in the western US "are expected to suffer adverse impacts, like . . . increased competition for water, while the benefits [of the technology] accrue to other parts of the nation . . . "[15] In examining the net economic impact of the pipelines, however, the analysts did not take into account the regional costs associated with any use of scarce water resources. Instead they employed only a few of the easily quantifiable market costs related to the technology (e.g., pumping water for use in the pipelines) and ignored the more massive, qualitative costs to the region.[16] After having examined only a subset of the costs to the region, the authors of the report concluded: "slurry pipelines can, according to this analysis, transport coal more economically than can other modes [of transport]".[17] The obvious question is: 'More economically for whom?'

Likewise, in a recent report on liquefied natural gas (LNG) transport technology, the OTA authors did no analysis of the problem of regional equity. Citizens living near LNG facilities are especially concerned about this problem, because the federal government (through the Federal Power Commission) has the 'right' to force a LNG terminal on an unwilling community. Because of the tendency of the gas to vaporize, flame, and explode over great distances, residents of ocean ports with LNG facilities obviously bear a disproportionate, and often involuntarily imposed, cost of the technology. Yet, owing to liability limitations, those injured by a LNG accident are left with little or no effective compensation. With regard to these difficulties of equity, the assessors merely noted that the federal government had the *legal right* to overrule the state on siting decisions, and that insurance problems following LNG accidents "are not greatly different" from, and are consistent with, those consequent upon other catastrophes, such as nuclear accidents. Obviously, however, consistency is not a sufficient condition for determining the ethical justifiability of a particular policy. If it is wrong to deprive a certain community of the rights to collect damages after a technology-related accident, this action is not rendered just merely because some other communities face the same problem from other technologies. More importantly, however, the LNG assessors concluded that the technology and US Coast Guard standards were cost-effective in insuring safety.[18] Again, the obvious question is: 'Cost-effective for whom? And for whose safety?'

3. PROBLEMS WITH GEOGRAPHICAL DISTRIBUTION OF IMPACTS

As the coal slurry and liquefied-natural-gas examples reveal, there is a basic question of political and ethical philosophy underlying the problem of geographical equity. When technological and environmental projects, especially ones allegedly undertaken in the national interest, place disproportionate costs on the citizens of a given state or locale, ought the federal government to preempt state or local control of those projects? If the federal government has this right, under what circumstances ought it to be exercised? And under what conditions, if any, ought the region in question to be compensated for the costs it bears?[19]

3.1. *The Dilemma of Federalism*

These questions are problematic because the context within which federal decisions are made is often radically different from that of local decisions. The federal government may be in favor of nuclear generation of electricity, for example, because it wishes to reduce oil imports and improve US welfare regarding foreign relations, the balance of payments, and national security. A particular state government, on the other hand, may be opposed to nuclear production of energy because it is concerned with the safety of its citizens living near the plant and with the effects of a major accident on the state's economy and land values.

In such situations, say a number of political, legal, and ethical philosophers, the federal government ought to have controlling power. They claim that centralized decisionmaking is necessary in order to:

1. protect the environment and to avoid 'the tragedy of the commons';[20]
2. gain national economies of scale;[21]
3. avoid regional disparities in effective representation of all sides to a dispute;[22]
4. compensate the victims of one region for spillovers from another locale;[23]
5. facilitate 'the politics of sacrifice' by imposing equal burdens on all areas.[24]

Although historically American political philosophy has been based on the presumption of decentralized decisionmaking, arguments such as these five have recently led to congressional legislation overriding the presumption.[25] Largely within the last decade, responsibility for environmental policy, for example, has shifted from states and communites to the federal government. The consensus of legal opinion appears to be that this shift has been necessary because the states have been unable to check environmental degradation.[26] The theory is that since states and regions have not zoned and planned adequately for technological and environmental projects, much decisionmaking authority has to be given to the federal government in lieu of its remaining in the hands of powerful vested interests dominating local policies.[27]

The increase of federal authority over environmental and technological projects, however, has been a mixed blessing. In attempting to equalize the inequities caused by technological development and to achieve consistent national environmental standards, the federal government has threatened local autonomy and created new problems. States are prevented, for example, from strengthening current federal radiation standards for nuclear plants within their borders, even though any amount of radiation is known to be carcinogenic, mutagenic, and teratogenic.[28] As a consequence of state difficulties like this one, federalism is being challenged by those who maintain that local policymaking is more desirable. At least six arguments are typically used to support grass-roots control.

1. Local policymaking is said to promote *diversity*, because it is better able to reflect geographical variations in preferences for goods. For example, a community may decide to license an electrical generating plant if it is needed for a new subway system but not if it is used to support resort development following an Olympics.[29]
2. It is claimed to offer a more flexible vehicle for experimenting with government laws and regulations (such as those concerning the industry vs. environment tradeoff) and for promoting the utility and *self-determination* of the local community.[30]
3. Regional control is alleged to enhance citizens' autonomy and liberty by giving them the capacity to satisfy their tastes for specific conditions of work/residence/recreation.[31]
4. It is said to encourage fraternity among citizens through participation and self-education in governmental decisionmaking.[32]
5. Local policymaking is claimed to enable communities to avoid inequitable, federally imposed sacrifices for the sake of national goals.[33]
6. Finally, regional control is alleged to lead to an increase of equality among persons and to protection against violations of rights.[34]

Without knowing the various cases to which the arguments for and against federalism may be applied, it is difficult to determine, *a priori*, whether or not decisionmaking ought to be centralized and whether or not the federal government ought to have the right to impose technological and environmental burdens on certain locales or regions. Maximizing local autonomy and grass-roots control is a desirable democratic goal. At the same time, however, there are obvious instances when (for the sake of everyone's survival) federal policies ought to preempt all others, e.g., in wartime.

3.1.1. *Federal Supremacy and the War Power.* Throughout US history, there have been at least three, and perhaps four, classes of cases in which federal control has legally superseded that of state and local authorities. These deal with the war power, preemption, interstate commerce, and eminent domain.[35]

The *war power* presents a clear instance in which federalism, legitimately applied, is necessary for national security and unity in a time of stress. What is peculiar to its application in technological and environmental matters, however, is that it is often invoked when there is neither a war nor imminent threat of one. The war power has been used in peacetime, for example, to push nuclear power plants on unwilling states.[36] And in *Ashwander v. Tennessee Valley Authority*, the Supreme Court allowed the construction of a dam and electrical generating facility on the basis of the war power. The court agreed that the construction was necessary to 'national security', even though it took place during peacetime.[37] The problematic issue, in both these typical cases, is what constitutes 'national security'. Given the extent to which electrical generating facilities are responsible for resource depletion and pollution, it could just as well be argued, I think, that the *Ashwander* construction threatened local, national, and global security and safety. Apart from whether the war power ought to be invoked in a particular case, what is clear is that spurious claims of 'national security' could be used to expand federal authority and to impose technological and environmental burdens on unwilling communities.[38]

3.1.2. *Federal Supremacy and Preemption.* A second justification for federal authority to impose such burdens is *preemption.* The basis for federal preemption, of local power or control over various projects, rests with the supremacy clause of the United States *Constitution* (article VI, clause 2). The doctrine provides that the *Constitution* and the laws of the US shall be the supreme law of the land. Where a "state law stands as 'an obstacle to the accomplishment and execution of the full purposes and objectives of an Act of Congress' the federal statute prevails and the state law is invalidated".[39]

The general criteria for the courts' finding preemption are easily stated. Its application to particular cases, however, has been highly variable and imprecise. The "decisions display inconsistent treatment of the degree of conflict necessary to support a finding of preemption and reliance on a broad range of presumptive factors".[40] This inconsistency is particulary apparent in cases dealing with technological development and pollution control. In numerous instances, the states have been granted the right to develop environmental standards more stringent than federal guidelines, on the grounds that they have primary responsibility for the health and safety of their citizens. In several selective, but highly significant, classes of cases (e.g., those involving radioactive pollution), the states have been denied this right and federal preemption has been upheld.[41] Federal preemption has also been used successfully, for example, to prevent the states from challenging federally imposed liability limits in the case of a nuclear accident.[42]

While most persons would probably agree that invoking the preemption doctrine is often necessary, e.g., to invalidate state laws which are racist or sexist, other applications of it are problematic. Most obviously the doctrine impedes those who are correct in challenging an unwise or unjust federal law. Also, in the case of technological and environmental projects, preemption could easily threaten the rights of local citizens to due process. Prior to beginning such projects, government hearings are held to insure that residents of the affected area are given the right to have their opinions heard. In the case of construction of a nuclear power plant, for example, the hearings rarely serve their purpose of due process. If a citizen, or indeed a whole state, disagrees with the federal government that radiation standards are safe, that nuclear liability ought to be limited, or that the emergency core cooling system does not need to be tested at full scale, then those views are discounted. Local or state laws in these matters are preempted by the relevant federal agency, the Nuclear Regulatory Commission,[43] even though the local citizens bear most of the costs and risks, and often receive none of the benefits, of the technology.

3.1.3. *Federal Supremacy and the Interstate Commerce Clause.* As happened in the case with nuclear energy, the federal government's historical power over interstate commerce often has been the main justification given for preemption of state laws dealing with technological development of environmental standards. Protecting *interstate commerce*, a third justification often given for federal authority over states, arose out of the US government's early concern about protecting the rights of private property. John Marshall, for example, declared that the federal government would stand between private corporations and any acts of the states which threatened them. Decades later, when the Grange agitated for state control of rail abuses in the 1860's, the Supreme Court responded by ruling that the states lacked the power to regulate interstate commerce.[44] This political and legal development, plus the fact that corporations have been defined as persons under the Fourteenth Amendment to the US *Constitution*, have permitted US industry to operate largely without local restrictions.[45]

Although many states are challenging federal regulation of commerce and seeking to regulate the imposition of environmental burdens such as atmospheric pollution, noise, and nuclear waste,[46] "private transport of pollutants between states constitutes interstate commerce".[47] Even the pipelines used for crude oil, gas, and natural gas, for example, are under federal (Interstate Commerce Commission) jurisdiction.[48] In one recent OTA study of coal-slurry pipelines, the authors said quite bluntly that "any state prohibition [even in western US where water is scarce] or unusual restriction on the use of water for coal slurry may be attacked as an unconstitutional discrimination against interstate commerce in coal".[49] Likewise any state restriction on radioactive emissions may be attacked as an unconstitutional discrimination against interstate commerce in nuclear-generated electricity. In the landmark case of *First Iowa Hydro-Electric Cooperative v. Federal Power Commission*, the Supreme Court ruled that where there is a national plan to promote interstate commerce, decisions must be made "by the federal government . . . on behalf of the people of all the states"[50]

Several doubtful assumptions appear to be built into policy on this issue. One is that *laissez-faire* interstate commerce is desirable. If any state restriction designed to guard the health and safety of citizens, to protect the environment, or to preserve nonrenewable resources can be declared illegal on the grounds that it is "an unconstitutional discrimination against interstate commerce", then commercial-industrial-technological interests are guaranteed a virtual free rein in whatever they wish to do. Laissez-faire commerce and technology, however, appear tolerable only if it can be shown that they are in the public interest.[51] This brings us to another questionable presupposition, viz., that promoting interstate commerce is always in the public interest. Clearly this is not the case, as conflicts between maximizing the quality of life and maximizing economic growth ought not always to be resolved in favor of economic growth. Moreover, in certain cases, promoting interstate commerce could easily be shown to be contrary to the public interest. Again take the example of nuclear technology, where the lion's share of federal-state conflicts have occurred in recent years. Nearly all attempted state restrictions on nuclear-power-plant emissions, sitings, liability, and waste disposal have been preempted by the federal government on the grounds that the states have attempted to interfere with interstate commerce.

Presumably the federal government has preempted state control over nuclear power because it believes that untrammeled development of 'interstate commerce in nuclear energy' is in the public interest. However this belief would be true only if atomic energy were the only or, in a given situation, the cheapest, safest energy option. If there is an alternative to nuclear energy, such as solar power, and if one wishes to encourage interstate commerce in this technology, then it is not necessarily in the national interest to promote atomic power.[52] Moreover why should the interstate-commerce principle be interpreted so as to force all states to use a particular energy technology? Couldn't a state argue that it wished to employ another means of generating electricity? It would seem possible, for example, for a state to argue that, on the grounds of the (federal) Energy Reorganization Act of 1974, a state is obliged to develop *all* energy resources, and not just nuclear power.[53] For all these reasons, it is not clear that current interpretations of federal authority over interstate commerce justify the exclusion of state and local decisionmaking.

3.1.4. *Federal Supremacy and the Law of Eminent Domain*. A fourth means often used to justify federal preemption of state or local authority is the law of *eminent domain*. It stipulates that the government has the power to purchase land to be used for some public purpose, such as a freeway. Just compensation is an essential part of the law, since the Fifth Amendment to the US *Constitution* prohibits the taking of private property without such compensation. The landowner is required to sell his property in this manner unless it can be shown that the government's action was arbitrary, or that the land was not used for a public purpose.[54] Buying land for coal-slurry pipelines is one example of how the government could use eminent domain to obtain property needed for a public purpose. As is sometimes the case when federal preemption is employed to promote interstate commerce, however, it is conceivable that use of land for certain technological projects might not necessarily be in the authentic public interest.

Consideration of the power of the federal government to overrule local autonomy by means of the *war power*, the *preemption doctrine*, the *interstate-commerce clause*, and *eminent domain* reveals that federalism could result in some inconsistent and perhaps inequitable policies and in a misperception of the common good or the national interest. Because of the seriousness of the problems, especially for citizens in regions directly affected by them, it is imperative that technology assessments and environmental-impact statements analyze the geographical distribution of costs and benefits for the projects they evaluate.

3.2. *The Ethics of Equality*

Admittedly, however, to argue that federalism can have undesirable consequences on given regions or communities, and that these consequences ought to be avoided, is to presuppose that equitable geographical distribution of technological and environmental costs and benefits is desirable. Is there reason to believe that equality among regions is a justifiable goal? Even more basically, is there reason to believe that any equality of treatment among persons or groups is an ethically defensible end?

This question is far more complex than it appears. The notion of equality has a long history and it is one of the oldest elements in liberal thought.[55] Surprisingly, a number of persons deny that there is any sense in which equality is a principle of justice.[56] Part of the difficulty is that there are many formulations of egalitarian principles, and the core concept is quite vague.[57]

3.2.1. *The Principle of Political Equality*. Most people appear to agree that some inequalities among persons (e.g., in educational achievement) are based on natural capacities, while others (e.g., inequalities in wealth) are a result of social roles or socialization, and that only inequalities of the latter type ought to be questioned.[58] Insofar as equality is a product of socialization, it is often construed in at least two important ways, as *political equality* (equality of treatment under the law or equality of civil rights) and as *economic equality*. Economic equality is often taken to mean either equality in the distribution of wealth or merely equality of opportunity. Political equality and economic equality are closely related, however, since real political equality often requires economic equality; otherwise political power is likely to be the handmaid of economic power. For example, numerous studies have shown that the greater a defendant's wealth, the less likely it is that he will be found guilty of the crime with which he is charged.[59] Moreover, factual inequalities of condition and differences in the distribution of wealth seem to militate against both political equality and equality of opportunity.[60]

From an ethical point of view, sanctioning at least a principle of *policical equality* appears eminently reasonable, although there is controversy over whether adoption of a principle of *economic equality* can be defined adequately and, if so, what form it ought to take.[61] Even the principle of political equality is subject to a number of interpretational difficulties, despite its general acceptance. This widespread acceptance is typically based on one or more of the following four considerations.

1. The comparison class is all humans, and all humans have the same capacity for a happy life.[62]
2. Free, informed, rational people would agree to the principle.[63]

3. This principle provides the basic justification for other important concepts of ethics; it is a presupposition of all schemes involving justice, fairness, rights, and autonomy.[64]
4. Equality of treatment is presupposed by the idea of law: "law itself embodies an ideal of equal treatment for persons similary situated".[65]

Perhaps the most significant of these considerations are (3) and (4). They amount to the claim that accepting a principle of political equality is necessary in order to insure fairness and consistency. The main interpretational question raised by the principle, however, is 'what sort of equality is required as a basis for fairness and consistency?'. Presumably one is talking about equal *treatment* under the law. But does equality of treatment mean giving everyone the *same* treatment?

3.2.1.1. *Equality or Sameness of Treatment?* It is not clear that there are always morally relevant reasons why everyone ought to receive the same treatment. Genuinely *equal* treatment (proportional to one's merits or to the strength of his claims to it) might require that treatment for all individuals not be the *same*.

Several cases in which, given appropriate conditions, one might justifiably permit/give treatment which is not the same for all individuals come to mind. Although they are not universalizable, these include circumstances in which different treatment is given/allowed:

1. as a reward for merit or virture;
2. as a recompense for past deeds;
3. as an incentive for future actions which are desirable; or
4. as a provision for special needs.

For example, if one wishes society to have the benefit of the services of the President of the US, then one ought to permit the President to have better police protection than everyone else receives. Following circumstances (3) and (4) above, this 'better' treatment might be justified on the grounds of incentives for those who might serve as future presidents and on the grounds of the special needs of the person holding this office. This means that there can be no universal claim that all persons ought to receive the same treatment. Hence to say that all persons ought to receive equal treatment under the law does not mean that the treatment must be the same.

3.2.1.2. *Equality of Treatment or Equality of Respect?* But if treatment ought not always be the same, how is it that the principle of political equality is necessary for consistent treatment of persons? What is it that is consistent? Legal philosopher Ronald Dworkin maintains that what is meant is that everyone ought to receive the same, or consistent, *concern* and *respect* in the political decision about how goods, treatment, and opportunities are to be distributed.[66] His point, as well as mine, is not that anyone's rights may ever be ignored. Rather, the point is that one person's interests may be outweighed by another's. For example, in certain circumstances, protecting the President of the US may outweigh protecting a particular citizen. If so, then one chooses the particular policy (e.g., regarding geographical distribution of environmental/technological risks, costs and benefits) which gives all persons/regions the same concern or respect, but which provides an equitable basis for deciding when one person's/region's interests outweigh another's.

If this reasoning about sameness/equality and treatment/respect is correct, then the federal government's imposing different environmental-technological risks, costs and benefits on various regions cannot be said to be unethical merely because this imposition does not give persons in all regions the *same* treatment. To establish that the distribution is unethical, one would have to argue either that there are no morally relevant reasons for different treatment, or that some person's/region's interests were erroneously judged to outweigh those of another. In other words, proponents of geographical equity ought not to make too simplistic an appeal for the 'same treatment'. Rather, they ought to argue their cause on a case-by-case basis and attempt to show, *in a particular* instance, that good reasons do/do not support giving the same cost-benefit distribution to all regions involved. Later in the essay, I will discuss whether there appear to be (morally) good reasons for discrimination in most geographical-equity cases and whether, in these cases, technology assessors and environmental-impact analysts tend to judge correctly that one region's interests outweigh those of another.

3.2.2. *The Principle of Prima-Facie Political Equality*. To argue, as has just been done, that a principle of political equality is reasonable but that, under some circumstances, 'good reasons' may justify giving persons/regions different treatment under the law, is to argue for a principle of *prima-facie* political equality. On this view, equality (sameness) of treatment of persons and communities needs no justification; it is presumed defensible, and only unequal treatment requires justification.[67] This means that the burden of proof is placed on the person who wishes to 'discriminate' through unequal treatment. Not to put this burden on the possible discriminator would be to encourage behavior under the law to be determined by influence and power rather than by consistency and fairness. Moreover, placing the burden of proof on the possible discriminator is consistent with the four considerations, given early in this section, in favor of a principle of political equality.

Despite the apparent plausibility of the principle of *prima-facie* political equality, not all persons accept it. One of the strongest arguments against formally adopting the principle comes from utilitarians. Their rejection of it is worth considering, since many scholars blame this utilitarian position for many of the ills besetting TA and EIA.

Utilitarians argue, generally, that the moral goal of all human actions is not to maximize equality or justice among persons and groups. Following the principle of utility, they claim that this goal is to achieve the greatest possible balance of good over bad for mankind as a whole. Principles of equality, such as the principle of *prima-facie* political equality, are recognized by utilitarians only to the extent that doing so will lead to the greatest good of the majority. As a consequence, application of the tenets of utilitarianism can lead to numerous violations of equal justice at the same time as the good is allegedly maximized for mankind as a whole. Although utilitarians admit that they sacrifice individual rights (e.g., to equality under the law) to the alleged common good, they claim that such violations of equality minimize human suffering and maximize social improvement more than would egalitarian schemes.[68] Hence they give *prima facie* assent to no rights or ethical principles of equality. Instead, in a given situation, they follow whatever *rule* or perform whatever *act* (they judge) will lead to the best consequences for the majority of people.

Because (both rule- and act-) utilitarians give no assent to any principle of equality, including a principle of *prima-facie* political equality (see note 70), a number of scholars have argued that acceptance of a utilitarian methodology to evaluate technological and environmental costs and benefits is the reason why distributional inequities are usually ignored in assessments and impact statements. More generally, they have argued that the 'real issue' in nearly all environment- and technology-related controversies is whether to maximize equity or efficiency.[69]

For purposes of analyzing regional equity in technology assessments and environmental-impact analyses, however, I am not sure that pursuing the debate between utilitarians and egalitarians is helpful. The real issue does not appear to be so simple. It does not seem to be the case merely that adherence to the principle of utility, rather than to some principle of equality, has allowed analysts to avoid consideration of the distribution of technological or environmental risks, costs and benefits. The previous arguments make this clear; equality of treatment is presupposed by our legal system, our notions of fairness and rationality, and by our concepts of right and justice. Hence, in some sense, assessors appear likely to subscribe to a principle of equality. Simple failure to adhere to some principle of political equality or equal distribution of the costs and benefits of a project appears unlikely to explain assessment omissions. Moreover, although on the rule-utilitarian scheme,[70] the goal of pursuing equality is theoretically subservient to that of maximizing the quantity of welfare for the majority, it is not clear that this ordering is used much in practice. As Brandt points out, "most utilitarians think that inequalities of distribution tend to reduce the total welfare"; hence, he says, they favor equal distributions of costs and benefits "except as there are special reasons to the contrary".[71] Inequalities lead to something less than the maximum net expectable utility per person, to resentment, snobbishness, competition, and to the rich (or those receiving inordinate benefits) losing the values of hard work and social concern. Equality of distribution (e.g., of environmental and technological costs and benefits) also makes sense because of the declining marginal utility of income.

Not only utilitarians, however, appear to accept in practice some principle of equality. It seems that most rational persons would do so, in part because equal consideration of persons is a presupposition of virtually any system of law or justice. As such, it is really a principle of impartiality and consistency.[72] The history of discrimination, paradoxically, also suggests that most persons accept some principle of equality, such as *prima-facie* political equality. As Pennock points out, most persons who have defended racism or antisemitism have claimed to accept a principle of equality, but have argued that certain facts justified their favoring unequal treatment in a given situation. Such 'facts' have included claims of conspiracy and of the moral inferiority of Blacks.[73] Proponents of apartheid clearly invoke the Aristotelian principle of equality; they argue that Blacks require different treatment to attain their different (from those of Whites) goals.[74] Likewise when persons fought against women's suffrage at the turn of the century in the US, most neither argued against a principle of equal concern or respect for equal beings nor required women to defend it. Rather they argued that a person's sex was a relevant basis for discrimination, that women were by nature unable to exercise political power, and that they were not ready to do so.[75]

3.3. *Relevant Bases for Justifying Inequality of Treatment*

If these brief comments on discrimination are correct, then they suggest that 'closet utilitarians', influencing assessment methodologies, are not necessarily the culprits in analyses where the risks, costs and benefits of regional inequities are ignored. Instead it appears that the authors of TA's and EIA's would very likely argue that, although they adhere to a principle of equality, there are *good reasons* for one to allow unequal distributions of technological and environmental costs and benefits, and *good reasons* for assessors not to consider various regional impacts of these distributions. If this is true, and if a principle of *absolute* sameness of treatment (see section 3.2.1.1 earlier) "is not seriously held" by any thinkers, even liberals or socialists,[76] then the two critical philosophical questions are these:

1. What constitutes *good reasons* for allowing unequal distributions of technological and environmental costs and benefits?
2. What constitutes *good reasons* for assessors not to consider the various regional impacts of these distributions?

Another way of asking for good reasons to discriminate is to ask, 'what constitutes relevant and irrelevant *differences* when we are applying the principle that similar cases ought to be treated similarly?'.[77] We have come to believe, for example, that the color of a person's skin does not constitute a relevant difference, but that severe mental illness might be such a difference. For example, we know that mental illness might be a good reason for discriminating against a person regarding his right to bear arms.

Traditionally one of the most common (although not one of the most justifiable) of the alleged good reasons for discriminating among so-called equals has been that this discrimination supposedly serves a higher interest, that of freedom.[78] The man who wants segregation in the schools, for example, says that integration, as an attempt to provide equal treatment and equal opportunity to all races, has resulted in a violation of his freedom of association.[79]

In environmental and technological matters, such as the question of pollution control, some proponents of regional *freedom* often argue that federal standards for certain industrial emissions are so strict that they do not allow for unhampered economic development in the area; likewise many proponents of *equality* of treatment for all regions argue that no locale ought to bear a greater health or safety risk for the sake of economic gain.[80] Apart from who is right in such controversies, one thing seems clear. The fact that a law, action, or policy discriminates among persons or regions does not necessarily make it unjust; the fact that it discriminates badly does.[81] Bad laws, actions, or policies, as well as 'bad' technology and environmental-impact assessments, ought to be shown to be so on the latter ground, that they discriminate badly. I hope to do this by answering questions (1) and (2) above.

3.3.1. *Does Inequality Serve Long-Term Equality*? What are the grounds on which authorities appear to sanction inequitable distributions of technological and environmental costs and benefits? John Rawls says that inequalities ought to be allowed only if there is reason to believe that the practice involving the inequity will work for the advantage of everyone.[82] If we discount unethical or political motivations, then it seems reasonable to assume that assessors and policymakers believe that permitting inequitable distributions of technological and environmental costs and benefits will work for the advantage of everyone. (Most assessors, however, do not state explicitly the ethical and econometric assumptions made in their analyses. That is one reason why evaluation of them is sometimes difficult.) Given assessors' tendencies to discount safety risks or pollution hazards on the grounds that "the economy needs" the risky technology,[83] or that certain pollution-control standards are not cost-effective and beneficial to industry and therefore to our national well-being,[84] it does not seem presumptuous to suggest how analysts might justify their discounting the importance of distributive, especially regional, inequities.

If we put the most favorable interpretation possible on the failure to evaluate distributive inequities, assessors of technological and environmental impacts might be said to justify these inequalities as "required for the promotion of equality in the long run". (Almost any other defense would be questionable and likely to be open to the charge that it was built upon using some humans as means, rather than treating them as ends in themselves.)[85] But if certain technology- and environment-related inequities are most justifiable on the grounds that they serve the interests of long-term equality, then this is probably said to be the case because it is believed that technological progress and the resultant economic growth help everyone, but especially the disenfranchised. As one prominent science editor put it: "if the industrial economies of these [developed] countries were not encouraged to grow", they would not be able "to provide the materials necessary for removing the disparity between nations. Technology can make a direct contribution to the improvement of the lot of developing nations".[86]

Technology and economic growth have also been said to be necessary to remove long-term inequalities among communities or regions of a given industrialized nation. Peter Drucker, for example, defends such a position. He argues that technologically induced increases in production and massive economic growth are necessary to help low-skilled Blacks and other minority group members gain employment and, ultimately, live as well as the rest of us.[87] On his view, "the environmental crisis [with its attendant ethical problems, such as distribution of risks, costs and benefits] is the result of success . . . success in raising farm output . . . by insecticides, pesticides, and chemical fertilizers", for example. Inequities have occurred because we have tried "to bring a decent living at the lowest possible cost to the largest possible number of people".[88] In other words, he says, technology-induced inequalities are justifiable because they are part of the long-term process of distributing a decent living to more and more people and part of the task of narrowing the gap between rich and poor. Margaret Maxey makes much the same point, but in a more emotive manner, in arguing for acceptance of the inequitable distribution of the risks and benefits of radioactive waste. She claims that curbing technological or environmental hazards and inequities would amount to curbing technology and economic growth; this would be wrong, she says, because it would return us to a primitive style of life where injustices and dangers were even more troublesome than they are today.[89]

The basic problem with using this sort of argument to justify distributive inequalities, however, is that it contains a highly questionable factual premise, that promoting technology, economic expansion, and increased production will lead to greater equality of treatment among persons in the long term. There are several reasons for doubting this premise. Perhaps the most obvious one is historical. Given past experience with technological and economic growth, there is little basis for believing that it will help to promote a more equitable distribution of wealth, which is often necessary for authentic equality of treatment. In the US, for example, in the last thirty-five years, although there has been an absolute increase in the standard of living, the relative shares of US wealth held by various groups have not changed. The poorest 20 percent of persons still receives 5 percent of the wealth, while the richest 20 percent still receives 41 percent; the shares of the middle three quintiles have remained just as constant.[90] But if economic and technological growth have not helped to promote distributional economic equality in the US, then (because of the close relationship between wealth and the ability to attain political equality and to utilize equal opportunities)[91] it is unlikely that technological and economic growth has promoted equal treatment in a political sense either. This is ultimately because, as Mishan put it, the poor rarely share in the growth of real wealth; they are "isolated from economic growth".[92] Hence only redistribution, achieved through political means, will bring about a more egalitarian society; economic and technological 'progress' makes inequities even wider.[93]

One reason why technological expansion does not ordinarily help to create a more egalitarian society is that technology generally eliminates jobs; it does not create them. For the last 40 years, for example, the total employment in the manufacturing sector of the US economy has declined; goods-producing industries have sought to use fewer workers and to increase output per worker. As a consequence, "the productivity index is really an automation index", an indicator of the degree to which energy and technology have been substituted for jobs.[94] What new jobs have become available, especially in the last 30 years, have not necessarily been the consequence of technological or economic growth, but instead the result of an expansion of the service sector of the economy. Between 1947 and 1977, for example, employment in the service areas increased 95 percent, more than in any other sector.[95] All this suggests that an expanding economy and increasing use of technology might not necessarily help employment. And if they do not, then there are also probably little grounds for arguing that they help to equalize opportunities and ultimately to equalize treatment of all persons within the political system.

If anything, the plight of the poor and the unskilled is exacerbated by technological progress because they must compete more frantically for scarcer jobs. Moreover because a larger proportion of the indigent are unemployable, progress makes little or no immediate impact on the problem of hard-core poverty, despite over-all increases in the standard of living.[96]

One of the most direct reasons, however, why technological progress heightens the plight of the poor, and why it can be argued that progress is unlikely to remove distributive inequities resulting from technology, is that the poor bear the brunt of adverse environmental impacts. The whole question of regional or geographical equity revolves around the issue of *whether persons in one locale ought to be involuntarily exposed to uncompensated health hazards or economic risks in order that the general or national well-being might be served*. This question really comes down to whether the poor ought to be exposed to these inequities since, in practice, technology- and environment-related geographical discrimination often amounts to increasing existent discrimination against the poor.

It is well known that most environmental policies "distribute the costs of controls in a regressive pattern while providing disproportionate benefits for the educated and wealthy, who can better afford to indulge an acquired taste for environmental quality, than the poor, who have more pressing needs and fewer resources with which to satisfy them".[97] This means that whatever environmental quality cannot be paid for cannot be had. Even when technological growth has brought increased employment opportunities, this often has been at the expense of the poor. The poor usually live near to technological facilities presenting a health hazard; they often cannot afford to move away. It is well known, for example, that a disproportionate number of deaths, among nonwhites and in low socioeconomic groups, occurs as a result of urban air pollution from sources such as asbestos, sulfur dioxide, and benzpyrene. Various studies have shown that "those square miles populated by nonwhites and by all low socioeconomic groups were the areas of highest pollution levels".[98] In fact, various adverse environmental and technological impacts are visited disproportionately upon the poor, while the rich receive the bulk of the benefits.[99] This suggests that uncompensated regional or geographical distributive inequities are especially noxious because they harm those who already bear many of society's adverse impacts.

3.3.2. *Morally Relevant Reasons Do Not Support Geographical Discrimination.* Owing to their poverty, those disproportionately burdened with environmental hazards are in a position of virtual helplessness. Their helplessness, however, is the key to arguing that geographical discrimination is a 'bad' discrimination and that good reasons do not support it. As Hans Jonas expressed it, one has a moral obligation to protect the utterly helpless. Absolute helplessness demands absolute protection.[100] If Jonas is correct, and I think he is, then unequal geographical distributions are likely to be wrong, not only because they do not protect the more helpless members of society, but also because they appear to inflict even greater, disproportionate burdens on them.

To the extent that policymakers or assessors sanction the belief, either that current inequities will be dispelled by technological progress in the long term, or that it is permissible to discriminate against the poor via distribution of environmental costs and benefits, then to that degree they probably do not have *good reasons* for allowing, or for failing to assess, technology- and environment-related geographical inequities. Moreover it appears that geographical considerations, alone, are not morally relevant grounds for determining who ought to receive disproportionate costs or benefits of technology. If all persons deserve equal concern or respect in the political decision about how to distribute costs and benefits, then allowing one particular uncompensated group of individuals to bear most of a technology's burden is arbitrary discrimination. In this sense, discrimination on the basis of geography is arbitrary. There is no morally relevant reason (e.g., merit, need) why where a person lives is grounds for discriminating against him.

The earlier discussion of federal protection of interstate commerce also suggests that geographical discrimination serves the interests of expendiency. In the absence of morally relevant grounds for it, the discrimination is likely to result in humans being used as *means* to some commercial or industrial *end*. Moreover, there appear to be no grounds for arguing (see section 3.2.1.2) that national interests outweigh the interests of regions subjected to disproportionate environmental or technological costs and benefits. A necessary condition for an ethically justifiable claim that the federal interest outweighs the regional interest (in a particular case) is that the national and regional costs and benefits are calculated and weighed. As was explained earlier (see section 1), however, analysis of the geographical costs and benefits of a project is rarely done in TA and EIA. If such analysis is not regularly done, and if it is a necessary condition for justifying grographical discrimination, then geographical discrimination cannot be shown to be ethically defensible. Hence, in the absence of such TA/EIA analysis, good reasons cannot be said to support geographical discrimination.

3.3.3. *A Rejoinder: Geographical Inequalities Are Justifiable Because They Cannot Be Measured/Avoided.* In response to the claims that morally relevant reasons do not support geographical discriminations and that, in the absence of TA/EIA analysis of regional costs and benefits, geographical discrimination cannot be said to be ethically defensible, technology assessors and environmental-impact analysts are likely to make several rebuttals. One is that they must concentrate on evaluating measurable parameters, and that the social costs of geographical inequities are not measurable. They may believe that evaluating such 'subjective' social impacts would compromise the alleged objectivity, accuracy, and nonpartisan character of technology assessment and environmental-impact analysis.[101] In other words, they may believe that, since they do not have the requisite objective data to measure distributive impacts, they are best ignored.

There are two obvious replies to this claim. *First*, to concentrate only on measurable quantities is to beg the question as to what impacts ought to be evaluated.[102] This *petitio* predetermines the conclusions likely to be reached by the assessor. *Second*, there are several quantitative ways in which to measure adverse geographical impacts. For example, property values often decrease in regions of high pollution or scarce resources. Likewise, in certain locales, prices may be placed on the 'value' of life and the hazards of premature death.[103]

Another reason why assessors may not adequately evaluate the impacts of geographical distributions is that they may not see them as involving questions of justice. As one author put it, for example, "no issue of justice is involved in the question whether a new highway should be built. This is purely a question of utility . . . whether the benefits of it would outweigh the cost. This is no more a question of what justice requires than is the question whether one should buy his wife a new coat".[104] For this analyst as, I suspect, for many others, technology assessment and environmental-impact analysis is purely a matter of cost-benefit calculation, and no distributional issues are considered.[105] He has *defined* the problem as not involving distributive equity and therefore justice. However, questions of the distribution of costs and benefits obviously are issues of equal treatment, and issues of equal treatment clearly involve problems of justice. To presuppose otherwise is dangerous. The potential for sanctioning unjust social impacts appears greater in proportion as assessors fail to recognize the ethical complexity of the problems with which they deal. (For further discussion of this point, see Chapters 3 and 4 this volume.)

4. A CASE STUDY

To illustrate this potential for harm when distributive geographical impacts are not evaluated adequately, consider the effects of offshore oil and gas development. In November, 1976, the OTA completed its assessment of these technologies. The study, *Coastal Effects of Offshore Energy Systems*, was generally excellent, except for some flaws concerning distributive regional impacts.[106] A brief look at the history of policy regarding offshore energy technologies and at the assessment methodology will help to explain these shortcomings.

Distribution of the costs and benefits of coastline oil development poses a particularly interesting question of ethics. Should roughly half of all Americans, those who live or work within 50 miles of a beach, bear the economic and environmental costs of offshore oil technology while virtually all citizens receive the benefits? By what methodological procedures does one weigh the relative importance of negative impacts upon the region, as compared to allegedly positive consequences for the nation as a whole? In examining this case study, I hope to answer these questions. I also will illustrate how regional distributive costs of the technology were either ignored or underestimated by the authors of the impact analysis. Because of these assessment flaws, I will argue that public policy regarding technology and the environment is likely to be misguided on several counts.

4.1. *Regulation of Offshore-Energy Technology*

Oil is now produced from developments off the coasts of Louisiana, Texas, California, and Alaska, with exploratory drilling off many other coastlines.[107] Jurisdiction over these offshore oil and gas deposits has been subject to dispute for more than 20 years in the US. By the Outer Continental Shelf (OCS) Lands Act of 1953, Congress and the federal government have exclusive control of these lands (that is, those beyond the three-mile limit), their deposits, leases to them, and pipeline corridors within a state's territorial waters. Given the fact that the USGS has estimated that one-third of all US oil reserves could lie in the OCS regions, and fact that roughly 50 percent of all US oil is imported from foreign countries upon whom the US is dependent,[108] there has been great pressure to develop OCS resources. In fact, Congress passed the Coastal Zone Management Act (CZMA) of 1972 because "state and local arrangements for regulating coastal development were inadequate to meet the energy demand and to evaluate all national interests".[109]

The CZMA deals with all coastal areas and all land within three miles of the shore. Although the states theoretically have control over these regions, the CZMA and its 1976 amendments prescribe the conditions according to which coastal development related to OCS oil activity may take place. The act provides for matching grants (with the federal government paying up to 80 percent of the cost) to coastal states. The purpose of the grants is to plan coastal development, e.g., how to locate "energy facilities, such as terminals and refineries related to OCS development".[110] The Commerce Department is responsible for making such grants, and all plans must be approved formally by the Secretary of Commerce. The approval is contingent upon the state's taking "adequate consideration of the national interest", and establishing ways to coordinate state and local agencies to implement the plan.[111]

According to present procedures, the states and the general public may formally participate in decisionmaking (regarding leasing OCS lands) at only one point in the process. This is after the release of the draft environmental-impact statement. The public may present comments at hearings on the draft document. Citizens or states may also go to court if they disagree with a plan for development of offshore-oil technology. They are allowed to challenge only the procedures by which the OCS decision was made, however, and not the substance of the decision itself.[112]

4.2. *Three Uncompensated Regional Costs of Offshore Technology*

Often, however, the substance of technology- and environment-related decisions, and the federal regulations governing them, are at the heart of the problem of geographical equity. When coastal residents oppose development of offshore energy technologies, for example, it is rarely because they believe some decisionmaking procedure was not followed properly. Rather, they disagree with the relevant laws and procedures themselves, even when they are followed perfectly.

What are some of these substantive areas of disagreement? For one thing, coastal residents often believe it is unfair that they have no decisionmaking power regarding whether to allow OCS oil development, while they must bear the uncompensated costs of oil spills. Moreover, insofar as oil-spill losses are calculated at all, they do not include damages to the public or funds necessary to handle liability claims. Government assessors maintain that oil spills from offshore developments annually cause about $3 to $4 million in losses, yet they admit that this figure includes only the value of the product lost and the cleanup cost, not damage to the public or the environment.[113] So long as these latter data are not included, distributional effects of the technology, on everyone from coastal motel owners to fishermen, are not adequately calculated. In the recent OTA study under consideration, data on how OCS development might expand employment and provide tax revenue were provided, while the distributive costs of spills were ignored.[114] Both these methodological shortcomings, incomplete calculation of oil-spill damages and failure to analyze the social impact of inadequate liability coverage, tend to skew the analysis in favor of the technology. Obviously if all negative impacts are not considered, then offshore development appears more desirable than might be the case. This bias is particularly troublesome to coastal residents because, although distributional costs of spills have not been calculated, citizens know they are likely to be adversely affected.

Using statistics from the past ten years of offshore oil development in the Gulf of Mexico, the OTA estimated, for example, that in one Atlantic Coast region, the Baltimore Canyon, 18 spills (releasing about 40,000 barrels of oil) could be expected over the next 30 years. They also noted that *no* offshore spill to date "has been contained and cleaned up on site", and that "there is no assurance that the technology utilized . . . would be adequate for oil-spill surveillance, containment, and cleanup".[115] In fact, if a spill occurred as far as 50 miles at sea, the government calculated that the odds are at least 1 in 10 that the oil slick would reach the Atlantic Coast.[116] Comparatively speaking, within the region out to 50 miles off the New Jersey and Delaware shores, for example, OCS developments are likely to spill more oil than small tanker operations.[117]

Perhaps one reason why the various distributive effects of oil spills on the public were not calculated in the government technology assessment is that "under existing law, damaged parties lack protection against economic losses that may result from oil reaching shore".[118] This lack of protection is a consequence of several facts. For one thing, the spill laws apply to vessels and not to development risks. *Second*, there is no provision [in current law] for compensating parties for damages that the cleanup effort cannot prevent, and insurance against spills is difficult to obtain. Also offshore operators are not required to demonstrate financial responsibility. This means that companies "which could not assume current required liability expenses would [nevertheless] be permitted to operate . . .".[119] In other words, as the assessment team admits, "existing laws are not adequate either to assign liability or compensate individuals or institutions for damages from oil spills resulting from exploration, development, or production . . .".[120]

In addition to the regional costs associated with development-related oil spills and limited liability coverage, coastal residents face several other adverse impacts from oil and gas technologies. One is that onshore facilities related to offshore drilling "may be a financial burden on the state and local communities".[121] The assessors clearly point out that "localized fiscal problems" will arise because of the development technology.[122] They note that "particular localities within a state will experience net adverse budgetary impacts during the course of OCS development, since there is little reason to expect that the tax-revenue-producing onshore facilities would be located in the tax jurisdiction of the communities that must provide public services and facilities for the population supporting offshore exploration and development".[123]

4.3. *Assessment Failure to Calculate Regional Costs*

Since assessors have made admissions that localized negative impacts are likely to occur as a result of OCS technology, it is paradoxical that no attempt was made to quantify or estimate these adverse consequences. This omission is even more significant in light of the fact that the authors use dollar amounts for employment benefits, per-capita tax revenues, and capital expenditures, all of which are allegedly positive impacts of the technology upon the local communities.[124] Because no comparable figures were calculated for adverse economic impacts on the region, it is not surprising that the assessment conclusions are largely in favor of the technology. Brief mention was made of negative onshore fiscal impacts,[125] but they were apparently discounted because they were not put in quantitative terms. This is consistent with

Gresham's Law, which states that quantitative information (here, data favoring the technology) drives out qualitative (in this case, information leading one to oppose the technology). This seems to happen, in part, because assessors do not know how to evaluate the qualitative material.

Although I believe a sound case can be made for quantification of all assessment parameters on the grounds that it would enable analysts to avoid the Gresham problem, this is not the place to present such an argument.[126] (For this argument, see Chapter Six.) Suffice it to say that qualitative data may be more easily interpreted in a number of radically different ways than may quantitative information. Moreover without a common quantitative basis for comparing diverse impacts, it is unclear that the most rational assessment possible could take place.

In the case of the OTA study of oil-development technology, failure to quantify the costs of significant distributive, geographical inequalities appears to have biased the evaluation in favor of offshore development. Although the assessment mentions the conflicts between state and local authorities and between state and federal goals regarding how (or whether) to develop the technology, no quantification of the various regional costs, under state versus federal policy, was attempted. Likewise the assessors noted that the oil and gas facilities would have negative consequences of regional air and water quality, but they included no quantification of these impacts.[127] (Much the same procedure was followed in a recent study of offshore oil technology completed by the Environmental Protection Agency. The authors noted various forms of water pollution resulting from OCS development (caused by freon extractables, fecal coliform and chlorine residual, floating solids, heavy metals, total dissolved solids, chlorides, oxygen-demand parameters, and phenolic compounds). Despite this fact, they cited no costs of the pollution, such as onshore effects or reduction of the fishing catches. The discussion was presented in a purely qualitative manner.[128]) The OTA authors similarly admitted that their study of biological impacts was "qualitative in its findings"; that there were "basic uncertainties about environmental and economic impacts" of the technology; and that "the effects of pollutants which may be discharged during OCS operations cannot presently be determined with any accuracy".[129] They also mentioned the fact that good water quality is essential to the tourist, fishing, and sport industries of the area.[130] Nonetheless they appear to have ignored these uncertainties and then drawn several conclusions which are little more than unsubstantiated value judgments.

4.4. *Value Judgments About Negative Impacts*

One such conclusion is that none of the alternatives for supplying "equivalent amounts of energy" offers "clear social, economic, or environmental advantages" over offshore oil development.[131] It is not evident, however, how this determination was made. Quantitative data seem necessary to support it, but these were not given. At least, it is necessary to have some explanation of the basis on which the qualitative remarks on negative regional impacts appear to have been discounted. Since these adverse consequences were not quantified, and since there was no evident procedure for determining that the positive impacts outweighed the negative geographical impacts, the OTA conclusion appears to be little more than an unsubstantiated value judgment.

Another conclusion of the assessors also appears vulnerable to similar lines of questioning. They judge that the net fiscal benefit of the offshore technological development outweighs the "localized fiscal problems".[132] Like the previous conclusion this one was drawn without apparent analysis of either quantitative data or the issues of distributive equity. Both judgments fail to answer the questions: 'Social, environmental, and economic advantages *for whom*? Net fiscal benefit *for whom*?'

Although the grounds for the two value judgments were not stated by the assessors, several facts appear certain. The authors apparently did not believe that adverse fiscal and environmental impacts to the various coastal regions were outweighed (and hence justified) by lower regional energy prices as a result of the technology. No such tradeoff seems conceived of, since they clearly note that "dramatic changes in regional energy prices should not be expected to follow OCS development".[133] The only key to these value judgments appears to be the report's emphases on the national need for oil and on the necessity for avoiding dependence on foreign energy suppliers.[134] Both emphases suggest that, as we discussed earlier,[135] 'national security' or 'the war power' might have been used to justify the assessment conclusions in favor of development of the offshore technology.

4.4.1. *Sanctioning Laissez-Faire Technology.* There is also some reason to believe that the OTA assessors see offshore oil technology as being permitted to operate in a *laissez-faire* fashion, and perhaps that they agree with this mode of operation. After all, the Department of Commerce oversees approval of state coastal plans, in order to insure that they are consistent with OCS development. It is conceivable that the commerce clause and a 'progress' ethic have been used (perhaps implicitly) to sanction the failure to assess a number of negative regional impacts. Evidence for the *laissez-faire* interpretation is provided by several facts, all of which have the potential for severe regional damage:

1. "The federal government does not set definitive standards for the industry to follow in carrying out its responsibility to provide cleanup equipment in the event of a major oil spill. The USGS does not inspect cleanup equipment but relies on industry to make its own inspections".[136]
2. "When the BPTCA [Best Practical [Pollution] Technology Currently Available] limitations were derived, it was concluded that they should be based on what was [already] being achieved by all [industrial] facilities . . .".[137]
3. The required 'environmental baseline study' for the Mid-Atlantic area was not scheduled for completion until 6 months after the lease sale of OCS lands for offshore development in the area. Moreover "there is no requirement that the information gathered [in the environmental baseline study] be used in the decisionmaking process for the sale of offshore lands and subsequent operations".[138]
4. There are no precise federal regulations with regard to construction of offshore platforms or pipelines,[139] "no standards that cleanup and containment equipment . . . must meet, and no assurance that a major oil spill actually could be confined, and removed from the water even if the best equipment is available".[140]

Despite this apparent evidence for approval of unrestricted development of offshore oil technology, and despite the OTA authors' admissions (noted earlier) that there were "basic uncertainties about the environmental and economic impacts of the technology",[141] the assessors nevertheless concluded that "no significant damage to the environment or changes in patterns of life" are anticipated during operation of the offshore technology.[142] This conclusion is surprising for several reasons. For one thing, it is an argument from ignorance and, as such, is beset with all the difficulties associated with such a logical fallacy. For assessors to have drawn conclusions favoring offshore oil technology, after they admitted that they were in ignorance about its economic and environmental impacts, also suggests another methodological problem. Perhaps the OTA assessors are as methodologically committed to the desirability of laissez-faire technology, despite its human costs, as federal policymakers appear to be committed to laissez-faire energy development, regardless of the medical and environmental costs it imposes.

4.4.2. *Begging the Question of the Importance of Regional Costs.* My quarrel, however is not with conclusions which are pro-technology or pro-commerce. In fact, in the case of offshore oil development, the assessors' essentially positive conclusions about the technology may well be correct, at least in certain cases, even after a proper geographical and distributional analysis is taken into account. One reason why they may be correct, for example, is that some of the regions bearing OCS oil costs already receive disproportionate oil-related benefits. For instance, although some costal residents in northeastern US, for example, would bear higher regional environmental and economic costs as a result of offshore development, it could be argued that they ought to bear the costs. Since the Northeast is an importer of oil and gas, it might now be said to receive disproportionate benefits which ought to be balanced by the costs of offshore development. After all, goes the argument, the Northeast consumes 26% of the nation's petroleum products, but has only 9% of the total refinery capacity. Perhaps this region ought not eschew the costs of energy production (as with OCS development) and, at the same time, claim disproportionate benefits (26% of petroleum use).[143]

Apart from whether the assessors' conclusions are substantively correct, my claim is that there is no methodological justification for evaluating a technology in such a way that problems of the distribution of negative regional impacts are ignored, especially when the assessment conclusion is in favor of the technology. If one were unable to calculate these distributive impacts, then at best he ought not to draw a conclusion regarding the overall impact of the technology without noting the methodological problems limiting its validity. This means that where value judgments (e.g., that progress is desirable, that technological growth helps the poor, that energy technologies ought to operate in a laissez-faire fashion) influence assessment conclusions, the evaluative presuppositions ought to be explicitly noted. As it is, in a technical report interspersed with quantitative data, one is likely to infer, wrongly, that conclusions favoring a particular technology are more objective than they are in reality.

In presupposing that negative regional impacts need not or cannot be quantified, or are not significant, however, the authors of the offshore-oil study have very likely (unwittingly perhaps) run into substantive, as well as methodological problems. By begging the question of the importance of negative regional impacts, they may have predetermined their finding that the technology was more economically and environmentally desirable than others available. If a comprehensive analysis of the distributional regional impacts of all energy alternatives were completed, then a different conclusion might be justified. Moreover since all the regional costs were not quantified in the OTA offshore energy study, the assessment conclusion has at least very likely erred in being less negative (regarding the technology) than it otherwise might have been. In any case, the authors' presupposition, that negative geographical impacts need not or cannot be evaluated, has led to an apparent pro-technology bias. Without hard data to the contrary, the assessors appear to have sanctioned 'business as usual'.

Another aspect of this presupposition is that it may easily lead to policy or to assessment conclusions in which decisionmakers sanction the tyranny of the majority over the minority. So long as costs to the minority (those living near/on a coastline) are neither assessed adequately, quantified, nor somehow evaluated pricisely in light of the benefits to that same minority, then the costs may be ignored. And if they are ignored, then those bearing them effectively become a new minority against whom a new form of technology-induced discrimination has been practiced.

4.4.3. *Ignoring Equal Protection.* One of the most obvious ways in which the offshore oil technology appears to result in discrimination against a geographical minority has to do with equal protection. In sanctioning policy preventing oil-spill victims from obtaining coverage for their losses, decisionmakers have clearly allowed one set of persons (coastal residents/property owners) to receive unequal protection from the law. In neglecting to count this liability problem as a cost of the offshore technology and in failing to evaluate regional economic and environmental losses resulting from the oil development, the assessors appear (at least implicitly) to have sanctioned unequal protection. It seems puzzling that a shoreline motel owner, for example, could suffer economic losses (from an oil spill) for which he could not legally be compensated. What is more puzzling is that, for some reason, his and other regional losses were not evaluated in any clear fashion by the OTA assessors of the offshore technology. For this reason, it is not clear that the assessors have provided morally relevant grounds for failing to apply the principle of *prima facie* political equality (see section 3.2.2. earlier) in this case.

Everyone, including owners of coastal motels, has a right to equal protection and to due process. Admittedly national interests could, in some cases, be shown to outweigh the interests of these rights holders (see section 3.2.1.2 earlier). Before this could be shown, however, both the costs to the rights holders (of this judgment) and the ethical justification for so weighing them would have to be established. In this OTA study, however, assessors neither recognized the necessity of this ethical justification nor counted costs to regional residents as a negative social-political impact of the technology. For practical purposes, therefore, one wonders if the assessment has rendered the concept of 'right' as meaningless. As Daniel Callahan put it,

The concept of a 'right' becomes meaningless if rights are wholly subject to tests of economic, social, or demographic utility, to be given or withheld depending upon their effectiveness in serving social goals.[144]

In the case of offshore oil technology, in the absence of appropriate ethical analysis, one wonders whether the concepts of equal rights and equal protection have been withheld in order to serve the social goals of economic progress or interstate commerce.

Perhaps one reason, why the equal protection clause of the Fourteenth Amendment has not been employed to prevent regional inequities (such as those discussed in this technology case), is that the amendment has traditionally been interpreted as referring only to *state action* violative of claims to equal protection and due process.[145] Currently, however, there is disagreement as to whether the equal-protection clause applies only to the states or whether it also prohibits individuals from discriminating. In one of the most recent Supreme-Court interpretations of this statute, *Griffin v. Breckenridge*, the Court held that Congress intended the amendment to regulate private conduct.[146]

Such a broadened interpretation of the Fourteenth Amendment seems desirable in the light of the fact that constitutional changes might be required to help the individual cope with new technology and with environmental degradation. A number of environmentalists have suggested as much.[147] Even Thomas Jefferson, writing to Samuel Kercheval in 1816, noted that "laws and institutions must . . . become more developed, more enlightened . . . must advance also, and keep pace with the times".[148] In favor of this wider understanding of the Fourteenth Amendment, it can be argued that a broadening of the scope of the clause is in the best interests of the people of this country. Charles A. Reich, in *The Greening of America*, put it well when he wrote:

lawyers talk about the rationality and equality of the law, but they simply do not get outside the accepted assumptions to think about how the law operates as an instrument of one class in society against another.[149]

If my earlier remarks about how the poor bear a disproportionate number of negative technological impacts and environmental hazards (see section 3.3.1) are correct, then the current, limited interpretations of the equal protection clause in effect discriminate against the poor. So long as law does not effectively regulate environment- and technology-related inequalities in protection, then Reich could be right. Law might well operate as the instrument of one class against another. Moreover, apart from the problem of further debilitating members of poorer classes, failure either to guarantee equal protection from technology-induced hazards, or to provide morally relevant reasons for justifying discrimination with respect to equality, can militate against the success of democratic government. Abraham Lincoln made a related point when he spoke to the Young Men's Lyceum of Springfield, Illinois. He said that

> if [citizens'] rights to be secure in their persons and property, are held by no better tenure than the caprice of a mob, the alienation of their affections from Government is the natural consequence; and to that, sooner or later, it must come.[150]

4.5. *Further Consequences of Ignoring Regional Inequalities*

If I am correct in arguing that the failure to calculate or to adequately assess regional inequalities in technological impacts can lead to ethically unjustified violations of equal protection, then this methodological difficulty of technology assessors probably also signals other undesirable consequences. For one thing, the authors' inattention to evaluation of certain regional costs reveals that, at least with respect to *distributive justice*, technology may well be 'out of control', or autonomous, if those delegated to monitor it fail to do so comprehensively.[151]

Assessment inattention to distributive impacts, an inattention which contributes to technological autonomy, suggests that policy analysts have persisted in 'frontier' values. They have not moved toward a 'spaceship' attitude, and they have not examined the second- and third-order consequences of so-called 'frontier' values.[152] Had they done so, they might have realized that virtually no technological developments result in isolated consequences on some distant frontier. Instead they have complex distributive impacts requiring analysis and evaluation.

If technology is 'out of control' with respect to distributive impacts, then it will likely be more difficult for society to move toward the goal of equal concern for all persons. One authority noted recently that economic inequality, often a cause of political inequality, is presently on the increase in most western industrial societies, certainly in the United States. "What I see", he said, "is the emergence of an affluent majority, the hardening of its attitude toward the poor, and the imposition of a majorial tyranny in which the poor are increasingly ghettoized".[153]

Assessors' and policymakers' ignoring evaluations of regional distributive impacts leads to other problems besides encouraging greater economic and political inequalities among persons. It also contributes to a loss of freedom, especially among those who bear the disproportionate geographical costs of technology. If one's fishing business is threatened by OCS oil spills, for example, or if one's property values fall because compensation cannot legally be obtained for all spill damage to shorelines, then his freedom is quite limited. If freedom involves both the opportunity to choose among genuine alternatives and ready access to knowledge that will make the selection an informed one,[154] then the OCS assessment of geographical impacts, and public policy regarding those impacts, contribute toward a limit on freedom The policy imposes limits because the coastal property owner, for example, is not free to obtain compensation for spill damage or to move and thereby receive a just price for his land. The assessment methodology imposes limits because omitting calculation and evaluation of geographical impacts deprives citizens of access to knowledge that could be used to argue for more equitable public policy. These constraints, in turn, might curb social progress. Social progress is, at least in part, defined in terms of the increase of options available to individuals and societies.[155] If both public policy and assessment methodology regarding technology- and environment-related impacts limit these options for some people, then progress may be thwarted. In other words, if it causes consequences destructive of opportunities for future choice, then a particular technology or environmental impact might be helping to create a 'closed', rather than an 'open', society.[156] Perhaps this is one reason why Ellul believes that the price of technological power is loss of freedom. Whether he is completely correct is debatable. What appears evident, however, at least on the basis of the OCS technology assessment, is that Ellul is right about one thing. He maintains that humans rarely ask what they will have to pay, *in toto*, for their power in technology.[157]

The fact that certain technologies and modes of assessing them threaten freedom also suggests that they threaten democracy as well. One government agency representative said recently that "technology assessment is performed almost secretly and outside the usual framework of the democratic process".[158] Given the limited provisions for public participation in OCS-related decisionmaking,[159] and the inattention of assessors to calculating the distributive social impacts of offshore-oil technology of various regions, it might well be said that this assessment was done outside the democratic process. What appears to have happened is that, given the powerlessness of affected state and local groups to control OCS policy, a select group of federal assessors has determined that policy. In doing so, they have prepared the required impact statement but have taken advantage of their position in order to omit evaluation of regional distributive impacts. This means that they have both ignored the concerns of local groups and that they have taken advantage of their disenfranchisement. Just such a situation appears to have been envisioned by Dwight D. Eisenhower. He spoke of the "danger that public policy could itself become the captive of a scientific technological elite".[160]

Scientific elites are often responsible for assessment work, and hence ultimately responsible for policy, because many questions of technological impacts are seen predominately as issues of scientific fact, not social policy. This characterization of assessment issues, however, seems largely misguided. (See Chapters Three and Four earlier in this work.) Key regulatory questions appear to involve social policy and not merely scientific expertise. This is because, although factual information is essential, informed public decision-making cannot be accomplished on a purely factual basis. As one author put it, the "central question . . . is what society really wants".[161] In the case of offshore-oil technology, for example, the issue is not merely what the probability of a spill is, but whether the public is willing to take that risk, whatever it is.

So long as technology and environmental-impact assessment is conceived of as involving analysis of largely factual issues, then ethical problems, such as that of geographical equity, will never be handled adequately in the evaluation. This is because no amount of scientific information can resolve a question of values. Likewise assessors are wrong to attempt to adjudicate a *political* issue (what people really want) in a quasi-*judicial* framework (preparing an assessment statement, holding technical hearings on the draft, revising the original statement of the basis of the hearings). Political and ethical problems need to be handled as political and ethical problems, not merely as legal or scientific ones.[162] Otherwise, democracy suffers; when one is dealing with purely technical (legal or scientific) questions, there is no framework for the non-expert or the impacted layman to speak about how the policy affects him. Just as guilt or innocence, in a democracy, is normally determined by *trial* by one's *peers*, not by a *decision* by psychologists or psychiatrists, so also, whether a technology is desirable or undesirable ought usually to be determined by those affected by it, not merely by scientists or technologists.

Apart from many authors' tendencies to handle political and ethical questions in a scientific or legal way, current assessment methodologies threaten democracy in a number of other ways. One of the most obvious problems is that a majority of American scientists "are (or wish they were) consultants to corporations".[163] Hence, even though they may not realize it, they very likely have a tendency to view a particular technology as desirable. Pro-technology bias is also probable since many decisions are "skewed in favor of well-organized and well-financed" interests,[164] and the geographical minorities subjected to technological costs are likely to have both poorer organizations and poorer finances than those who promote the technology.

5. AN OUTLINE FOR REFORM

How can assessment methodologies be improved so as to deal better with some of these problems? Assessors' tendencies to ignore evaluation of local distributive impacts and to treat political and ethical questions (e.g., What do citizens of this region really want?) as scientific or legal ones might be overcome in several ways. One solution might be to allow interested parties, particularly those receiving disproportionate geographical or distributive impacts, to do their own analysis of impacts affecting them. They might receive federal funding for this purpose, and they could place their own costs on items usually only handled qualitatively, if at all, by assessors.[165] In other words TA and EIA conclusions might be more sensitive to ethical constraints if some form of 'adversary' assessment were used. (See Chapter Nine later in this volume.)

Another means of better handling issues of geographical equity would be to require assessors both to state and defend their presuppositions (e.g., that technological progress increases political equality among persons) and to use the best available methods of social-systems accounting. Instead of employing outdated judgments about the desirability of an increasing GNP, for example, assessors need to focus on social-impact analysis, and on defining a good quality of life.[166] They ought to attempt to justify their recurrent assumption that increased levels of either energy consumption or energy supplies, for example, raise the quality of life. In this way their implicit, often unanalyzed, value judgments could be made more explicit and hence subject to evaluation by the public. Any one of a number of methodological devices could be used to render such presuppositions explicit. One means might be B. M. Gross' method for analyzing 'system performance'. On his scheme, technologies or environmental projects would have to be evaluated according to how well they promoted beauty, creativity, security, freedom and other goals, as well as how they serve traditional economic interests.[167]

One of the biggest blocks to adequate social-impact accounting, whether in technology assessment, environmental-impact analysis, or some other area, is the existence of externalities, also known as disamenities, spillovers, or diseconomies.[168] So long as the externalities, or social costs, of a technology are not internalized and 'priced' within economic accounting, then the number of persons who bear uncompensated social costs of technology is likely to increase. This was part of my point in discussing liability for damages induced by OCS technology. So long as assessors fail to quantify all factors involved in spill damage to the environment, for example, then that damage is likely to increase, to be ignored by assessors, and to go uncompensated. (See Chapter Six earlier in this volume.)

With an adequate system of pricing externalities, assessors could provide a clear framework according to which technological impacts could be equalized. Regional 'losers' could be compensated by regional 'gainers'.[169] Such a system might be both methodologically superior to those now used,[170] and it ought to be more just.[171] Boulding makes the point nicely:

> the fundamental principle that we should count all costs, whether easily countable or not, and evaluate all rewards, however hard they are to evaluate, is one which emerges squarely out of economics and one which is at least a preliminary guide line in the formation of moral judgment in what might be called the 'economic ethic'.[172]

Of course what might happen in pricing externalities is that the social costs associated with certain environmental and technological projects might be found to be so high that they are prohibitive. In this case, calculation of spillovers might support policy based on prohibitive law rather than merely on permissive and compensatory law. Such a situation, however, might be desirable from the point of view of equal concern for all persons. As welfare economist Ezra Mishan points out, equity among persons/regions is usually better served by prohibitive laws regarding externalities. This is because, with a prohibitive law, the burden is put on the person who damages the interests of others; with a permissive/compensatory law, he argues, the victims are unable to limit in advance the degree to which their welfare is hurt.[173]

Apart from these methodological suggestions involving using an adversary means of assessment, internalizing externalities, and improving schemes of social-systems accounting, there are several other ways in which assessment procedures might be modified so as to better serve analysis of problems of geographical equity. One solution might be to provide ethically weighted TA's and EIA's. (This proposal is discussed later, in Chapter Eight). Also, instead of merely attempting to list the total costs and benefits consequent upon a particular technological or environmental project, analysts might make more use of the 'scenario' approach.[174] That is, they might find it useful to engage in morphological analysis by

1. using a few main features to summarize the problem under study;
2. listing the main possible values of each feature;
3. enumerating all possible combinations of all modalities, one for each feature; and
4. evaluating the risks, costs and benefits of each plausible scenario derived from (3).

By using the scenario approach, impact analysts would not be limited to assessing risks, costs and benefits merely under current policies. They could examine costs of regional inequality with various scenarios, each of which presupposed a different regulatory framework. In the case of OCS-development technology, for example, assessors might examine two different scenarios for social costs, one in which states charged a petroleum transport fee (as Maine now does), and one in which there are no such fees.[175] Or, they might evaluate and compare regional and national benefits and costs within two other scenarios: one in which the federal government has complete control over environmental standards for oil-spill cleanup, and another in which the federal government was allowed only to set minimum standards for environmental quality in this regard.[176] Likewise assessors might examine the effects of allowing, versus prohibiting, state veto of technological or environmental projects within their borders,[177] or the consequences of requiring, versus not requiring, that local areas make zoning decisions far in advance of any proposal to use land for a particular technological purpose.[178] Once various scenarios were articulated, then citizens would have a better idea, not only of whether they were willing to accept certain impacts, but also of the conditions under which those impacts might be heightened or lessened.

If reforms such as these discussed were implemented, then there is reason to believe that both impact assessment and public policy would pay closer attention to ethical analysis of egalitarian principles central to most conceptions of distributive justice. Without such reforms, as the case study of OCS-development technology reveals, geographical minorities will continue to bear disproportionate risks from hazards such as oil spills, limited liability, and onshore fiscal burdens. Analogous costs, in fact, are likely to be borne by numerous local groups who are inequitably affected by the environmental or technological impacts of other projects. To ignore these regional impacts is not only to run the risk of sanctioning studies which are likely to be built on presuppositions about the argument from ignorance, *laissez-faire* technology, and the misapplication of 'national security' arguments. It is also to ignore the degree to which assessment conclusions might promote more rational, equitable, and democratic decisionmaking about technology. Once considerations such as geographical equity are evaluated, then environmental-impact analysis and technology assessment will have moved a long way toward bridging the gap between technological *utility* and environmental *justice*. After all, as Cicero pointed out long ago:

> nothing is *genuinely* utile that is not at the same time just, and nothing just that is not at the same time utile; . . . no greater curse has ever assailed human life than the doctrine of those who have separated these two.[179]

NOTES

[1] See M. A. Boroush, K. Chen, and A. N. Christakis, *Technology Assessment: Creative Futures*, North Holland, New York, 1980, p. 268; hereafter cited as: Boroush *et al.*, *TA*. Here they point out, for example, that the arid lands of the western US have benefitted from publicly funded irrigation diversion programs at the expense of the rest of the nation, and particularly of the South, whose agriculture has suffered from the competition. On the other hand, they note, the southern states have benefitted disproportionately from military and space activities. Urban renewal programs have provided construction funds to cities primarily in the Northeast, while farm subsidies have driven numbers of people to these cities and swelled their welfare rolls. Boroush *et al.* also point out that railroad rate regulation has benefitted eastern and western US at the expense of the South. Likewise if representatives of western states are correct, their area would be adversely affected by current plans for massive strip-mining which would deplete their non-renewable resources and impair western community life in return for a domestic energy supply to meet national demands.

[2] See D. R. Godschalk and D. J. Brower, 'Beyond the City Limits: Regional Equity as an Emerging Issue', *Land Use and Environment Law Review–1979* (ed. F. A. Strom), Clark Boardman, New York, 1979, pp. 450, 457–459; hereafter cited as: Godschalk and Brower, Equity.

[3] See Boroush *et al., TA*, pp. 266–268, 362–363, for examples of this point. See also H. P. Green, 'The Adversary Process in Technology Assessment', in *Technology Assessment: Understanding the Social Consequences of Technological Applications* (ed. R. G. Kasper), Praeger, New York, 1972, p. 50; hereafter cited as: Green, Process, in Kasper, *Technology*. For discussion of the historical origins, types and characteristics of environmental-impact analyses (EIA) and technology assessments (TA), see A. L. Porter, F. A. Rossini, S. R. Carpenter, and A. T. Roper, *A Guidebook for Technology Assessment and Impact Analysis*, North Holland, New York, 1980, pp. 51–63; hereafter

cited as: Porter *et al., Guidebook*. See also S. R. Carpenter, 'Philosophical Issues in Technology Assessment', *Philosophy of Science* 44 (4), (December 1977), 574 ff.; J. P. Martino, *Technological Forecasting for Decisionmaking*, Elsevier, New York, 1978 (hereafter cited as: *Forecasting*); R. T. Taylor, 'NEPA Pre-emption Legislation: Decisionmaking Alternative for Crucial Federal Projects', *Land Use and Environment Law Review–1979* (ed. F. A. Strom), Clark Boardman, New York, 1979, pp. 98–100 (hereafter cited as: NEPA); K. M. Murchison, 'Waiver of Intergovernmental Immunity in Federal Environmental Statutes', *Environment Law Review–1977* (ed. H. F. Sherrod), Clark Boardman, New York, 1979, p. 653 (hereafter cited as: Waiver); and Congress of the United States, Office of Technology Assessment, *Technology Assessment Activities in the Industrial, Academic, and Governmental Communities*, Hearings Before the Technology Assessment Board of the Office of Technology Assessment, 94th Congress, Second Session, June 8–10, 14, 1976, US Government Printing Office, Washington, D.C., 1976, p. 76; hereafter cited as: Congress, *TA in IAG*. For distinctions between EIA and TA, see Porter *et al., Guidebook*, pp. 3–4, 48–51, and Congress, *TA in IAG*, pp. 75, 236. and K. S. Shrader-Frechette, 'Technology Assessment as Applied Philosophy of Science', *Science, Technology, and Human Values* 6 (33), (Fall 1980), 33–50; hereafter cited as: Shrader-Frechette, TA.

[4] See R. B. Stewart, 'Pyramids of Sacrifice? Problems of Federalism in Mandating State Implementation of National Environmental Policy', *Land Use and Environment Law Review–1978* (ed. By F. A. Strom), Clark Boardman, New York, 1978, p. 166; hereafter cited as: Stewart, Pyramids. See also, for example, Meteorology Group, Teknekron, Inc., Berkeley, California, *Results of Multiscale Air Quality Impact Assessment for the Ohio River Basin Energy Study*, Environmental Technology Applications Group, Waltham, Massachusetts, 1978, pp. 1, 3, 5–6, 12; for an excellent bibliography of materials related to air-quality impact assessment, see p. 15.

[5] This example is used by W. G. Murray and C. J. Seneker, 'Industrial Siting: Allocating the Burden of Pollution', *Land Use and Environment Law Review–1979* (note 2), p. 433; hereafter cited as: Industrial.

[6] Martino, *Forecasting*, p. 516.

[7] The same suggestion is made by Robert Coburn, 'Technology Assessment, Human Good, and Freedom', in *Ethics and the Problems of the 21st Century* (ed. by K. Goodpaster and K. Sayre), University of Notre Dame Press, Notre Dame, 1979, pp. 106–121, esp. pp. 116–117; hereafter cited as: Coburn, TA, in Goodpaster and Sayre, *Ethics*. Coburn's proposal relies heavily on the work of John Rawls.

[8] Boroush *et al., TA*, p. 241; see also pp. 240 ff. and 363.

[9] For discussion of the assumption of aggregation, see Sergio Koreisha and Robert Stobaugh, 'Appendix: Limits to Models', in *Energy Future: Report of the Energy Project at the Harvard Business School* (ed. by R. Stobaugh and D. Yergin), Random House, New York, 1979, pp. 237–240; hereafter cited as: *EF*. See also M. Dogan and S. Rokkan, *Quantitative Ecological Analysis in the Social Sciences*, MIT Press, Cambridge, 1969; hereafter cited as: *QEA*.

[10] For an excellent example of this point, see Oskar Morgenstern, *On the Accuracy of Economic Observations*, Princeton University Press, Princeton, 1963, p. 37; hereafter cited as: *Accuracy*.

[11] This point is also made by Gail Kennedy, 'Social Choice and Policy Formation', in *Human Values and Economic Policy* (ed. by Sidney Hook), New York University Press, New York, 1967, pp. 140–149, esp. p. 148; hereafter cited as: *HV and EP*.

[12] H. R. Bowen, Chairperson, National Commission on Technology, Automation, and Economic Progress, *Applying Technology to Unmet Needs*, Vol. 5, US Government Printing Office, Washington, D.C., 1966, p. V-240. See also E. J. Mishan, *Cost-Benefit Analysis*, Praeger, New York, 1976, p. 407 (hereafter cited as: *Cost-Benefit*), who criticizes what he calls 'the fallacy of aggregation'.

[13] See, for example, Porter *et al.*, Guidebook, pp. 318–319; Boroush *et al., TA*, pp. 268, 363; and Stobaugh and Yergin, 'The End of Easy Oil', in Stobaugh and Yergin, *EF*, pp. 3 ff. Some assessors, however, neither argue for consideration of distributional impacts nor provide such analyses in their own work. See C. Starr, 'Social Benefit Versus Technological Risk', *Technology and Society*, (ed. by N. de Nevers), Addison-Wesley, London, 1972, pp. 214–217, esp. p. 214; hereafter cited as: Starr, Benefit.

[14] Congress of the US, Office of Technology Assessment, *Annual Report to the Congress for 1978*, US Government Printing Office, Washington, D.C., 1978, pp. 73–74. All annual reports of the OTA are hereafter cited as: *AR 1978, AR 1979*, etc.

[15] Congress of the US, Office of Technology Assessment, *A Technology Assessment of Coal Slurry Pipelines*, US Government Printing Office, Washington, D.C., 1978, p. 6; see also p. 80. Hereafter cited as: Congress, OTA, *Coal Slurry*.

[16] Congress, OTA, *Coal Slurry*, pp. 84, 99. For discussion of the problem of pricing natural resources, see M. A. Lutz and K. Lux, *The Challenge of Humanistic Economics*, Benjamin/Cummings, London, 1979, pp. 297–308, esp. pp. 305–307; hereafter cited as: *Challenge*.

[17] Congress, OTA, *Coal Slurry*, p. 15.

[18] Congress of the US, Office of Technology Assessment, *Transportation of Liquefied Natural Gas*, US Government Printing Office, Washington, D.C., 1977, p. 42; see also pp. 55, 68, 70. Hereafter cited as: Congress, OTA, *LNG*.

[19] For an excellent discussion of whether equity should play a role in environmental considerations, see A. M. Freeman, 'Distribution of Environmental Quality', in *Environmental Quality Analysis*, (ed. by A. V. Kneese and B. T. Bower), Johns Hopkins, Baltimore, 1972, pp. 243–278; hereafter cited as: Freeman, DEQ, and Kneese and Bower, *EQA*.

[20] See Stewart, Pyramids, pp. 162–163, 216.

[21] See Stewart, Pyramids, pp. 162–163; see also R. B. Stewart, 'Paradoxes of Liberty, Integrity, and Fraternity: the Collective Nature of Environmental Quality and Judicial Review of Administrative Action', *Environmental Law* 7 (3), (Spring 1977), 472; hereafter cited as: Paradoxes. Another variant of this argument is that the common interests of all regions, and the system as a whole, are better served by policies of federalism. See M. Markovic, 'The Relationship between Equality and Local Autonomy', in *Equality and Social Policy* (ed. by W. Feinberg), University of Illinois Press, Urbana, 1978, p. 96; hereafter cited as: Markovic, Relationship, and Feinberg, *ESP*.

[22] See Stewart, Pyramids, pp. 164–165, and Stewart, Paradoxes, p. 473. According to this argument, racist and sexist discrimination, for example, is more easily outlawed by federal policies. (See Godschalk and Brower, Equity, p. 468, and note 24 below.) Another variant of the argument is that decisons must be made federally because inadequate technical expertise exists at the state and local level. See K. S. Shrader-Frechette, *Nuclear Power and Public Policy*, pp. 40–43, for discussion of a case in which this argument was made. See also N. Notis-McConarty, 'Federal Accountability: Delegation of Responsibility by HUD under NEPA', *Environmental Affairs* 5 (1), (Winter 1976), 136–137; hereafter cited as: Federal.

[23] See Stewart, Pyramids, pp. 166–168, and Stewart, Paradoxes, p. 473.
[24] See Stewart, Pyramids, pp. 168–170, and Stewart, Paradoxes, p. 473. In this regard, authors have also claimed that federalism helps to avoid egoistic autonomy, parochialism, and laissez-faire regionalism. See Markovic, Relationship, p. 96, and O. Patterson, 'Inequality, Freedom, and the Equal Opportunity Doctrine', in W. Feinberg, *ESP*, pp. 32–33; hereafter cited as: Inequality. Patterson argues that centralized decisionmaking enhances personal freedom, equality, and welfare-state spending. He shows why decentralized control usually results in widespread deprivation of various human rights. Following a similar line of reasoning, Notis-McConarty (Federal, p. 135) notes that NEPA has been abused by various community groups attempting to block the entry of low-income citizens and minorities into their neighborhoods.
[25] J. L. Huffman, 'Individual Liberty and Enviromental Regulation: Can We Protect People While Preserving the Environment?' *Environmental Law* 7 (3), (Spring 1977), 435; hereafter cited as: Liberty. See also Stewart, Paradoxes, pp. 472–475, and Stewart, Pyramids, p. 162.
[26] Stewart, Pyramids, p. 147; see also pp. 148–153, 178–181. Admittedly, however, a number of authors have argued in favor of federal, rather than state or local, authority because they wish to avoid environment-motivated delays or prohibitions regarding technological and industrial projects. They believe that crucial federal projects will be less likely to be challenged for environmental reasons if the federal government controls them. See Taylor, NEPA, pp. 110–111 and Murray and Seneker, Industrial, pp. 445–446.
[27] See Murray and Seneker, Industrial, p. 424. According to Stewart, Pyramids, p. 153, many state and local officials shun strict environmental standards because they are afraid they will curtail economic development of their communities. On this view, environmental quality is being bought at the price of economic freedom. (See Huffman, Liberty, p. 431 and Godschalk and Brower, Equity, p. 468.)
[28] In this regard, see Dorothy Nelkin, 'Science, Technology, and Political Conflict: Analyzing the Issues', in *Controversy: Politics of Technical Decisions* (ed. by D. Nelkin), Sage, Beverly Hills, 1979, pp. 9–10; 12–14; hereafter cited as: *Controversy*. For discussion of the low-level radiation case, see Shrader-Frechette, *Nuclear Power and Public Policy*, D. Reidel, Boston, 1983, pp. 39–43; hereafter cited as: *Nuclear Power*. See also A. W. Murphy and D. B. LaPierre, 'Nuclear "Moratorium" Legislation in the States and the Supremacy Clause: A Case of Express Preemption', *Environment Law Review 1977* (ed. by H. F. Sherrod), Clark Boardman, New York, 1977, p. 445; hereafter cited as: Moratorium.
[29] E. D. Muchnicki, 'The Proper Role of the Public in Nuclear Power Plant Licensing Decisions', *Atomic Energy Law Journal* 15 (1), (Spring 1973), 46–47 (hereafter cited as: Role), uses this example. The same point is made by Stewart, Pyramids, p. 161 and in Stewart, Paradoxes, p. 472.
[30] This point is also made by Stewart, Pyramids, p. 161, and Stewart, Paradoxes, p. 472. See note 28.
[31] For a discussion of the problem of local autonomy and equality, see Markovic, Relationship, pp. 85–92. A point similar to this one is made by Stewart, Pyramids, p. 161, and in Paradoxes, p. 472.
[32] Stewart, Pyramids, p. 161, and Stewart, Paradoxes, p. 472, makes this same point. See also Stewart, Pyramids, pp. 170–173.

[33] See Stewart, Pyramids, p. 148. Markovic, Relationship, p. 97, points out in a similar vein that even a majority vote undergirding federal policies can be very undemocratic; the larger part of society could always impose its will on the smaller. See J. R. Brydon, 'Slaying the Nuclear Giants', *Pacific Law Journal* 8 (2), (July 1977), 767 (hereafter cited as: Giants), who gives a pertinent example. The city of Long Beach, California, he says, was powerless to exclude the operation of a federally licensed radioactive waste disposal facility within its borders.

[34] Markovic, Relationship, pp. 85, 98, makes this point. He argues that local autonomy promotes equality because persons can only be soverign in smaller communities. See note 31.

[35] J. Mills and R. D. Woodson, 'Energy Policy: A Test for Federalism', *Land Use and Environment Law Review–1978* (ed. by F. A. Strom), Clark Boardman, New York, 1978, pp. 291–377 (hereafter cited as: Energy), argue that the war power, preemption, and interstate commerce are the three main ways that the federal government overrides the states in regulating technology. Eminent domain may also be a means, according to B. Blinderman, 'The Marshland Issue: Legislative Compensation for Losses Resulting from Governmental Land Use Regulations – From Inverse Condemnation to Productive Use', *Land Use and Environment Law Review* (ed. F. A. Strom), Clark Boardman, New York, 1978, pp. 32–34; hereafter cited as: Marshland. See also Congress, OTA, *Coal Slurry*, pp. 133–137.

[36] Mills and Woodson, Energy, pp. 300–302, 331.

[37] Mills and Woodson, Energy, p. 299, gives this example.

[38] Mills and Woodson, Energy, p. 331 makes this same point.

[39] Murphy and LaPierre, Moratorium p. 437. See also Mills and Woodson, Energy, pp. 293–294.

[40] Murphy and LaPierre, Moratorium p. 438; see also p. 439.

[41] For an analysis of the preemption doctrine and of the inconsistency between radiation standards and other environmental guidelines, see Shrader-Frechette, *Nuclear Power*, pp. 39–43. See also G. B. Karpinski, 'Federal Preemption of State Laws Controlling Nuclear Power', *Georgetown Law Journal* 64 (6), (July 1976), 1341; hereafter cited as: Preemption.

[42] See Shrader-Frechette, *Nuclear Power*, Chapter Four, and Karpinski, Preemption, pp. 1334–1338.

[43] See J. Lieberman, 'Generic Hearings: Preparation for the Future', *Atomic Energy Law Journal* 16 (2), (Summer 1974), 147; hereafter cited as: Hearings.

[44] B. R. Beede and J. A. Sigler, *The Legal Sources of Public Policy*, D. C. Heath, Lexington, Massachusetts, 1977, pp. 89–90; hereafter cited as: *Legal.*

[45] For discussion of this point, see S. Novick, *The Electric War*, Sierra, San Francisco, 1976, pp. 50–69, 72–74, 79–84, 89–91, 98–101, 240; hereafter cited as: *Electric*. See also Mills and Woodson, Energy, pp. 296–298.

[46] Beede and Sigler, *Legal*, p. 104.

[47] Stewart, Pyramids, p. 173.

[48] Beede and Sigler, *Legal*, pp. 96–98.

[49] Congress, OTA, *Coal Slurry*, p. 131.

[50] Murphy and LaPierre, Moratorium p. 450; see also pp. 448–449.

[51] See F. P. Huddle, 'The Social Function of Technology Assessment', in Kasper, *Technology*, p. 163, who makes a similar point.

[52] According to a recent OTA study of solar technology, by the middle 1980's solar energy could supply half this country's energy needs at competitive prices. See Congress of the US, Office of Technology Assessment, *Application of Solar Technology to Today's Energy Needs*, 2 vols., US Government Printing Office, Washington, DC, September 1978.

[53] Karpinski, Preemption, p. 1337. See also note 50.

[54] See Blinderman, Marshland, pp. 32–34, for a discussion of eminent domain.

[55] See Markovic, Relationship, pp. 82–85. See also Isaiah Berlin, 'Equality', in *The Concept of Equality* (ed. by W. T. Blackstone), Burgess, Minneapolis, 1969, pp. 14–34; hereafter cited as: Berlin, Equality, in Blackstone, *Equality*.

[56] See, for example, Richard Taylor, 'Justice and the Common Good', in *Law and Philosophy* (ed. by Sidney Hook), New York University Press, New York, pp. 86–87 ff.; hereafter cited as: Taylor, Justice, and Hook, *Law*.

[57] See H. A. Bedau, 'Egalitarianism and the Idea of Equality', in *Equality* (ed. by J. R. Pennock and J. W. Chapman), Nomos IX, Yearbook of the American Society for Political and Legal Philosophy, Atherton Press, New York, 1967, p. 26; hereafter cited as: Bedau, Egalitarianism, in Pennock and Chapman, *Equality*. See also Berlin, Equality, p. 14.

[58] See Markovic, Relationship, pp. 87–88.

[59] See Patterson, Inequality, pp. 33–34, and Harold Laski, 'Liberty and Equality', in Blackstone, *Equality*, pp. 170, 173; hereafter cited as: Laski, Liberty, in Blackstone, *Equality*.

[60] Markovic, *Relationship*, p. 85. See John Rees, *Equality*, Praeger, New York, 1971, pp. 61–79, and H. J. Gans, 'The Costs of Inequality', in *Small Comforts for Hard Times* (ed. by M. Mooney and F. Stuber), Columbia University Press, New York, 1977, pp. 50–51; hereafter cited as: Mooney and Stuber, *Comforts*.

[61] Some persons have argued that equality of opportunity, for example, is not a desirable goal because it results in a meritocracy, which in turn causes more inequality among persons. They claim that the equal-opportunity doctrine promotes competition and oligarchy and presupposes that these, rather than democracy, are good for society. Moreover, they say, equalization of opportunity is usually attempted by means of education, and education fails to accomplish this goal. (For arguments to this effect see, for example, Patterson, Inequality, pp. 21–30, and Bernard Williams, 'The Idea of Equality', in Blackstone, *Equality*, pp. 49–53; J. H. Schaar, 'Equality of Opportunity, and Beyond', in Pennock and Chapman, *Equality*, pp. 231–240; hereafter cited as: Schaar, Equality. See also J. R. Pennock, *Democratic Political Theory*, Princeton University Press, Princeton, 1979, pp. 36–37; hereafter cited as: *DPT*.)

[62] W. T. Blackstone, 'On the Meaning and Justification of the Equality Principle', in Blackstone, *Equality*, p. 121, uses this argument, as does Frankena; see note 61.

[63] See note 62. John Rawls, 'Justice as Fairness', in Feinberg and Gross, *POL*, p. 284, also makes this point.

[64] For arguments to this effect, see M. C. Beardsley, 'Equality and Obedience to Law', in Hook, *Law*, pp. 35–36; hereafter cited as: Equality. See also Berlin, Equality, p. 33; Frankena, Beliefs, pp. 250–251; Markovic, Relationship, p. 93; John Rawls, 'Justice as Fairness', pp. 277, 280, 282; G. Vlastos, 'Justice and Equality', in *Social Justice* (ed. by R. B. Brandt), Prentice-Hall, Englewood Cliffs, N.J., 1962, pp. 50, 56; hereafter cited as: Vlastos, Justice, and Brandt, *Justice*.

[65] J. R. Pennock, 'Introduction', in *The Limits of Law*, (ed. by J. R. Pennock and

J. W. Chapman) Nomos XV, The Yearbook of the American Society for Political and Legal Philosophy, Lieber-Atherton, New York, 1974, pp. 2, 6; hereafter cited as: Pennock and Chapman, *LL*.

[66] R. Dworkin, *Taking Rights Seriously*, Harvard University, Cambridge, 1977, p. 273; hereafter cited as: *Rights*.

[67] For an excellent defense of this position, see W. K. Frankena, 'Some Beliefs about Justice', in *Philosophy of Law* (ed. by J. Feinberg and H. Cross), Dickenson, Encino, California, 1975, pp. 252–257; hereafter cited as: Frankena, Beliefs, in Feinberg and Gross, *POL*. The position, described here as '*prima facie* political egalitanism', appears to be close to what Frankena defends as 'procedural egalitarianism'. For Frankena, procedural egalitarians are to be distinguished from substantive egalitarians, who believe that there is some factual respect in which all human beings are equal. Procedural egalitarians deny that there is some such factual respect.

[68] For discussions of Utilitarianism, see J. S. Mill, *Utilitarianism, Liberty, and Representative Government*, E. P. Dutton, New York, 1910, esp. pp. 6–24, 38–60. See also Jeremy Bentham, *The Utilitarians: An Introduction to the Principles of Morals and Legislation*, Doubleday, Garden City, New York, 1961, esp. pp. 17–22; P. Nowell-Smith, *Ethics*, Penguin, Baltimore, 1954, p. 34; and J. J. C. Smart and B. Williams (eds.), *Utilitarianism: For and Against*, Cambridge University Press, Cambridge, 1973, esp. pp. 3–74. For a treatment of egalitarianism, see John Rawls, *A Theory of Justice*, Harvard University Press, Cambridge, 1971, pp. 14–15; hereafter cited as: Rawls, *Justice*. See also Charles Fried, *An Anatomy of Values*, Harvard University Press, Cambridge, 1970, pp. 42–43; Charles Fried, *Right and Wrong*, Harvard University Press, Cambridge, 1978, pp. 116–117, 126–127; and Alan Donagan, *The Theory of Morality*, University of Chicago Press, Chicago, 1977, pp. 221–239.

[69] See, for example, K. Sayre and K. Goodpaster, 'An Ethical Analysis of Power Company Decisionmaking', in *Values in the Electric Power Industry* (ed. by K. Sayre and K. Goodpaster), University of Notre Dame Press, Notre Dame, 1977, pp. 266–279; hereafter cited as: Sayre, *Values*. See also Alasdair MacIntyre, 'Utilitarianism and Cost-Benefit Analysis: An Essay on the Relevance of Moral Philosophy to Bureaucratic Theory', in Sayre, *Values*, pp. 219–224; Starr, Benefit, pp. 214–217; Stewart, Pyramids, p. 187; Porter *et al., Guidebook*, p. 358; Taylor, Justice, pp. 90–91; and Shrader-Frechette, *Nuclear Power*, pp. 31–35.

[70] See Nelkin, *Controversy*, pp. 23–83. For the distinction between rule and act utilitarianism, see J. J. C. Smart, 'Utilitarianism', *The Encyclopedia of Philosophy* (ed. by Paul Edwards), vol. 7, Macmillan, New York, 1967, pp. 206–212.

[71] R. B. Brandt, *Ethical Theory*, Prentice-Hall, Englewood Cliffs, N.J., 1959, pp. 415–420; hereafter cited as: *Ethical*. See Taylor, Justice; Beardsley, Equality p. 193; and Pennock, *DPT*, p. 143, who agree with this point.

[72] For John Rawls' view on this point, see notes 63 and 68.

[73] Pennock, *DPT*, pp. 148–149. See note 74.

[74] Rees, *Equality*, pp. 118–120.

[75] Beardsley, Equality, p. 193.

[76] This point is also made by Rees, *Equality*, p. 122.

[77] For analysis of this question, see W. K. Frankena, 'The Concept of Social Justice', in Brandt, *Justice*, pp. 10, 14; hereafter cited as: Concept. See also Taylor, Justice, pp. 94–97.

[78] For discussion of this point, see Rees, *Equality*, pp. 116–117, 120 and Stewart,

Paradoxes, pp. 474–476; Pennock, *DPT*, pp. 16–58, and Patterson, Inequality, p. 31.

79 Pennock, *DPT*, p. 38, uses this example.

80 For arguments to this effect, see section 3.1 of this essay. See also notes 19–23 and 29–34.

81 R. A. Wasserstrom, 'Equity', in Feinberg and Gross, *POL*, p. 246, also makes this point. Even the Fourteenth Amendment, under the equal-protection clause, does not prohibit all discrimination, but merely whatever is 'arbitrary'. In this regard, see N. Dorsen, 'A Lawyer's Look at Egalitarianism and Equality', in Pennock and Chapman, *Equality*, p. 33; hereafter cited as: Look.

82 See Rawls, *Justice*. See also note 63 and S. I. Benn, 'Egalitarianism and the Equal Consideration of Interests', in Pennock and Chapman, *Equality*, pp. 75–76.

83 In this regard, see Hans Bethe, 'The Necessity of Fission Power', *Scientific American* 234 (1), (January 1976), 26 ff., who makes such an argument.

84 This argument is made, for example, by assessors in the employ of the Nuclear Regulatory Commission. See Shrader-Frechette, *Nuclear Power*, p. 29.

85 Frankena, Concept, p. 15, uses this argument. He offers it as a sound (and apparently the only) basis for justifying inequalities and differences in treatment among persons.

86 John Maddox, *The Doomsday Syndrome*, Macmillan, London, 1972, p. 213.

87 Peter Drucker, 'Saving the Crusade', in *Environmental Ethics* (ed. by K. Shrader-Frechette), Boxwood Press, Pacific Grove, 1980, p. 102; see also p. 200. Hereafter cited as: *EE*.

88 Drucker, 'Saving the Crusade', in Shrader-Frechette, *EE*, p. 103.

89 See, for example, M. M. Maxey, 'Radwastes and Public Ethics', *Health Physics* 34 (2), (February 1978), 129–135, esp. 132.

90 Andrew Larkin, 'The Ethical Problem of Economic Growth Vs. Environmental Degradation', in Shrader-Frechette, *EE*, p. 212, gives these Census Bureau statistics; Larkin hereafter cited as: Ethical. See also D. C. North and R. L. Miller, *The Economics of Public Issues*, Harper and Row, New York, 1971, p. 151, who substantiate this same point. Similar statistics for England are cited by Rees, *Equality*, pp. 30–32. Patterson, Inequality, p. 36, notes that individuals "will accept inequalities if they perceive their society to be changing in such a way that there is an absolute increase in the standard of living". For this reason, he says, there is social stability in rapidly industrialized countries.

91 See section 3.2 of this chapter and notes 59–61. See also J. P. Plamenatz, 'Equality of Opportunity', in Blackstone, *Equality*, p. 88; hereafter cited as: Equality.

92 E. J. Mishan, *21 Popular Economic Fallacies*, Praeger, New York, 1969, p. 236; hereafter cited as: *Fallacies*. See also Shrader-Frechette, *Nuclear Power*, pp. 123 ff.

93 See Mishan, *Fallacies*, pp. 232–233, 245 ff.; Rees, *Equality*, p. 36. See also Plamenatz, Equality, and Larkin, Ethical (notes 90 and 91).

94 R. Grossman and G. Daneker, *Jobs and Energy*, Environmentalists for Full Employment, Washington, D.C. 1977, pp. 1–2.

95 See note 94.

96 Mishan, *Fallacies*, p. 237.

97 Stewart, Pyramids, p. 172. Numerous detailed economic analyses support this point. See, for example, A. M. Freeman, 'Distribution of Environmental Quality', in *Environmental Quality Analysis* (ed. by A. V. Kneese and B. T. Bower), Johns Hopkins,

Baltimore, 1972, pp. 271–275; hereafter cited as: Freeman, DEQ, and Kneese and Bower, *EQA*. See also A. V. Kneese and C. L. Schultze, *Pollution, Prices, and Public Policy*, Brookings Institution, Washington, D.C., 1975, p. 28.

[98] Virginia Brodine, 'A Special Burden', *Environment* 13 (2), (March 1971), 24. See D. N. Dane, 'Bad Air for Children', *Environment* 18 (9), (November 1976), 26–34. See also A. M. Freeman, 'Income Distribution and Environmental Quality', in *Pollution, Resources, and the Environment* (ed. by A. C. Enthoven and A. M. Freeman), W. W. Norton, New York, 1973, p. 101 (hereafter cited as: Freeman, IDEQ) and Enthoven and Freeman, *PRE*, make the same point, regarding air pollution, as do Kneese and Haveman (A. V. Kneese, 'Economics and the Quality of the Environment', in Enthoven and Freeman, *PRE*, pp. 74–79; hereafter cited as: Kneese, *EQE*. See also A. M. Freeman, R. H. Haveman, A. V. Kneese, *The Economics of Environmental Policy*, John Wiley, New York, 1973, p. 143; hereafter cited as: Freeman, Haveman, and Kneese, *EEP*. See also P. Asch and J. J. Seneca, 'Some Evidence on the Distribution of Air Quality', *Land Economics* 54 (3), (August 1978), 278–297, and D. D. Ramsey, 'A Note on Air Pollution, Property Values, and Fiscal Variables', *Land Economics* 52 (2), (May 1976), 230–234.

[99] See, for example, Jane Stein, 'Water for the Wealthy', *Environment* 19 (4), (May 1977), 6–14. The point is documented well by Freeman (DEQ, p. 275), who argues that distribution of environmental pollution is a consequence "of the broader distribution forces at work in the economic system". Although there are minor exceptions to this rule, says Freeman, the poor bear a disproportionate burden of environmental hazards, and pollution is not "the great leveler", since the wealthy have "the means to protect themselves" from environmental insults. Even though the issue of who benefits most from pollution controls is complex (Freeman, DEQ, pp. 271–273; Freeman, IDEQ, pp. 101–104; Freeman, Haveman, Kneese, *EEP*, pp. 144–145; Kneese, EQE, pp. 78–80), Freeman, Haveman, Kneese and other economists (see *EEP*, pp. 143–144) conclude: "on balance, that the improvement would be pro poor". In any case, there are several means whereby the costs of pollution control can be shifted from the poor and middle class to members of higher income groups (see Freeman, IDEQ, pp. 104–105; and Freeman, Haveman, and Kneese, *EEP*, pp. 145–148).

[100] Hans Jonas, 'Philosophical Reflections on Experimenting with Human Subjects', in *Ethics in Perspective* (ed. by K. J. Struhl and P. R. Struhl), Random House, New York, 1975, pp. 242–353.

[101] Assessments are widely held to be objective, accurate, and nonpartisan. See Congress, *TA in IAG*, pp. 220–226. Congress, OTA, *AR 1978*, p. 7; Congress, OTA, *AR 1977*, p. 4; Congress, OTA, *Technology Assessment in Business and Government*, US Government Printing Office, Washington, D.C., 1977, p. 9; hereafter cited as: *TA in B and G*. For an excellent discussion of economists' reasons for ignoring considerations of distributive equity, see Nicholas Rescher, *Distributive Justice*, Bobbs Merrill, New York, 1966, pp. 106–107.

[102] Patterson, Inequality, p. 17, makes this same point.

[103] See, for example, M. W. Jones-Lee, *The Value of Life: An Economic Analysis*, University of Chicago Press, Chicago, 1976 (hereafter cited as: *Value*) for an economic approach to these issues.

[104] Taylor, Justice, pp. 86–97.

[105] This point is also made in Congress, OTA, *TA in B and G*, p. 13; Congress, OTA,

AR 1976, pp. 66–67; Congress, *TA in IAG*, p. 27; Kasper, *Technology*, p. 4; J. R. Ravetz, *Scientific Knowledge and Its Social Problems*, Clarendon Press, Oxford, 1971, pp. 369–370, 396; S. Koreisha and R. Stobaugh, 'Appendix', in Stobaugh and Yergin, *EF*, p. 234. See also E. F. Schumacher, *Small Is Beautiful*, Harper, New York, 1973, p. 38; J. K. Galbraith, *The New Industrial State*, Houghton Mifflin, Boston, 1967, p. 408, and E. J. Mishan, *Welfare Economics*, Random House, New York, 1969, p. 5; hereafter cited as: *WE*.

[106] US Congress, Office of Technology Assessment, *Coastal Effects of Offshore Energy Systems*, 2 vols., US Government Printing Office, Washington, D.C., November 1976; hereafter cited as: USC, OTA, *Coastal-1* or *Coastal-2*. According to this study (USC, OTA, *Coastal-1*, p. 16), federal-state conflicts are "the major issues in the question of whether to develop offshore energy resources". Statistics on the proportion of Americans living near/on coastlines are given in USC, OTA, *Coastal-1*, p. 3.

[107] US Environmental Protection Agency, *Development Document for Interim Final Effluent Limitations Guidelines and New Source Performance Standards for the Offshore Segment of the Oil and Gas Extraction Point Source Category*, EPA 440/1–75/055, Group II, US EPA, Washington, D.C., September 1975, pp. 24–25; hereafter cited as: US EPA, *Offshore*.

[108] USC, OTA, *Coastal-1*, p. 3; see also Mills and Woodson, Energy, pp. 303–304.

[109] M. Gendler, 'Toward Better Use of Coastal Resources: Coordinated State and Federal Planning under the Coastal Zone Management Act', *Land Use and Environment Law Review–1978* (ed. F. A. Strom), Clark Boardman, New York, 1978, pp. 225, 228; hereafter cited as: Toward. See also Mills and Woodson, Energy, pp. 291– 292.

[110] USC, OTA, *Coastal-2*, Part I, p. II-1.

[111] USC, OTA, *Coastal-2*, Part I, p. II-2.

[112] USC, OTA, *Coastal-2*, Part I, p. II-1. As one recent technology assessment concluded: "The State role at present is little more than that of commentator." (USC, OTA, *Coastal-1*, p. 17.)

[113] USC, OTA, *Coastal-1*, p. 169.

[114] USC, OTA, *Coastal-1*, pp. 157–160; see also p. 57.

[115] USC, OTA, *Coastal-1*, pp. 16–17, 57, 58–59.

[116] USC, OTA, *Coastal-1*, p. 15.

[117] USC, OTA, *Coastal-2*, Part III, WP-3, p. 2.

[118] USC, OTA, *Coastal-1*, p. 51.

[119] USC, OTA, *Coastal-1*, pp. 52–53.

[120] USC, OTA, *Coastal-1*, p. 16; see pp. 51–56. In 1976, however, the Supreme Court did uphold Florida legislation demanding strict liability for oil-spill damage; the court ruled that federal and state laws in this regard were "parts of an integrated whole". (Mills and Woodson, Energy, p. 294.)

[121] USC, OTA, *Coastal-1*, p. 272.

[122] USC, OTA, *Coastal-1*, p. 19.

[123] USC, OTA, *Coastal-1*, p. 16. See also Gendler, Toward, pp. 233–234.

[124] USC, OTA, *Coastal-1*, p. 16.

[125] See note 124.

[126] B. M. Gross, 'The State of the Nation: Social Systems Accounting', in *Social Indicators* (ed. by R. A. Bauer), MIT Press, Cambridge, 1966, p. 222. See also p. 260 where Gross discusses related problems when one attempts to forego quantitative measures of parameters. Hereafter cited as: Gross, State.

[127] USC, OTA, *Coastal-2*, Part IV.
[128] US EPA, *Offshore*, pp. 53–60. Ignoring the economic effect of major oxygen demand is especially significant because dissolved oxygen is needed to keep organisms living and to sustain species reproduction.
[129] USC, OTA, *Coastal-2*, p. I-1 and II-1; see Part V. See also USC, OTA, *Coastal-1*, p. 13.
[130] See preceding note.
[131] USC, OTA, *Coastal-1*, p. 12.
[132] USC, OTA, *Coastal-1*, p. 19.
[133] USC, OTA, *Coastal-1*, p. 16.
[134] See, for example, USC, OTA, *Coastal-1*, p. 13.
[135] See section 3.1 of this chapter and notes 35–38. See also section 4.1.
[136] USC, OTA, *Coastal-1*, p. 19.
[137] US EPA, *Offshore*, p. 134.
[138] USC, OTA, *Coastal-1*, pp. 17, 60–62, 134.
[139] USC, OTA, *Coastal-1*, pp. 18, 47–50, 152–155.
[140] USC, OTA, *Coastal-1*, pp. 19, 57–59, 166–169.
[141] USC, OTA, *Coastal-1*, p. 13. The assessors noted, for example, that "the effects of pollutants which may be discharged during OCS operations cannot presently be determined with any accuracy . . ." (p. 17; see pp. 67–69).
[142] USC, OTA, *Coastal-1*, p. 11.
[143] Statistics on petroleum use and refinery capacity are taken from USC, OTA, *Coastal-2*, Part V, pp. i–ii.
[144] Daniel Callahan, 'Ethics and Population Limitation', in *Philosophical Problems of Science and Technology* (ed. A. M. Michalos), Allyn and Bacon, Boston, 1974, p. 560; hereafter cited as: Ethics.
[145] Stewart, Pyramids, p. 197, for example, argues that it is not plausible that *due process* secures the right to environmental quality.
[146] See, for example, Norman Dorsen, 'A Lawyer's Look at Egalitarianism and Equality', in Pennock and Chapman, *Equality*, p. 36.
[147] See, for example, W. H. Ferry, 'Must We Rewrite the Constitution To Control Technology?' in *Technology, Society, and Man* (ed. by R. C. Dorf) Boyd and Fraser, San Francisco, 1974, pp. 18–19.
[148] Cited by L. K. Caldwell, 'The Coming Polity of Spaceship Earth', in *Environment and Society* (ed. by R. T. Roelofs, J. N. Crowley, D. L. Hardesty), Prentice-Hall, Englewood Cliffs, New Jersey, 1974, p. 177.
[149] Charles A. Reich, *The Greening of America*, Random House, New York, 1970, p. 73.
[150] Quoted by J. L. Huffman, 'Individual Liberty and Environmental Regulation: Can We Protect People While Preserving the Environment?' *Environmental Law* 7 (3), (Spring 1977), 436; hereafter cited as: Liberty.
[151] Langdon Winner, *Autonomous Technology: Technics-Out-of-Control As a Theme in Political Thought*, The MIT Press, Cambridge, Massachusetts, 1977, esp. pp. 13–43; hereafter cited as: Winner, *AT*.
[152] For a discussion of 'frontier' versus 'spaceship' values, see Shrader-Frechette, *EE*, Chapter 2.
[153] Patterson, Inequality, p. 35; see also p. 34.
[154] This definition is used by M. C. Tool, *The Discretionary Economy: A Normative*

Theory of Political Economy, Goodyear, Santa Monica, 1979, p. 321; hereafter cited as: Tool, *DE*. See also Callahan, Ethics, p. 546, for a discussion of freedom and the extent to which policy regarding technology limits freedom.

[155] F. Pohl, 'Keynote Address', in *Technology and Social Progress – Synergism or Conflict* (ed. by P. K. Echman), American Astronautical Society, Washington, D.C., 1969, p. 8, uses this definition.

[156] See Tool, *DE*, p. 324.

[157] Winner, *AT*, p. 187.

[158] Quoted in C. H. Danhof, 'Assessment Information Systems', in Kasper, *TA*, p. 26.

[159] See section 4.1 of this essay.

[160] Winner, *AT*, p. 148, quotes him.

[161] Muchnicki, Role, p. 59.

[162] See Muchnicki, Rile, p. 48.

[163] D. K. Price, *The Scientific Estate*, Harvard University Press, Cambridge, 1965, p. 10.

[164] Stewart, Paradoxes, p. 479.

[165] Boroush *et al., TA*, p. 363, also makes this suggestion.

[166] For a discussion of the distinction between a higher GNP and a higher quality of life, see (for example) E. J. Mishan, *The Costs of Economic Growth*, Praeger, New York, 1967, pp. 109–137; hereafter cited as: *CEG*. See also J. A. Hobson, *Confessions of an Economic Heretic*, Harvester Press, Sussex, England, 1976, pp. 171, 208–209. This point is also emphasized by A. V. Kneese and C. L. Schultze, *Pollution, Prices, and Public Policy*, Brookings Institution, Washington, 1975, p. 109; hereafter cited as: *PPPP*.

[167] Gross, State, pp. 213–255.

[168] For a discussion of diseconomies, see Mishan, *Cost-Benefit*, pp. 127–132; Mishan, *CEG*, pp. 57–73; and D. C. North, 'Political Economy and Environmental Policies', *Environmental Law* 7 (3), (Spring 1977), 449–462; hereafter cited as: Political.

[169] This suggestion is also made by North, Political, pp. 450–451. See also Mishan, *Cost-Benefit*, pp. 127–132, and D. Jordan, 'The Town Dilemma', *Environment* 19 (2), (March 1977), 6–15; Brandt, *Ethical*, pp. 422–424; K. Boulding, *Economics as a Science*, McGraw-Hill, New York, 1970, p. 130.

[170] Many externalities are already priced, notably those related to health costs. See, for example, Jones-Lee, *Value*. What I am calling for is an expanded use of the notion of internalizing externalities.

[171] Brandt, *Ethical*, pp. 422–424, for example, believes that this is the case.

[172] K. Boulding, *Economics as a Science*, McGraw-Hill, New York, 1970, p. 130.

[173] Mishan, *Cost-Benefit*, pp. 446–448; see also pp. 217–232; and Mishan, *Welfare Economics*, pp. 253–260.

[174] See Boroush *et al;. TA*, pp. 221–232.

[175] Huffman, Liberty, p. 431.

[176] For discussion of the minimum-standard view, see Karpinski, Preemption, p. 1341.

[177] For discussion of this point, see Muchnicki, Role, p. 51; Karpinski, Preemption, pp. 1335–1336; and Novick, *Electric*, p. 222.

[178] For discussion of the zoning issue, see Muchnicki, Role, p. 52.

[179] Cited by Nicholas Rescher, *Distributive Justice*, Bobbs Merrill, New York, 1966, p. vii.

PART IV

STEPS TOWARDS SOLUTIONS

CHAPTER EIGHT

ETHICALLY WEIGHTED RISK-COST-BENEFIT ANALYSIS

1. INTRODUCTION

If Chapters Three and Four of this volume are correct, then technology assessors and environmental-impact analysts have attempted, unsuccessfully, to avoid the normative dimensions of their tasks. This avoidance is, in part, the result of adherence to a positivist philosophy of science and the consequence of our failure as a society to rethink our ethical and social commitments.

Although assessors have attempted to avoid explicitly normative studies, because of their adherence to certain methodological tenets of risk-cost-benefit analysis (RCBA), their conclusions have tended to give implicit support to various ethical assumptions. As Chapters Two, Five, and Seven revealed, their analyses typically sanction problematic utilitarian doctrines doctrines and attendant violations of distributive equity.

One of the ways to avoid the tendency to interpret the RCBA calculus in a utilitarian manner would be to weight each of the risks, costs, and benefits in the analysis according to alternative ethical criteria and then to examine the effects of various weightings on the RCBA conclusions. For example, when the people of London were trying to decide, a decade ago, whether to build a new airport, many citizens argued vehemently that the various parameters ought to be weighted according to ethical considerations. The Roskill Commission, the group performing the RCBA, dismissed their arguments and instead produced a classical market analysis. The claims of the public, nevertheless, raise an interesting point. Why, for example, should the homes of those displaced by the proposed airport be valued at market price? Aren't there some incommensurable values, like remaining in the home in which one was born and raised his children? Doesn't one's home have some value not reflected by market considerations, and couldn't this value be addressed, as the people suggested, by numerically weighting the market costs of the houses?[1]

2. REASONS FOR ADOPTING A SYSTEM OF ETHICAL WEIGHTS FOR RCBA

Pursuing the same insight as that of the Londoners discussing the analyses of airport options, a few economists have suggested that practitioners of

RCBA ought to weight the various risks, costs, and benefits within their calculations so that the numerical assignments will more closely reflect the actual risks, costs and benefits which are experienced by the public. My position is somewhat different; I have argued elsewhere that RCBA ought to be weighted by various ethical parameters, but that teams of analysts, including ethicians and members of various public interest groups, ought to be responsible for devising alternative, ethically weighted RCBA's. Once these were completed, normal democratic procedures could be used to amend them and to choose the most reasonable RCBA. Before going into that proposal, however, it makes sense to explain why any ethical weighting at all appears a desirable policy move. I can think of at least four reasons.

First, since no necessary connection exists between Pareto Optimality, the central concept of RCBA, and socially desirable policy,[2] but since there is a great tendency (discussed earlier) to assume that RCBA alone reveals socially desirable policy, it would be helpful if there were some way to avoid the tendency to accept this erroneous assumption. If RCBA conclusions are misinterpreted or misapplied in this manner, then one means of avoiding such misuse would be to provide a number of alternative, ethically weighted RCBA's. They would enable persons to see that desirable social policy is not only a matter of economic calculations but also a question of ethical analysis. If people follow Gresham's Law and thus persist in their tendency to accord primacy to quantitative results, such as those of RCBA, then using ethically weighted RCBA's might keep them from ignoring the ethical parameters normally not represented in a quantitative analysis. Ethical weights also might be one way to avoid the tendency to identify *un*weighted RCBA results with a prescription for desirable social policy. (An alternative way of avoiding this problem would be to re-educate policymakers and the public as to the significance of quantitative results; for a variety of reasons, this does not appear to be a practical option, regardless of how desirable it might be.)

A second reason for using a number of ethically weighted RCBA's in policy analysis is that they would provide a more helpful framework for democratic decisionmaking. According to Pearce, policy analysis should show how the chosen measures of social risks, costs, and benefits respond "to changed value assumptions".[3] One good way to do this is actually to carry out the calculations and to see how the relationships among the risks, costs, and benefits change as they are weighted in terms of different ethical considerations. If the real goal of policy analysis is to clarify decision options, and if a major part of decision options are their ethical assumptions and consequences, then one good way of clarifying this normative dimension for the public would be to show the effects of RCBA's done under different value constraints. Such weighted options would also aid democratic decision-making in the sense that the ethical aspects of policy analysis would not be left merely to the vagaries of the political process. Instead they could be clarified through alternative, ethically weighted RCBA's.

Finally, ethically weighted RCBA's might aid public decisionmaking because they could explicitly bring values into policy considerations at a very early stage of the process, rather than wait until the RCBA conclusions were completed. In this way, citizens might be able to exercise more direct control over the values to which policy gives assent. If ethical values should determine our social choices, rather than that our choices should determine our ethical values, then it makes sense for those values to enter the policy process explicitly and early, as they would if a number of alternative ethical weights were attached to RCBA parameters.

Even were a system of ethical weights not used, of course, RCBA would still embody a number of implicit ethical presuppositions. As Gunnar Myrdal put it, "a disinterested social science has never existed and never will exist. For logical reasons it is impossible ... Valuations enter into the choice of approach, the selection of problems, the definition of concepts, and the gathering of data".[4] Acceptance of the willingness-to-pay measure, for example, implies acceptance of the existing scheme of property rights and income distribution, since calculations of risks, costs and benefits is done in terms of existing market prices and conditions.[5] This means that if there is something ethically suspect about the existing distribution of income and goods, then there is also something ethically suspect about the calculations in the RCBA which presupposes it. Suspect or not, all unweighted RCBA calculations contain implicit normative presuppositions which, correct or incorrect, ought to be made explicit so that they may be evaluated by the public. To employ a system of alternative, ethically weighted RCBA's, among which policymakers and the public can decide, would be to assent to the thesis that existing RCBA's already contain ethical weights (which are probably unrecognized). To employ alternative, ethically weighted RCBA's would also be to assent to the thesis that proponents of the ethics implicit in current analyses ought to be required to plead their cases, along with advocates of other ethical presuppositions, in the public court of human reason. In a word, since implicit ethics already affects decisionmaking based on RCBA, then it would be desirable to make the ethical presuppositions in RCBA explicit and open to evaluation.

Another way of expressing this third reason for using alternative, ethically weighted RCBA's is to argue that weighting is already done anyway, both by means of implicit presuppositions of economists who practice RCBA and by means of the political process by means of which the public responds to RCBA. Given that weighting is done anyway, it makes sense to bring as much clarity, ethical precision, and objectivity to it as is possible. For this reason, it makes sense to do it explicitly by weighting the RCBA parameters. Such reasoning is especially appealing in light of a recent statement by the economist Weisbrod. He maintained that all public projects which were adopted, despite their failure to be vindicated by classical (not ethically weighted) RCBA, were implemented because of an implicit set of utility *weights* attached by the political process.[6] If this is true, then it makes sense to inform the political process by supplying it with a number of ethically weighted RCBA's among which the public can decide.

A fourth reason for using weighted RCBA is that it appears more desirable than the alternatives likely to be adopted in its place. One alternative to such weights is to use the current market approach toward valuing RCBA parameters. As has already been seen, this commits one to a value bias regarding current income distribution and property rights. Moreover it requires that one assign no quantitative value to nonmarket items such as air and water, and to externalities such as pollution and resource depletion. As I have argued elsewhere, failure to represent such parameters quantitatively most often results in their not being considered at all, since policymakers tend to discount qualitative information.[7]

The main alternative to following market assignments for risks, costs and benefits is to use what economists and risk assessors call "revealed preferences".[8] On this scheme, society's current revealed preferences are determined on the basis of inductive inferences about what was allegedly preferred in the past. This means, for example, that if, in the past, society "accepted" X number of automobile fatalities per 100,000 miles driven, then (all things being equal, which they never are) society will accept the same level of fatalities now. Hence, on this theory, society 'reveals' its preferences by means of past behavior which it tolerated. Using the theory of revealed preferences, practitioners of RCBA can assign weights to current risks, costs, and benefits on the basis of society's past preferences for certain risks, costs, and benefits.

Of course, there are numerous problems with weighting RCBA parameters by means of a system of revealed preferences. Such a framework requires one to assume, for example, that because risk levels were of a certain magnitude in the past, society actually accepted them; that social preferences regarding risk aversion do not change over time; and that qualitatively diverse risks, costs, and benefits may be compared solely on the basis of their magnitude.[9] Many of these questionable assumptions could be avoided if, instead of employing a system of revealed preferences for weighting RCBA parameters, one followed the practice of preparing a number of alternative, ethically weighted RCBA's. For example, one might avoid the assumption that qualitatively diverse risks, costs, and benefits are comparable, solely on the basis of their magnitudes, by preparing RCBA's weighted according to the equity with which risks, costs, and benefits are distributed or weighted according to the involuntariness with which they are imposed.

3. KNEESE'S SUGGESTION FOR ETHICALLY WEIGHTING RCBA

These four reasons for employing ethically weighted RCBA's all stem from the consideration that some risks, costs, and benefits are of such overarching significance that they are really different *in kind* from other RCBA parameters. In fact, there is evidence that Mill himself believed that some sort of weighting of certain consequences was necessary in order to deal with non-utilitarian value schemes. In his discussion of justice, Mill says:

> Our notion, therefore, of the claim we have on our fellow creatures to join in making safe for us the very groundwork of our existence gathers feelings around it so much more intense than those concerned in any of the more common cases of utility, that the difference in degree becomes a real difference in kind. The claim assumes that character of absoluteness, that apparent infinity, and incommensurability with all other considerations, which constitute the distinction between the feeling of right and wrong and that of ordinary expediency and inexpediency.[10]

Following Mill's lead, several economists have proposed that RCBA attempt to take account of the distinction between *right* and *expediency* by means of using a system of weights. Two of the best known suggestions in this regard are those of Weisbrod and the United Nations Industrial Development Organization.[11] Weisbrod's approach has been widely criticized because it is built on the assumption that efficiency and distribution are the only two criteria for project choice.[12] Obviously many other ethical criteria could be used to evaluate a particular RCBA proposal. The UNIDO approach likewise is suspect, both because it is a revealed-preferences model and because it allows RCBA practitioners to disregard political constraints.[13] Also, both the Weisbrod and UNIDO approaches fail to discuss the *ethical* dimensions of the weighting problem in any comprehensive or sophisticated way. For all these reasons, these two approaches seem unlikely to yield a suitable framework for investigating a system of ethical weights for use in RCBA.

One more recent approach, however, appears to offer more hope. Developed by economists Allen Kneese, Shaul Ben-David, and William Schulze, working under National Science Foundation funding, this approach is based on reweighting risks, costs, and benefits by means of alternative ethical criteria. On this scheme, each ethical system is represented as a general, transitive criterion for individual or social behavior. For example, one criterion might be "do unto others as you would have them do unto you." Kneese *et al.* stipulate that the general criteria representing alternative ethical systems be *transitive* so as to avoid the consequences of Arrow's famous Impossibility Theorem. Further, they suggest that each ethical system be represented by a general criterion rather than by a list of *rules*, such as the Ten Commandments.[14]

Kneese's, Ben David's, and Schulze's reasons for avoiding representing ethical systems as lists of rules are (1) that the rules could conflict (requiring a hierarchical ordering), (2) that they might fail to cover certain situations, and (3) that they would likely be more difficult to treat than would general criteria. The difficulty arises from the fact that a list of rules would have to be treated as a set of mathematically specified constraints on the outcomes of a particular RCBA.[15]

According to Kneese *et al.*, the requirement that an ethical system be represented as a transitive criterion for individual or social behavior leaves at least four ethical systems (and probably more) for use in reweighting RCBA parameters. These are Benthamite utilitarianism, Rawlsian egalitarianism, Nietzschean elitism and Paretian libertarianism. On this scheme, the Benthamite criterion is that one ought to maximize the sum of the cardinal utilities of all individuals in a society. The Rawlsian criterion is that one ought to try to maximize the utility of the individual with the minimum utility, so long as he remains the worst off. According to the Nietzschean weighting scheme, one ought to maximize the utility of the individual who can attain the greatest utility. Finally, according to the Paretian criterion, says Kneese *et al.*, one ought to act in such a way that no one is harmed and, if possible, that the well-being of some individuals is improved.[16]

The merit of the ethical weighting criteria described by Kneese *et al.* is not that they tell policymakers what is right or wrong. Rather, their value is that they could theoretically provide a unique sort of nonmarket information. For example, one could ask an individual how much he would be willing to pay for redistributing income to less fortunate members of society. In this way, at least to a limited extent, ethical beliefs could be brought into RCBA. The purpose of bringing them in, again, would not be to provide a prescription for policy, but to allow the public and policymakers to explore the consequences of weighting risks, benefits, and costs with a variety of ethical criteria, even though they are admittedly quite simplistic. In this way persons could see, for example, how different assumptions about the desirability of given distributions change the overall ratio of costs to benefits. They would be able to examine alternative, ethically weighted options, and not merely marginal costs or benefits across opportunities.

The work of Kneese *et al.* illustrates dramatically that what is said to be feasible or unfeasible in terms of RCBA can change extensively when different ethical weighting criteria are employed. Using case studies on helium storage, nuclear fission, and automobile-emission standards, Kneese, Ben-David, and Schulze showed how alternative ethical weights can be used to generate contradictory RCBA conclusions.

In the first case study, the authors show that if a Benthamite criterion is used, RCBA dictates that helium should be stored for the future. After providing measures of the welfare of various generations, both with and without helium storage, Kneese *et al.* employ the Benthamite criterion to weight all generations the same. In so doing, they counteract typical tendencies to undervalue the welfare of our descendants. (Conventional RCBA weights the future by a discounting factor determined in part by the time preferences of the present generation.) Calculating on the basis of weighted costs and benefits both with and without storage, they show that a Benthamite ought to store helium whenever the cost of extraction from the air is greater than the cost of separating and storing helium from natural gas. Since the cost of extraction from air is greater, they conclude that a consistent Benthamite ought to store helium.

According to the Paretian criterion, however, helium should not be stored. This is because, on Paretian grounds, helium should be stored only if it makes one generation or group better off without making anyone else (another generation) worse off. Since those who receive the benefits of helium storage are in the future and have little possibility of compensating members of the present generation for their losses, and since storage of helium is unfeasible on Paretian grounds, unless the earlier generation is completely compensated, Kneese *et al.* conclude that a consistent Paretian ought not to store helium.

The other two ethical criteria, the Rawlsian and Nietzschean, each sanction storage under certain conditions but proscribe it in others; these conditions have to do with the relative wealth of present and future generations. Since the Nietzschean criterion calls for maximization of the generation with the greatest utility, while the Rawlsian criterion requires that the generation with the lowest utility should be made better off, Kneese *et al.* conclude that the benefits and costs of helium storage may be secondary to the relative levels of income of various generations. This is because, under different scenarios of future welfare, contrasting conclusions may be drawn. The authors show that, if one expects the future to be "poorer than the present", then a consistent Rawlsian ought to store helium for future benefit. In this same situation (i.e., members of future generations are poor, relative to present individuals), they show that a consistent Nietzschean ought to leave no helium. On the other hand, claim Kneese *et al.*, if members of future generations are thought to be wealthier than present persons, then a consistent Nietzschean ought to call for storage of helium while a consistent Rawlsian (in the same circumstances) ought to call for no storage.[17]

With respect to the second case study, involving nuclear power, Kneese *et al.* show that, when used with a Benthamite weighting criterion, RCBA reveals that using nuclear fission for generation of electricity is not feasible. On either the Paretian, Rawlsian, or Nietzschean criterion, nuclear power may be said to be feasible or unfeasible, depending on the value attributed to a number of factors. Some of these factors include the value of compensation for damages and the utility attributed to various future generations and to the present.[18]

Using ethically weighted RCBA parameters, Kneese *et al.* also drew some interesting conclusions as to the desirability of stricter air pollution control standards for automobiles. According to traditional RCBA methods, the stricter standards could not be said to be efficient. However, risks, costs, and benefits of an air-emissions-standards program are not distributed evenly among income classes. This means that, when either Nietzschean, Benthamite, or Paretian weightings are employed, the strict standards can be shown to be unfeasible, but when Rawlsian weightings are used, then stricter standards can be shown to be feasible.[19]

4. WEIGHTING RCBA'S BY MEANS OF LEXICOGRAPHIC ORDERING OF CLAIMS

Admittedly, the weighting scheme outlined by Kneese *et al.* has a number of limitations. Its most obvious deficiency is that in employing simple criteria, it fails to capture the complexity of ethical systems. It is unable to represent a priority weighting of different ethical claims within the same ethical system.

For example, as the earlier discussion revealed, Kneese *et al.* translate the Rawlsian criterion as trying to maximize the utility of the individual with the minimum utility, so long as he remains the worst off. In reality, however, the sophisticated Rawlsian ethics hardly begins to be captured by such a rule. Rawls maintains, for instance, that there are at least two central principles of justice, and that they ought to be lexicographically ordered, such that the claims of the first are to be satisfied first. The *first* principle is that each person is to have an equal right to the most extensive total system of equal basic liberties compatible with a similar system of liberty for all. The *second* principle is that social and economic inequalities are to be arranged so that they are both (a) to the greatest benefit of the least advantaged, consistent with the just-savings principle, and (b) attached to offices and positions open to all under conditions of fair equality of opportunity. Thus, although Rawls assents to a maximin criterion (the second principle) which Kneese *et al.* identify with Rawlsian ethics, this rule alone does not capture Rawls' position and, indeed, may misrepresent it, since the first principle is supposed to be satisfied prior to the second, or maximin rule. A Rawlsian criterion appears to demand, at least, a lexicographically ordered list of rules. A lexicographic version of the second principle, for example, might prescribe that one maximize the welfare of the second worst-off person (i.e., rank the distributions according to his welfare level), in case the worst-off persons are equally well off in two or more different distributions. If the second worst-off persons are also equally well off, then the rule prescribes that one maximize the welfare level of the third worst-off person, and so on.[20]

As the Rawls example reveals, many ethical nuances of decision situations cannot be accounted for in the Kneese framework. This is why, despite its complexity, a more desirable weighting option appears to be to use a list of transitive, lexicographically ordered rules to represent the priority weightings of different ethical claims within a given ethical system, and then to prepare alternative RCBA's each weighted in terms of a different ethical system, among which the public could decide. (A lexicographic ordering is one in which the attributes of alternatives are ordered by importance, and the alternative is chosen with the best value on the most important attribute. If there are two alternatives of equal value, then this procedure is repeated for those alternatives in terms of the next attribute, and so on until a unique alternative emerges or until all alternatives have been considered. The lexicographic ordering on a set of tuples $(x_1, \ldots, x_n)$, where x_i is an element of

X_i and X_i is simply ordered by $\leqq$, is defined as follows. $(a_1, \ldots, a_n) \leqq (b_1, \ldots, b_n)$ if and only if

(i) $a_1 \leqq b_1$, or
(ii) $(a_1 = b_1$ and $a_2 \leqq b_2)$, or
. . .
(n) $a_1 = b_1$ and . . . and $a_{n-1} = b_{n-1}$ and $a_n \leqq b_n$.

A simple ordering on a set X is a relation (which may be denoted $\leqq$) having the properties

(i) for every a and b in X, either $a \leqq b$ or $b \leqq a$ (comparability);
(ii) for every a, b, and c in X, if $a \leqq b$ and $b \leqq c$, then $a \leqq c$ (transitivity).[21]

As Kneese recognizes, a list of transitive, lexicographically ordered rules would have to be treated as a set of mathematically specified constraints on the outcomes of a particular RCBA.

Admittedly, any weighting system based on *ordering* rules to represent the priority of ethical claims will fall victim to the objection that the resultant individual utility functions and social-welfare function do not satisfy the von Neumann-Morgenstern conditions. That is, they do not possess the expected-utility property for probability distributions over the possible outcomes. Despite the importance of this objection, let us put it aside for the time being, and then discuss it later, in section five of this chapter, after the weighting system has been outlined.

Later, I will suggest what a list of transitive, lexicographically ordered rules might be like. First, I want to discuss some of the conditions for the possibility of using such a system and thereby clear some philosophical ground for later system builders. The big advantage of such a system, it seems to me, is that it would allow one to order social situations on the basis of the relative priority of ethical claims, without requiring that all the different types of considerations relevant to a choice be collapsed into one dimension. This means that, insofar as the proponents of the deficiency argument reject (see Chapter Two) RCBA because they do not wish all dimensions of a social choice to be reduced to one, the lexicographic-rules system could defuse such an objection. This is a significant advantage, since most critics of RCBA appear to reject it because they do not wish all dimensions of a social choice to be reduced to one.

Peter Self and Gunnar Myrdal, for example, both emphasize the fact that all the dimensions of social choice cannot be collapsed into one because such a collapse would be insensitive to the different *kinds* of claims and parameters at stake. Lovins, Mishan, Georgescu-Roegen, Rotwein, Shackle, and others make the same point: there is no common denominator among the ethical claims and social factors which are at issue; hence they cannot be 'reduced' by means of RCBA quantification. Still other opponents of the RCBA 'reduction', such as MacLean, Hampshire, and Wolff argue that the attempt to collapse all the dimensions of social choice into one fails because it is dehumanizing and extends quantification (RCBA) into qualitative areas. Coburn, Gewirth, and MacIntyre all make much the same objection, formulated similar to classic arguments against utilitarianism: a social choice cannot be collapsed into one dimension because many relevant considerations are not commensurable in terms of a single dimension.[22]

Because my proposal for a lexicographic ordering prevents the collapse of allegedly different ethical dimensions into a single one, it could defuse all these major objections to RCBA. This is because the proposal takes account of the possibility that there are factors of choice which are incommensurable. In a two-rule, lexicographically ordered ethical system, for example, the rules might not be commensurable. (Suppose the rules were, for example, to maximize the overall utility of all persons in society and to maximize the utility of the worst-off persons.) In this case, a real, order-preserving function could be used to assign numbers in a plane, and the plane could be lexicographically ordered. A lexicographic ordering of 3 rules could be represented in 3-space, an ordering of 4 rules in 4-space, and so on.

It might be objected, however, that this lexicographic ordering prevents a collapse of various ethical dimensions into one by imposing a ranking of dimensions by absolute priority. Rawls himself notes the implausibility (in a finished theory) of such a strong requirement (absolute priority). To support this objection, one might give the example of a representation of each possible state of a three-person society by an ordered triple in which the n-th member ($n = 1, 2, 3$) is the utility of that state to the n-th person. For instance, we might have: S1 = (9, 9, 9); S2 = (27, 1, 0); S3 = (0, 0, 27) and, using our rules, we might conclude that S2 was superior to S1 (greater utility) and S1 to S3 (equal utility, higher minimum). Would such an objection raise the question of the plausibility of representing ethical criteria by lexicographically ordered rules?

While this objection to a ranking of dimensions by absolute priority is well taken, there are at least three related reasons why it does not necessarily count against using lexicographic ordering to represent ethical claims. *First*, the target of the objection is not lexicographic ordering *per se*, but only applications of it which employ such a strong requirement of absolute priority. The strong requirement is not built into lexicographic ordering; one can put whatever conditions he wishes on a particular ordering. For example, in a lexicographic ordering of four rules, one might specify no absolute priority among rules. Instead one might say that, when condition *A* obtained, the priority ranking of rules would be one, two, three, four; when condition *B* obtained, the priority ranking of the rules would be two, one, three, four; when condition *C* obtained, the priority ranking would be two, one, four, three, and so on . . . *Second*, the objection does not count against lexicographic ordering, *per se*, but only against simplistic or overly rigid interpretations of the ethics which the ordering is taken to model. Marvin Minsky once said that the real problem with computer representations of mind is not that the computer models are wrong or simplistic, but that we don't understand mind in the first place. If we did, then we could model it. Likewise, in this case, the real problem with using lexicographic orderings to represent ethical claims does not appear to be that the lexicographic orderings are in-principle inadequate, but that we don't understand the highly complex ethical claims in the first place. If we did, then we would know precisely under which conditions a particular lexicographic ordering of claims ought to obtain. *Third*, lexicographic ordering doesn't teach us about ethics; it just gives us a vehicle for representing: our ethical claims; their effects on various costs and benefits; and the degree to which alternative RCBA conclusions (regarding the same policy choices) are a function of different ethical weights. This being so, someone could always object that a particular lexicographic ordering was an inappropriate vehicle for representing a given set of ethical claims. Such an objection would not count against ordering *per se*, however, but only against the specific conceptualization of ethics which a particular ordering represented.

To my knowledge, no one has attempted such a lexicographic representation, although Steven Strasnick wrote a number of papers proposing that RCBA be weighted in terms of a lexicographic ordering of claims based on general needs. In his lexicographic system, Strasnick assumes that social situations cannot be ordered "on the basis of the relative priority of individual needs". For this reason, he postulates a basic need theory according to which all preferences of all individuals are directed at satisfying the same needs in the same hierarchical order.[23]

Georgescu-Roegen and Maslow notwithstanding, I find it difficult to believe that the preference orderings of all persons are based on an identical system of needs, hierarchically arranged. For this reason, I find Strasnick's proposal unconvincing. Investigating and amending a key assumption in his account, however, might provide some basis for believing that my proposal, to order the ethical claims of individuals or systems, is both possible and desirable. To see why this is so, let's attempt to see where Strasnick might have gone wrong, and how his most crucial assumption might be amended to help my theory of lexicographically ordered, ethical rules for weighting RCBA.

Why does Strasnick assume that one cannot order possible positions "on the basis of the relative priority of individual needs"? He maintains that this ordering cannot be accomplished because, more generally, no ordering of social situations "on the basis of the relative priority of individual claims" is possible. And he claims that no ordering of social situations on the basis of the relative priority of individual claims is possible because a preference ordering will not be representable by a social-welfare function that assigns a *real number* to represent the moral desirability of situations.[24]

But why is a preference ordering not capable of being represented by a function that assigns a real number to represent the moral desirability of situations? Strasnick reasons in this manner:

Suppose a group of individuals are competing for some scarce good and that we want to rank these individuals in terms of their relative priority. Each individual has two types of claims that he can make against this good, C^1 and C^2, with $C^1{}_i$, $C^2{}_i$ representing the strength of individual i's two claims and taking values between 0 and 1. Finally, suppose claims of the first type are viewed as taking precedence over claims of the second type, so that individual i will always be ranked higher in priority than individual j whenever $C^1{}_i > C^1{}_j$. If $C^1{}_i = C^1{}_j$, however, individual i will be ranked higher just in case $C^2{}_i > C^2{}_j$. Clearly, by using these priority rules we will be able to define a complete ranking of individual priority for all possible values of $C^1{}_i$ and $C^2{}_i$. *But while this is obviously a reasonable method of moral judgment*, given the priority of C^1 claims over C^2 claims, it is one that is not representable by a function which assigns to each individual a number, s, representing the strength of his respective claims and ranks one individual higher than another just in case his s-value is greater.[25]

To substantiate his last claim, Strasnick appeals to a mathematical note published thirty years ago:

Consider the lexicographic ordering of the plane: a point of coordinates (a', b') is better than the point (a, b) if "$a' > a$" or if "$a' = a$ and $b' > b$". Suppose that there exists a real order-preserving function $\alpha(a, b)$. Take two fixed numbers $b_1 < b_2$ and with a number a associate the two numbers, $\alpha_1(a) = \alpha(a, b_1)$ and $\alpha_2(a, b_2)$. [sic] To two different numbers a, a' correspond two disjoint intervals $[\alpha_1(a), \alpha_2(a)]$ and $[\alpha_1(a'), \alpha_2(a')]$. One obtains therefore a one-to-one correspondence between the set of real numbers (*noncountable*) and a set of *non-degenerate disjoint* intervals (countable).[26]

Although the note upon which Strasnick bases his argument reasons correctly, Strasnick misapplies it in a subtle way. Namely, he assumes that C^1_i and C^2_i can take on *any real* values between 0 and 1.[27] If one makes this assumption, then one is able to derive a contradiction. Why need one assume, however, that *real* numbers have to be possible values of this assignment? There is an infinity of rational numbers between 0 and 1 which should suffice. *If one uses the rationals, the contradiction does not follow; one does not obtain a one-to-one correspondence between a non-countable and a countable set of numbers.* (The rationals are countable because there is a 1–1 correspondence between them and 1, 2, 3, 4, ... The reals are not countable because they are greater in number than countable (1, 2, 3, 4, . . .) numbers; to see this, one assumes that there is a one-to-one correspondence between the reals and the counting numbers. Then one considers a number whose decimal representation is $\cdot a_1 a_2 \ldots$ and which has the property that the digit a_i is not the same as the ith digit in the ith real number. Since $\cdot a_1 a_2 \ldots$ is not among the reals in the assumed one-to-one correspondence, there can be no such correspondence.) This means that, unless Strasnick can show why the function for assigning numbers on the basis of a preference ordering must include all real numbers in a particular interval, e.g., 0 to 1, he has not proved that a claims-based ordering is not representable by a social welfare function.

Why might one believe that the rationals would not suffice and that one ought to assume that the values of C^1 and C^2 (the priority claims) must belong to the uncountable set of all real numbers? Most obviously, one might argue that using the rationals limits one to operations of addition, subtraction, multiplication, and division, but that, if one assumes that the reals are potential values for this ethically ordered assignment, then one is able to take roots, use logarithms, trigonometric functions, and so on. Admittedly this is a powerful objection. All things being equal, one ought to use the reals because this provides one with a greater variety of mathematical tools. Hence, *prima-facie* plausibility appears to be on the side of the reals. Why, however, in this application of welfare economics, does one need the reals? So long as the values of the lexicographic assignment are rationals, then they can be represented by a real, order-preserving function which in turn can be represented by a real social-welfare function. Let us see why this is so.

If the values of the items in the lexicographic ordering are rational pairs, or rational triples, and so on, or *if* the values of the items in the lexicographic ordering are from any countable set, *then* the following lemma guarantees that there exists a real, order-preserving function that maps the rational pairs, or rational triples, and so on, into the reals:

> Lemma II. Let X be a completely ordered set, $Z = (z_0, z_1, \ldots)$ a countable subset of X. If for every pair x, y of elements of X such that $x \leq y$, there is an element z_i of Z such that $x \leq z_i \leq y$, then there exists on X a real, order-preserving function, continuous in any natural topology.[28]

And *if* there is a real, order-preserving function that maps the rational pairs, or rational triples, and so on, into the reals, then we can define on them a real, order-preserving welfare function.

For example, suppose that four individuals, *A, B, C,* and *D*, are competing for some scarce good, and that a correct moral resolution of the conflict depends on a moral weighting of the competing claims of each person. If one individual's claim is judged to be of greater priority than any other claim of another person, perhaps on the basis of this individual's greater merit, for example, or because another individual is bound to him through obligation, then let us suppose that the fact of this greater priority will be sufficient to determine the correct moral outcome, apart from whether or not there is some measure of the relative utility or desirability of this outcome. If this view of moral evaluation is ever appropriate, then the model of morality based on the social-welfare function must be supplemented by another kind of model, one appropriate for representing a priority weighting of different claims. The lexicographic version of Rawls' difference principle is a type of claims-based moral system, as was already mentioned (near the beginning of section 4 in this chapter). If we assume that a measure of utility, for each individual, exists in various situations, then this measure can also be interpreted as the strength of the individual's claim for satisfaction. The ranking of the claims of the four individuals, *A, B, C,* and *D*, can thus be interpreted as a ranking of priority of individual claims for the realization of a given social situation. In the case of Rawls' difference principle, for example, since stronger claims must be satisfied first, one situation will be socially preferred to another if the strongest claim for the first situation is stronger than the strongest claim for the second. In case of ties, the next strongest claims are considered, and so on.

In the case of individuals *A, B, C,* and *D*, let us suppose further that each individual has three claims that he can make against this good, *M, N, S*, representing the claims of merit, need, and seniority, respectively. On this supposition, (M_A, N_A, S_A) is a rational triple representing the strength of individual A's three claims. Suppose further that M claims are viewed as taking precedence over N claims, and that N claims are viewed as taking precedence over S claims. This means that individual A will always be ranked highest in priority whenever $M_A > M_B$ and $M_A > M_C$ and $M_A > M_D$. If $M_A = M_B$, individual A will be ranked higher than B if $N_A > N_B$. If $M_A = M_C$, individual A will be ranked higher than C if $N_A > N_C$. If $M_A = M_D$, individual A will be ranked higher than D if $N_A > N_D$. But if $M_A = M_B$, then individual A will be ranked higher than B if $S_A > S_B$. And if $M_A = M_C$ and $N_A = N_C$, then the individual A will be ranked higher than C if $S_A > S_C$, and so on. By using these priority rules, we will be able to define a complete ranking of individual priority for all possible rational values of (M_A, N_A, S_A), (M_B, N_B, S_B), (M_C, N_C, S_C) and (M_D, N_D, S_D). We know that $\{(M_A, N_A, S_A), (M_B, N_B, S_B), (M_C, N_C, S_C), (M_D, N_D, S_D)\}$ is a completely ordered set (which we will call X) with a countable subset $\{(M_A, N_A, S_A), (M_B, N_B, S_B) \ldots\}$, which we will call Z. Since X is countable, it can play the role of Z in Lemma II, because it is a countable subset of itself. By Lemma II, we know that, for every pair of elements of X, for example, (M_A, N_A, S_A), (M_B, N_B, S_B), such that $(M_A, N_A, S_A) < (M_B, N_B, S_B)$, there is an element (M_A, N_A, S_A) of the subset such that $(M_A, N_A, S_A) \leqq (M_A, N_A, S_A) \leqq (M_B, N_B, S_B)$, then there exists on X a real, order-preserving function, continuous in any natural topology.

For example, if the rational triples (1, 2, 3), (4, 5, 6), (7, 8, 9), (1, 3, 4) represent the strength of the three claims, respectively, of individuals *A*, *B*, *C*, and *D*, then the lexicographic ordering of these individuals' claims, largest to smallest, will be (7, 8, 9), (4, 5, 6), (1, 3, 4), (1, 2, 3). By Lemma II, there is a real, continuous, order-preserving function which assigns values, for example, 11, 10, 9, and 8, respectively, to the triples which are lexicographically ordered. These values, measures of the strength of individuals *C*, *B*, *D*, and *A*'s claims for satisfaction, respectively, may also be assumed to be measures of the individual utilities of *C*, *B*, *D*, and *A*, respectively, in a particular situation. And these values, 11, 10, 9, 8 respectively, preserve the order of the triples and are representable by a real social-welfare function. The welfare function used in RCBA, $W(x)$, is merely the summation of the utilities (in this case, *C*, *B*, *D*, and *A*) in social state x. Hence, if one denotes individual *A*'s utility from state x as $U^A(x)$ or 8, then $W(x)$ in this case $= U^A(x) + U^B(x) + U^C(x) + U^D(x) = 8 + 10 + 11 + 9$, if one assumes that there are individuals *A*, *B*, *C*, *D* with respective utilities of 8, 10, 11, 9. In RCBA, social state y would be said to be superior to x only if $W(y) > W(x)$ or $\{U^A(y) - U^A(x)\} + \{U^B(y) - U^B(x)\} + \{U^C(y) - U^C(x)\} + \{U^D(y) -$

$U^D(x)\} > 0$.[29] In this result, social welfare is an additive function of measures of individual utility where the measures of individual utility are assumed to be measures of the strength of individual claims for satisfaction. Because the strength of individual claims for satisfaction is measured by means of lexicographic ordering, the resultant measure of social welfare is not purely utilitarian. It might more properly be called lexicographic utilitarianism, a combination of the utilitarian, aggregative view with the deontological, claims-based view of morality.

In fact, in situations in which one wished to guarantee a particular deontological outcome, e.g., recognition of a specific right, apart from the utilitarian values at stake, one could do so. For example, suppose there were three persons, *X, Y* and *Z*, who were competing for some scarce good. Following Dworkin's account of strong and weak rights, suppose further that person *Y* had *A* claims (claims based on an absolute right, e.g., the right not to be murdered) and *W* claims (claims based on a weak right, e.g., the right to drive a car) to this contested good, whereas persons *X* and *Z* had only *W* claims to this good. If we suppose further that the rational pairs (0, 5), (9999, 10), and (0, 10) represent the strength of the *A* claims and the *W* claims, respectively, of individuals *X, Y,* and *Z*, then the lexicographic ordering of these individuals' claims, largest to smallest, with *A* claims taking priority over *W* claims, would be (9999, 10), (0, 10), (0, 5). By Lemma II, there would be a real, continuous, order-preserving function which assigned values, for example, 10^{99}, 2, 1, respectively, to the pairs which were lexicographically ordered. In order to insure that the *A* claims of one or more individuals were not given second or lower priority behind the *W* claims of others, we would need to be certain that the real, continuous, order-preserving function assigned numbers (to the pairs representing *A* claims), each of which was larger than the maximum possible aggregate of all the numbers assigned to pairs at the next lower level (pairs representing only *W* claims). Let us call such a function, which assigns values in this way, a *D*-function (for 'deontological').

In the example just given, if we wished to insure that no person's *A* claims were violated in order to recognize the *W* claims of others, then we would need to be certain that the real, continuous, order-preserving function assigned numbers in a specific way. That is, the function would have to assign a number to the strength of *Y*'s claims, represented by (9999, 10), which was larger than the maximum possible aggregate of all the numbers assigned to the pairs at lower levels (pairs representing only *W* claims). Since the maximum possible aggregate of all the numbers assigned to the pairs ((0, 10) and (0, 5)) at the lower level is $2 + 1 = 3$ in the example just given, then the number assigned to represent the strength of *Y*'s claims would have to be greater than 3, in order to insure that no person's *A* claims were violated. Since 10^{99} is greater than 3, this condition is met. By thus fulfilling a similar condition in a given situation, we could always insure that the aggregate utilities of the majority did not outweigh the absolute rights of even one individual. Obviously, however, the extent to which such an assignment of values (to represent the strength of individuals' claims) mirrors utilitarian or deontological ethics would depend on the actual magnitudes associated with each *n*-tuple. The function would have to assign a number larger than the maximum possible aggregate of all the numbers assigned to the *n*-tuples at the lower levels. In other words, it would have to be a *D*-function.

As these examples illustrate, Lemma II enables one to use a welfare function to represent a lexicographic ordering, provided that the values of the items in the lexicographic ordering are rational pairs, or rational triples, and so on. (Note that Lemma II applies to a number of different cases. The one in which we are interested is the one in which *X* is a completely ordered set of rational triples and in which *Z* is equal to *X*.) *This means that, although numerous economists (see note 23) are correct in noting that, if a set is completely ordered by the preferences of some agent, it is not always possible to define on that set a real-valued, order-preserving, welfare function, there is a case in which a real-valued, order-preserving welfare function can be defined: If a set is completely ordered by the preferences of some agent, and if the values of the preferences are rationals, then it is always possible to define on that set a real-valued, order-preserving welfare function.*[30]

One economist with whom I discussed this problem conjectured that theorists required preference orderings to be mapped into the reals (rather than the rationals) so that they would have a continuous function. As the lemma shows, however, both the problem of a *continuous* function and that of a *real* function for representing lexicographically ordered preferences is resolved.

5. OBJECTIONS TO WEIGHTING RCBA'S BY MEANS OF LEXICOGRAPHIC ORDERING

In addition to the problems with continuity and with using a real social-welfare function, one can raise other objections to my proposal to weight RCBA parameters by means of lexicographic ordering. Hierarchical, transitive, ethical rules which were simple and clear enough to follow might be alleged either to represent a particular ethical position inadequately or to fail to cover certain circumstances. Other persons might claim that the ethical rules were too complicated and academic, or too difficult to apply. There would also be the problem of deciding who would set the relative weights or determine which ethical rules would be followed. Should the weighting scheme reflect democratic, meritocratic, or autocratic values?

One objection to my weighting scheme might be that ordering claims (A has greater merit than B so his claim is greater) is not ordering possible ways of dividing the good between A and B. The brunt of this response appears to be that ordering *claims* is not ordering ethical *outcomes*. This, of course, is true (which I never denied), but it is beside the point, for two reasons. *First*, ordering claims leads to the consequence that one orders outcomes, so the two orderings are closely related. *Second*, conventional economic wisdom has it that moral desirability is an additive function of measures of individual well being. If one wishes to challenge this conventional dictum, then so be it. My point is neither to defend nor to challenge it but simply to argue that, *if* one accepts this convention, *then* (contrary to what has been argued/believed) a preference ordering will be representable by a social-welfare function. If the point of this objection is to assert that 'ordering individual sets of claims' does not provide a means of 'ordering divisions of a good based on the claims of several individuals', then again the quarrel is not with my weighting scheme but with the aggregative aspects of economic theory itself. According to classical Pareto theory, optimal situations are defined in terms of aggregations of compensating variations (cv's), where each cv is determined on the basis of a subjective judgment. Hence the theory provides a way, however imperfect, of moving from individual judgments to decisions about the social good. While this move may be suspect, its discussion is neither the object of this chapter nor necessary in order to appreciate the significance of my principal result, viz., that a claims-based ordering is representable by a social-welfare function.

It could also be objected that the real, order-preserving function mapping ordered n-tuples of rationals into reals is not unique, and that the welfare ordering of two social states depends on the particular choice of function. Again, while true, this non-uniqueness is neither surprising nor problematic. Economists know that their conclusions about welfare are dependent on the particular welfare function used and on the specific cost and benefit parameters employed, just as philosophers know that their conclusions about welfare are dependent, for example, on particular assumptions about discount rates. Neither dependence invalidates, in general, economists' and philosophers' attempts to ascertain welfare and to make judgments about it. Nor does the observation, that the welfare function is not unique, undermine the main result of this chapter. Apart from whether or not the welfare ordering of social states depends on the choice of function, my main thesis is that, contrary to the received economic wisdom, a preference ordering can be represented by a social-welfare function.

One of the strongest and most astute objections to my proposal is that although the mapping is order-preserving, the resultant individual utility functions and social-welfare function do not satisfy the von Neumann-Morgenstern conditions: i.e., they do not possess the expected-utility property . . . for probability distributions over the possible outcomes. This objection is very well taken in that it is generally accepted that von Neumann-Morgenstern conclusively showed that the cardinals are needed to describe rational behavior under uncertainty.

While correct, nevertheless this objection counts less against my proposal than might be suspected. Why so? Any weighting system based on *ordering* rules to represent the priority of ethical claims would fall victim to this von Neumann-Morgenstern objection. As was explained earlier (in section 3), however, such ordering rules are necessary if ethical systems as complex as that of Rawls are to be accommodated within weighted RCBA. Because of the importance of representing this ethical complexity, it might be reasonable to put up with the von Neumann-Morgenstern problem. In other words, the ethical complexity of my ordering scheme has been bought at the price of broad applicability, since the ordering constraint precludes using the scheme in situations of choice under uncertainty. But if one chose to gain broad applicability, by foregoing ordering, then this would mean that one would lose the ability to accommodate some types of ethical complexity in one's weighting system. Hence it is not at all obvious that, in response to this von Neumann-Morgenstern objection, one ought to forego use of a lexicographic ordering scheme. Economists certainly have not foregone use of ordinal welfare functions simply because of this problem. Moreover, the three main points of this chapter are still intact, viz., (1) that it is possible to represent preference orderings by a social-welfare function; (2) that those who allege that such a representation is impossible are wrong; and (3) therefore, that those who allege that RCBA weighting schemes are impossible, *because such a representation is impossible*, are wrong. It may well be that RCBA weighting schemes are wrong *on other grounds*, perhaps on von Neumann-Morgenstern grounds. But that is not the issue here. The issue here is that the reason commonly alleged to prevent the use of a lexicographic-weighting scheme (viz., the impossibility of representing a preference ordering by a social-welfare function) is itself wrong.

Mishan's main objection to any sort of weighting scheme for RCBA is that it would be "at variance with the allocative principles by which the competitive economy is vindicated".[31] His claim appears to be that one ought to use RCBA principles which are consistent with those on which the competitive economy is based. Although it is not clear to which principles Mishan refers, policy decisions inconsistent with certain presuppositions underlying the competitive economy are often made and, I would claim, ought to be made. Whenever government takes account of an externality such as pollution and, as a consequence, imposes restrictive taxes or outright prohibitions, then it is clearly making decisions inconsistent with particular presuppositions underlying a purely competitive economy. Indeed, if it did not, then grave harm, such as pollution-induced deaths, could occur. Moreover, even RCBA itself is based on principles which are in part inconsistent with "the allocative principles by which the competitive economy is vindicated". Every time a practitioner of RCBA 'costs' a nonmarket item, e.g., human life, in terms of criteria other than discounted future earnings, for example, or every time he puts a dollar figure on the aesthetic value of undeveloped land, then he is using principles inconsistent with market allocation to determine how to aggregate risks, costs, and benefits. Also it is well known that welfare economists 'correct' their market parameters for risks, costs, and benefits. Hence not even they use only allocative principles consistent solely with a competitive economy. In fact, if they are to compensate for market imperfections, then they must rely also on other types of principles. Moreover, as economist Y. K. Ng argued, and as I stressed earlier in this chapter, if a Pareto Improvement is not a sufficient condition for a good change, then it cannot be used independently of other considerations, such as my weighting scheme.[32] If this is so, then better for the weighting to be explicit and careful, rather than to be left to the vagaries of the political process.

Another classic objection to the use of any weighting scheme in RCBA is that, as Culyer says, weightings ought to be left to politicians and the democratic process and not taken over by economists.[33] To some extent, this objection was already addressed earlier in the chapter: I discussed what an analytic method could add to the political-democratic process. More to the point, Culyer seems to assume that implementing a system of weighted RCBA's somehow means that economists, and not the people, will be making policy decisions. Clearly this is not the case. Admittedly, economists and ethicians might help to formulate the lexicographic rules and to weight RCBA parameters in terms of these rules, but these alleged experts, on my plan (see section 4 earlier in this chapter), would be responsible only for helping to formulate a number of ethically weighted analyses for a given project. The public and policymakers would be left to decide among the alternative RCBA's. The merit of having a number of ethical options, of course, would be that citizens could select the *values* by which they would choose social policy as well as the policy itself. One reason why my proposed weighting scheme would not lend itself to control by experts, who have no business dictating policy in a democracy, is that it calls for preparation of alternative RCBA's, each with different ethical weights. This amounts to employing some sort of adversary means of policy analysis. Since I defend adversary methods in the next chapter, I will not take the time to do so here.

While most of these criticisms of weighted RCBA are reasonable, especially the claims that it would be too difficult to accomplish and presupposes too simplistic a representation of an ethical system, none of the objections seems to me to address any in-principle difficulty with a modified RCBA model. For one thing, such a weighting scheme may be no more difficult to accomplish, econometrically, legally, and politically, than Mishan's proposed assimilation of externalities within the risk-cost-benefit framework. Yet, such an assimilation is widely acknowledged to be absolutely essential to accurate RCBA. Moreover, although economists have not been employing ethically weighted schemes in the precise ways just suggested, they have traditionally used a related method, a discounting procedure, in an attempt to weight future values, so that they are commensurate with present ones.[34] Economists usually justify the use of a discount rate on future benefits and costs (i.e., weighting them at less than current economic values) because a given amount of money invested now will yield a much larger amount later, given the rate of return over inflation. Kneese *et al.* say, for example, that if, in 50 years, we were to give a 'fair' compensation of $2500 to some future individual for effects of a carcinogen we are currently using, then we would only need to invest $558 now, if there were a 3 percent real rate of return over inflation in the next 50 years.[35]

There are numerous controversies surrounding both the justification for, and the application of, these traditional discounting schemes, especially as regards future public goods.[36] I do not wish to argue about their merit here. My point simply is that, given the complexity of using discounting methods, emplying a system of ethical weights is not in principle more difficult.

6. CONCLUSION

In light of all the reasons supporting lexicographically ordered systems of weights for RCBA, perhaps we ought at least to try to develop and use them. Moreover, to the degree that one accepts the claims of proponents of the deficiency argument, that RCBA falls victim to some of the same deficiencies as classical utilitarianism, then to that same extent ought one to be willing to avoid these alleged Benthamite flaws by means of some sort of weighting scheme.

I argued in Chapter Two that utilitarianism evidenced in many RCBA's is not essential to it but is rather the product of the belief systems of many of its practitioners. Regardless of the precise ways in which utilitarianism influences contemporary policy analyses, however, they should be delivered from implicit and unrecognized ethical and evaluative presuppositions. Whatever normative judgments exist should be made explicitly and with a view to pointing out their consequences. Deciding among alternative, ethically weighted RCBA's should help this goal.

NOTES

[1] Peter Self, *Econocrats and the Policy Process: The Politics and Philosophy of Cost-Benefit Analysis*, Macmillan, London, 1975, pp. 155–165 discusses the Roskill Report. Self's book is hereafter cited as: *PPCBA*.

[2] D. W. Pearce, *The Valuation of Social Cost*, George Allen and Unwin, 1978; hereafter cited as: Pearce, VSC. This point is defended in great detail in K. S. Shrader-Frechette, 'Technology Assessment as Applied Philosophy of Science', *Science, Technology, and Human Values* 6 (33), (1980), 34–41; hereafter cited as: Shrader-Frechette, TA.

[3] Pearce, VSC, p. 134.

[4] G. Myrdal, quoted by M. C. Tool, *The Discretionary Economy*, Goodyear, Santa Monica, 1979, p. 291.

[5] A. M. Freeman, 'Distribution of Environmental Quality', in *Environmental Quality Analysis* (ed. by A. V. Kneese and B. T. Bower), Johns Hopkins, Baltimore, 1972, pp. 247–248 makes this same point, as does Shrader-Frechette, TA, pp. 35–37.

[6] See K. Basu, *Revealed Preference of Government*, Cambridge University Press, Cambridge, 1980, p. 23; hereafter cited as: *RPG*.

[7] See this volume, Chapter 6, and K. S. Shrader-Frechette, 'Das Quantifizierungsproblem bei der Technikbewertung', in *Technikphilosophie in der Diskussion* (ed. by Friedrich Rapp and Paul Durbin), Vieweg, Wiesbaden, 1982, pp. 123–138. For discussion of Gresham's Law, see B. M. Gross, 'The State of the Nation', in *Social Indicators* (ed. R. A. Bauer), MIT Press, Cambridge, 1966, p. 222.

[8] Some of the main practitioners of the method of revealed preferences include C. Starr, C. Whipple, and D. Okrent. See, for example, Starr and Whipple, 'Risks of Risk Decisions', *Science* 208 (4448), (June 6, 1980), and D. Okrent, 'Comment on Societal Risk', *Science* 208 (4442), (1980), 374. See also C. Starr, *Current Issues in Energy*, Pergamon, New York, 1979, and D. Okrent and C. Whipple, *Approach to Societal Risk Acceptance Criteria and Risk Management*, PB-271-264, US Department of Commerce, Washington, D.C., 1977. Although other means (e.g., the method of expressed preferences) of assigning measures to RCBA parameters have been discussed, I treat

only the methods of market assignment of values and expressed preferences since these two dominate all current RCBA practice.

[9] For an economist's perspective on the problems with the method of revealed preferences, see Basu, *RPG*, especially Chapter 9.

[10] J. S. Mill, 'Utilitarianism', in *John Stuart Mill* (ed. by M. Warnock), Meridian, New York, 1962, p. 310; see also p. 321.

[11] B. A. Weisbrod, 'Income Redistribution Effects and Benefit-Cost Analysis', in *Problems in Public Expenditure Analysis* (ed. by S. B. Chase), Brookings Institution, Washington, D.C., pp. 177–208; hereafter cited as: Weisbrod, IRE, in Brookings, *Problems*. See also UNIDO, *Guidelines for Project Evaluation* (by P. Dasgupta, S. A. Marglin and A. K. Sen), United Nations Industrial Development Organization, Project Formulation and Evaluation Series, No. 2, United Nations, New York, 1972.

[12] See, for example, R. Haveman, 'Comment on the Weisbrod Model', Brookings, *Problems*, in pp. 209–222. See also the next note.

[13] See Basu, *PRG*, pp. 23–24. For other criticisms of the Weisbrod and UNIDO approaches, see note 12 and A. M. Freeman, 'Income Redistribution and Social Choice: A Pragmatic Approach', *Public Choice* 7 (Fall 1969), 3–22; E. J. Mishan, 'Flexibility and Consistency in Project Evaluation', *Economica* 41 (161), (1974), 81–96; and R. A. Musgrave, 'Cost-Benefit Analysis and the Theory of Public Finance', in *Cost-Benefit Analysis* (ed. by R. Layard), Penguin, Baltimore, 1972, pp. 101–116.

[14] A. V. Kneese, Shaul Ben-David, and W. D. Schulze, 'The Ethical Foundations of Benefit-Cost Analysis', in *Energy and the Future* (ed. by D. MacLean and P. G. Brown), Rowman and Littlefield, Totowa, N.J., 1982, 59–74; hereafter cited as: Foundations. A. V. Kneese, S. Ben-David, and W. Schulze, *A Study of the Ethical Foundations of Benefit-Cost Analysis Techniques*, unpublished report, done with funding from the National Science Foundation, Program in Ethics and Values in Science and Technology, August, 1979; hereafter cited as: *Study*.

[15] Kneese *et al.*, Foundations, pp. 62–63; Kneese *et al.*, *Study*, pp. 11–13.

[16] Kneese *et al.*, Foundations, pp. 63–65; Kneese *et al.*, *Study*, pp. 13–23.

[17] Kneese *et al.*, Foundations, pp. 65–73; Kneese *et al.*, *Study*, pp. 46–82.

[18] Kneese *et al.*, *Study*, pp. 83–119.

[19] Kneese *et al.*, *Study*, pp. 120–130.

[20] John Rawls, *A Theory of Justice*, Harvard University Press, Cambridge, 1971, pp. 244, 302. Lucian Kern, 'Comparative Distributive Ethics', in *Decision Theory and Social Ethics* (ed. by H. W. Gottinger and W. Leinfellner), Reidel, Boston, 1978, p. 189.

[21] See K. R. MacCrimmon and D. A. Wehrung, 'Trade-off Analysis', in *Conflicting Objectives in Decisions* (ed. by D. Bell, R. Keeney, and H. Raiffa), Wiley, New York, 1977, p. 143. See Kneese *et al.*, Foundations, pp. 62–63.

[22] See Self, PPCBA, p. 89; G. Myrdal, *Against the Stream*, Random House, New York, 1973, p. 168; A. Lovins, 'Cost-Risk-Benefit Assessment . . .' *George Washington Law Review* 45 (5), (August 1977), p. 927; N. Georgescu-Roegen, *Analytical Economics*, Harvard University Press, Cambridge, 1966, p. 196; E. Rotwein, 'Mathematical Economics', in *The Structure of Economic Science* (ed. by S. R. Krupp), Prentice-Hall, Englewood Cliffs, 1966, p. 102; G. Shackle, *Epistemics and Economics*, University Press, Cambridge, 1972, pp. 45–47. D. MacLean, 'Quantified Risk Assessment', in *Uncertain Power* (ed. by D. Zinberg), Pergamon, New York, 1983, section V; S. Hampshire, 'Morality and Pessimism', in *Public and Private Morality* (ed. by Hampshire), University

Press, Cambridge, 1978, p. 5; R. Wolff, 'The Derivation of the Minimal State', in *Reading Nozick*, Rowman and Littlefield, Totowa, N.J., 1981, pp. 99–101; R. Coburn, 'Technology Assessment, Human Good, and Freedom' in *Ethics and Problems of the 21st Century* (ed. K. Sayre and K. Goodpaster), University Press, Notre Dame, 1979, p. 108; A. Gewirth, 'Human Righrs and the Prevention of Cancer' in *Ethics and the Environment* (ed. D. Scherer and T. Attig), Prentice-Hall, Englewood Cliffs, 1983, p. 177; and A. MacIntyre, 'Utilitarianism and Cost-Benefit Analysis', in *Ethics and the Environment* (ed. Scherer and Attig), Prentice-Hall, Englewood Cliffs, 1983, pp. 139–142.

[23] Steven Strasnick, 'Neo-Utilitarian Ethics and the Ordinal Representation Assumption', in *Philosophy in Economics* (ed. by J. C. Pitt), Reidel, Boston, 1981, pp. 63–92; hereafter cited as: NU.

[24] Strasnick, NU, pp. 70, 84. Numerous economists hold this same position. Y. K. Ng, *Welfare Economics*, John Wiley, New York, 1980, p. 27 (hereafter cited as: *WE*), for instance, says that "the standard example where representation by a real-valued function is not possible is the so-called lexicographic order." See also Amartya Sen, *On Economic Inequality*, Norton, New York, 1973, pp. 2–3; hereafter cited as: Sen, *OEI*. Finally, see A. K. Dasgupta and D. W. Pearce, *Cost-Benefit Analysis*, Barnes and Noble, New York, 1972, p. 75.

[25] Strasnick, NU, pp. 69–70.

[26] Gerard Debreu, 'Representation of a Preference Ordering by a Numerical Function', in *Decision Processes* (ed. by R. M. Thrall, C. H. Coombs, and R. L. Davis), John Wiley, New York, 1954, p. 164; hereafter cited as: Representation. It appears that Debreu should have written "$\alpha_2(a) = \alpha(a, b_2)$" in the quoted passage and not "$\alpha_2(a, b_2)$".

[27] Strasnick, NU, p. 70.

[28] Debreu, Representation, p. 161.

[29] *Basu*, RPG, p. 7.

[30] One of the anonymous manuscript referees for Reidel Publishing Company noted, in response to this argument: "I fail to see the relevance of the fact that the irrational numbers are denser than the rationals. See Bolzano–Weierstrass Theorem on points of accumulation". However, my argument has absolutely nothing to do either with the concept of density or with the Bolzano–Weierstrass Theorem. Hence the objection is irrelevant.

[31] E. Mishan, *Economics for Social Decisions*, Praeger, New York, 1972, p. 23. See also E. Mishan, *Cost-Benefit Analysis*, Praeger, New York, 1976, pp. 403–415; hereafter cited as: *CBA*.

[32] Y. K. Ng, *Welfare Economics*, John Wiley, New York, 1980, pp. 68–72.

[33] A. J. Culyer, 'The Quality of Life and the Limits of Cost-Benefit Analysis', in *Public Economics and the Quality of Life* (ed. by L. Wingo and A. Evans), Johns Hopkins University Press, Baltimore, 1977, pp. 143, 150, 151.

[34] See Raymond F. Mikesell, *The Rate of Discount for Evaluating Public Projects*, American Enterprise Institute for Public Policy Research, Washington, D.C., 1977. See also Mishan, *CBA*, pp. 175–219, 408–410.

[35] Kneese *et al.*, *Study*, pp. 8–9. See also pp. 32–42.

[36] See, for example Talbot Page, *Conservation and Economic Efficiency*, Johns Hopkins, Baltimore, 1977.

CHAPTER NINE

ASSESSMENT THROUGH ADVERSARY PROCEEDINGS

1. INTRODUCTION

Many controversies over technology- and environment-related policy have been characterized by experts' disagreement over the relevant scientific facts. Prominent examples include conflict over nuclear fission, fluoridation, food additives, depletion of the ozone layer, and high-voltage transmission lines. Often the industrialists and technocrats who are party to such controversy blame the conflict on the fact that their opponents are paranoid neo-Luddites who are scientifically illiterate and uninformed about the relevant technical issues.[1] Likewise, often the radical environmentalists and consumer advocates involved in such a dispute blame the allegedly scientific controversy on the fact that their opponents are spokespersons for a 'technomanagerial elite' rather than advocates of disinterested scientific inquiry.[2] The existence and intensity of these controversies suggests that the present governmental bodies responsible for using the results of technology assessment (TA) and environmental-impact analysis (EIA) are unable to handle the interface between technological, environmental, and public policy components of contemporary decisionmaking. Perhaps one reason for this failure, as Chapters Three–Seven of this volume suggest, is that TA's and EIA's are often not responsive to the complex ethical and methodological issues which need to be addressed as part of comprehensive policy-making.

The most prominent of all proposals for improving the means whereby disputed scientific facts are employed in governmental decisionmaking is that of the 'science court'. Comprised of a panel of allegedly impartial, competent scientists from disciplines adjacent to, but not involved with, a particular dispute, the members of the court would hear testimony from both sides of the issue in question and would prepare a report summarizing their findings on the scientific facts at issue. Proponents of the science-court proposal maintain that it would help adjudicate technology-related controversy by helping to provide accurate scientific information; that it would promote more openness in policy formulation; that it would force environmentalists to recognize scientific facts; and that it would prevent scientists in the employ of industry from exercising inordinate power and from promoting value judgments under the guise of empirical data.

In the last six years since the 'Science Court Experiment' was proposed by a U.S. Task Force, it has been hotly debated by the scientific, environmental, governmental, and industrial communities. The proposal has not received extensive philosophical evaluation, however, despite its obvious relevance to philosophy of science and technology, applied ethics, and political philosophy. The single best philosophical analysis of the proposal, to date, is that of Alex Michalos, who supported the idea but argued that one of the major assumptions underlying it was epistemologically untenable. This assumption is that facts can clearly be separated from values, and that the court could deal only with the former. Denying the fact-value dichotomy, Michalos argued effectively that the court could hope not to avoid values, but only "to arrive at a *timely and authoritative agreement* about issues broadly classifiable as scientific". He reaffirmed all other basic science-court-proposal components, in particular the notion that scientists, not "ordinary citizens", should serve on the court.[3]

In this chapter, I would like to continue the philosophical analysis begun by Michalos, but to extend the debate so as to consider a more radical conclusion than that which he offers. Specifically, I will argue that epistemological analysis, ethical ideals, and political realities require that intelligent citizens, not scientists, serve on the court, which ought to be renamed, and that its focus be on issues of technology policy generally and not merely on scientific fact. My analysis consists of six main parts: (1) a consideration of the reasons for proposing the science-court experiment; (2) a presentation of the main components of the science-court proposal; (3) an evaluation of the problems not addressed by the proposal outlined in (2); (4) a sketch of an alternative proposal for such a court; (5) an overview of several arguments in favor of the populist alternative outlined in (4); and (6) a response to major arguments against the alternative, which I call the 'Technology Tribunal'.

2. REASONS FOR THE SCIENCE-COURT PROPOSAL

In order to understand the difficulties inherent in the science-court proposal and my reasons for suggesting an alternative to it, consider first the reasons why the court experiment has been recommended by a number of policy analysts. Three main reasons have advanced. The first, and perhaps the most obvious, is the need for accurate, scientific information to serve as a basis for deciding policy questions. Decisionmakers need to know, for example, how dangerous saccharin is as a food additive, and whether freon will damage the upper atmosphere. Without the accurate scientific information provided by a science court, its proponents claim that the public would continue to be subjected to a barrage of quasi-popular, media-interpreted claims and counterclaims about the relevant scientific facts. In the absence of this procedure, they charge, decisionmaking regarding technological matters is characterized by confusion and controlled by the loudest voices. As Michalos puts it, without a court, "the best hustlers might lead us to the worst calamities".[4] By providing accurate scientific information, such a court should be able to identify discredited claims when they occur in the course of public debate. Supporters maintain that this should enable the public to cut through the distortions and falsehoods promoted by various parties to technological controversies and to attain closure on scientific debates which have been kept controversial only because of misinformation.

A second reason for the court, say its proponents, is that it would help limit the inordinate power sometimes exercised by scientists. Without such a court, they believe, policymakers are often inclined to ask scientists for policy recommendations and not simply for the facts. To prevent the social views of scientists from being imposed under the guise of technical expertise, proponents maintain that the court would provide statements only on relevant scientific information. Moreover, they claim, a mechanism such as the court is particularly needed to provide a framework for adversary proceedings. Many scientists in a given field often owe their livelihood to particular policies regarding technology, both because of industrial employment and because of government and industry grants and research contracts.[5] By limiting court scientists to assessments only of technical points, proponents of the court believe that the policy process could be made more subject to public scrutiny and democratic control. They also maintain that it would provide a framework for airing all sides of an issue rather than merely the side advocated by a majority of the members of a particular scientific community.

The main benefit of adversary proceedings would be to extend opportunities for cross-examination of experts; one way to keep scientists honest would be to have them confront peers of equal competence. A good example of the effectiveness of an adversary format occurred in 1972, when Henry Kendall, an MIT physicist, and other intervenors challenged the US Atomic Energy Commission's (AEC's) nuclear reactor safety standards. At issue was the safety of the emergency core cooling system (ECCS). No full-scale working test of the system had ever been made, and critics questioned several key assumptions in the AEC's theoretical calculations of its operation under emergency conditions. The intervenors presented an 80-page critique of AEC standards for the system. Under the Freedom of Information Act, they obtained government documents showing that regulators were uncertain about the system's performance and had intimidated employees who questioned its performance. The intervenors' testimony stood up to two weeks of cross-examination by the AEC, whereas the AEC experts could provide virtually no evidence to support their assumptions that the cooling system would work. As a consequence, safety standards were modified, a new government safety director for nuclear power was appointed, and the ECCS was tested empirically.[6] Other recent examples of the effectiveness of an adversary format are the Lovins-Bethe debates and, earlier, the Pauling-Teller debates. Both these exchanges brought about a widespread discussion of complex scientific issues. Proponents of the science court believe that it would regularize such public debates and make science policy more accessible to democratic control.

A third reason often given for advocating the formation of a science court is that it would provide for more openness in policy decisions and would help eliminate the opportunity for regulators and politicians to hide policy decisions behind scientific conclusions. Proponents of the court claim that it would force policymakers to address and grapple with the value issues involved in a particular technology-related decision and to be clear about the particular value choices they espouse.

3. AN OUTLINE OF THE SCIENCE-COURT PROPOSAL

Formally, the science court exists only as a 1976 proposal of the Task Force of the Presidential Advisory Group on Anticipated Advances in Science and Technology. Their proposal was titled: "The Science Court Experiment: An Interim Report".[7] When Jimmy Carter took office, the Task Force was dissolved and the science court was never tried.

As outlined in a seminal article by the Task Force Chairman, Arthur Kantrowitz and as described in the Task Force report, the science court proceeding has three stages. At the first stage, the court identifies the significant questions of science and technology associated with the controversial issue in question. It addresses itself to these issues alone, and leaves other questions – the political, ethical, and policy ones – to subsequent consideration by other groups in the national decisionmaking process. The second step is an adversary proceeding which is presided over by a panel of impartial scientific judges. During this proceeding, scientist-advocates debate the technical questions that are in dispute. In addition to presenting their own cases, the debaters are able to cross-examine opponents and criticize their arguments. At the third and final stage of the court procedure, the panel of judges issues its decision as to the scientific facts relevant to the disputed technical questions. This decision is made public, unless national security dictates otherwise, and is designed to provide the scientific basis for reaching political decisions through the democratic process.[8]

4. PROBLEMS WITH THE SCIENCE-COURT PROPOSAL

In response to the proposal of the White House Task Force to create a science court, a number of its opponents have raised objections. Some of these are that good people would not be willing to interrupt their careers to work on such a court; that no one of stature is going to risk exposure as a loser; that all knowledgeable personnel will be committed to one side or the other of any live issue; that there are already too many special task forces and boards; that the court's findings will be used only if they are compatible with the views of those in power; that the adversary procedure will polarize issues and exclude intermediate ones; that the court will operate as an agent of whoever sponsors it; that it would be extraordinarily difficult to alter the decisions of the court; that the existence of a court would undermine efforts of scientists to be concerned with their social responsibilities; that the court will be used by those who wish to stall for time; that the proceedings will be biased by cleverness and determination rather than by adherence to the facts; that the court will reinforce the misleading impression that science is a field in which complete certainty is possible; and that, because of the prestige of the science court, people would become unwilling to serve on other important committees, e.g., of the National Academy of Sciences.

One need not be concerned with analyzing and evaluating all the objections to the science court, both because such a task would be lengthy and because Kantrowitz, the originator of the court idea, and Michalos, in his philosophical analysis of the proposal, have answered each of these objecttions.[9] There are, however, at least two other objections to the court which point to important ethical and political considerations and which need to be addressed here. One is the objection that it is epistemologically impossible for the court to deal only with matters of pure fact; the other objection is that the court will likely overemphasize the factual aspects of controversies and, as a consequence, will be dominated excessively by scientific experts. In the remainder of this chapter, I would like to consider both of these objections in detail and to argue that the alleged court needs to be modified somewhat in order to take account of these problems.

The first objection is considered by Michalos in his classic piece. This is that the science court cannot work because facts cannot be separated from values.[10] In discussing this objection, Michalos employs arguments used by philosophers ever since the post-1935 demise of the fact-value dichotomy in order to prove that the court cannot possibly deal only with matters of fact. He claims, rightly, that facts and values are not clearly distinguishable, and that every statement of what is good about anything presupposes some sort of evaluation. After correctly demolishing the fact-value dichotomy, however, Michalos argues that accepting the dichotomy "is not essential to the main purposes of the Court", which ought to focus on "what is important, whether it's overtly value-free or value-laden".[11] Having said this, Michalos sanctions the remaining assumptions and suggested procedures essential to the science court. Hence, as a consequence of the Michalos critique, the basic structure of the science court is left intact, as proposed, save that the procedure will deal with issues that are value-laden, not merely purely factual, and that it will ideally result in establishing some findings which are also value-laden and broadly classifiable as scientific, rather then in findings which settle some solely technical controversy.

It is at this point that one needs to consider a second important objection to the court, one which does not trouble Michalos. This objection is that the court would overemphasize facts and technical aspects of problems and, because of this, would be operated by and for recognized experts in various fields.[12]

In response to worries about overemphasis on facts and domination by experts, Michalos makes three main counterarguments:

(1) One ought not to worry about overemphasis on the facts because the science court will not stick only to technical matters. Realizing that there is no fact-value dichotomy, members of the court will be able to deal with issues that are value-laden and only "broadly classifiable as scientific".[13]

(2) One ought not to worry about overemphasis on the facts because civil servants and various groups within the governmental decision-making process will also be working on the problems addressed by various science courts. The court's concentration on facts does not mean that total societal emphasis will be on the facts alone, since other groups will be working through the evaluative aspects of these issues.[14]

(3) One ought not to fear that the court will become dangerously elitist, by virtue of scientists' making decisions, because all alternatives to this procedure are far less

desirable than the science court. Populist forms of technological decisionmaking are, in Michalos' words, "self-defeating. . . . For example," he asks, "how much light could an ordinary citizen be expected to shed on the question of whether or not saccharin is carcinogenic?" Unless only scientists were allowed to serve, the science court "could get some real duds".[15]

Let us consider each of these counterarguments and determine whether Michalos succeeds in allaying worries about overemphasis on facts and domination by technical experts. Michalos' first counterargument, that we need not worry about the court's overemphasizing facts, because it will deal with value-laden issues as well, is problematic in large part because its central assumption appears incompatible with a key assumption undergirding the third counterargument. If one makes the *first counterargument*, that the court can avoid overemphasis on facts by dealing with evaluative as well as technical issues, then (given acceptance of the rest of the science-court procedure) one must assume that scientists ought to decide issues which are ethical and political as well as scientific. However, if one makes the *third counterargument*, that ordinary citizens cannot be expected to shed light on purely technical considerations, and that the science-court deals with purely technical considerations, then one must assume that scientists ought to decide the purely scientific issues.

In his first counterargument, Michalos appeals to the ethical/evaluative dimension of the issues treated by the court, in order to answer the charge of overemphasis on the facts. In his third counterargument, he appeals to the purely scientific aspect of these issues in order to justify his reason for agreeing that only scientists ought to make the decisions. But Michalos cannot have it both ways. Either the issues are purely scientific or they are not. Michalos cannot argue as if they were, for purposes of his third counterargument, and argue as if they were not, for purposes of his first counterargument. If we agree with a preponderance of philosophers and with Michalos' earlier argument, that facts cannot be completely separated from values, then the issues treated by the science court must be value-laden. But if so, then Michalos' third counterargument rests on a false assumption. This leaves him with the assumption of his first counterargument, that scientists alone ought to adjudicate the value-laden issues addressed by the court. I will examine this assumption later in greater detail, since I believe it is the Achilles' heel of the current science-court proposal. First, however, let us consider Michalos' second counterargument. This is that one ought not to worry about the court's overemphasis on facts because other societal and governmental groups will be dealing with the evaluative aspects of the issues treated by the court.

The main problem with this second counterargument is that, if it is right, it leaves the court proposal open to an even greater objection. This objection is that the court cannot possibly achieve closure on certain science-related controversies, even though such closure was alleged as one of the three main reasons for proposing the court (this is the first of three reasons given in section two of this chapter). If the court emphasizes the facts, but other societal bodies play equally important policy roles, especially in dealing with evaluative aspects of the issue in question, then (by virtue of the power or influence of these other bodies) the court could not be said to achieve closure on a particular controversy. Rather, the court would be just one more group, competing for its share of media attention in a situation characterized by incompatible claims and counterclaims. Michalos has proposed to solve the problem of overemphasis on the facts by arguing that other bodies will address values; in so doing, he has undercut the court's claim to final authority. Moreover, if the court deals with issues that are evaluatively laden, then it is possible (indeed probable) that its findings will be incompatible with those of other groups, since different evaluative assumptions allow the same data to be interpreted differently. But if this is possible, then the court likely would add merely to the proliferation of opinion on an issue, rather than to a final conclusion regarding it. Hence Michalos appears to need some other strategy for avoiding the charge that the court would overemphasize the factual aspects of issues.

5. THE TECHNOLOGY TRIBUNAL

One obvious way to avoid possible overemphasis on the factual aspects of issues handled by the science court would be to stipulate that it address the major social-political-ethical-scientific facets of current controversies over technology, and not just the scientific aspects. Moreover, if Michalos and most other contemporary philosophers are right (and I think that they are) it would be impossible to deal with what is solely scientific anyway. Instead, one could make it a formal requirement of the court that it deal with the many disciplinary facets of a particular issue, and not just with the technical aspects. This seems a reasonable procedure, both because it would avoid overemphasis on largely technical matters and because it would enable the court to address what is on people's minds, namely the policy questions, and not merely the scientific questions. In other words, the court could address questions like "how safe is safe enough?" and not merely questions like "how safe is this particular technology?".

But if it is reasonable for the court to address the political and evaluative aspects of controversies, rather than merely the technical ones, then it likewise seems reasonable for intelligent citizens and not just scientists to act as juries after the adversary presentations and cross examinations. In other words, if it is epistemologically impossible for the court to avoid value issues, then it is not clear that only scientists should adjudicate the controversies in question.

Amended in these two ways, so as to focus broadly on science-and-technology-related policy questions and so as to allow for citizen adjudication and not just scientist determination, the 'science court' might do better with another name. My suggestion is 'technology tribunal'. This designation would correctly locate the issues to be dealt with as technological, rather than purely scientific, and it would avoid the authoritarian and antidemocratic connotations of the term, 'court'. Instead the label, 'tribunal', would serve to describe adversary proceedings in which scientists and other experts took part, but it would leave the final policy recommendation to some democratic procedure rather than solely to expert determination. Conceivably technical experts could present alternative positions on a technological issue on public television, and citizens, thus suitably educated, could play a larger role in policy formation regarding science and technology.[16]

6. ARGUMENTS IN FAVOR OF THE TECHNOLOGY TRIBUNAL AND AGAINST THE ASSUMPTION THAT SCIENTISTS OUGHT TO ADJUDICATE CASES HANDLED BY THE TRIBUNAL

The proposal to create a technology tribunal responsible both for coordinating adversary proceedings about issues of science and technology policy for democratic adjudication of these matters, however, rests upon a crucial assumption. This is that scientists alone ought not to be responsible for deciding crucial issues related to science and technology, because they have social, ethical, and political dimensions, as well as technical ones. Let us examine this assumption and the arguments and counterarguments which can be brought regarding it.

Most of the arguments in favor of this key assumption come down to five main claims which I call, respectively, (1) the argument from balance and objectivity; (2) the argument from democracy; (3) the argument from education; (4) the argument from political realism; and (5) the argument from G. E. Moore. Let us examine each of these.

6.1. *The Argument from Balance and Objectivity*

Perhaps the most basic reason for the technology tribunal, the argument from balance and objectivity rests on the premise that, while proponents of the science court go one step in the right direction of guaranteeing a hearing to all sides in a particular controversy, they do not go far enough in attempting to insure balance and objectivity in resolving the controversy. One of the key elements of the science-court proposal is to guarantee an adversary forum within which proponents of various positions on a particular technology-related issue would be provided with funds and with an equal footing to develop and make their cases effectively. In appealing to the argument from balance and objectivity to support the tribunal, I agree in this respect with advocates of the science court. They believe that developing a new "jurisprudence of science", and having public, adversary processes with adequate funding for all participants, could be a significant move toward rectifying the imbalances of political power following from the imbalances in economic resources in our society.[17] As Kantrowitz put it, such adversary proceedings, with funding for all sides, are necessary "to balance the politicization of science that has occurred in recent decades".[18]

Proponents of the science court do well to stipulate that the group of scientists-advocates participating in the adversary hearings be different from the group of scientists acting as judges of the hearings. Moreover, although they propose a second condition, that the judges be established experts in areas adjacent to, but not the same as, the dispute,[19] they appear to believe that these two conditions are sufficient to guarantee the objectivity and balance of the decisions reached. For many reasons, I think that this belief is false, and that the desired balance and objectivity could better be attained by having intelligent citizens or lawmakers act as judges. If one believes that scientist-judges in related fields are in a position to be more balanced and objective than intelligent citizens are about something, then he forgets that such scientists also must make their decisions within a political context. Granted, this political context is not one in which they decide the scientific merits of a proposal likely to provide them personally with employment, status, or income. However, the context is such that, because of their associations with 'big science', or 'big industry', they could be biased in favor of the proposed technology. One need only think, in this regard, of issues like the antiballistic missile or the supersonic transport or nuclear fission generation of electricity, all of which involve very large stakes for very powerful interest groups. Even if a scientist is not directly associated with such an interest group, it is highly unlikely that he would have no ties whatsoever. A recent issue of *Science*, for example, reported that the Air Force Office of Scientific Research was supporting the studies of more than 1000 doctoral candidates, all of whom rank at the top of the nation's younger generation of scientists; such financial incentives, noted the report, gave the scientists grounds for "developing their expertise in areas particularly relevant to Department of Defense interests".[20]

One way that such a scientist could be biased, even if there were no direct conflict of interest, might consist of being insensitive to consumer viewpoints and prone to underestimate risks and overestimate benefits associated with a particular protechnology decision. If an analytical chemist employed by a pesticide firm, for example, were called in to act as a judge in adversary proceedings regarding some controversy over radwaste disposal, it could well be that he had no direct conflict of interest, if he were not associated with the nuclear industry. Given the sorts of public health controversies faced both by the nuclear industry and by pesticide manufacturers, however, it would be likely that he shared a number of beliefs analogous to those of scientists employed by the nuclear industry, viz., less sensitivity to consumer issues and to questions of involuntarily imposed risks. The point is not that such scientists are immoral or knowingly biased. Rather the point is that scientists in analogous work or in similar industries often share common beliefs which might render them less effective than intelligent laymen might be as judges of adversary proceedings. As Jeffrey Martin put it, such scientists might tend to discount anomalous evidence or novel scientific explanations and instead to see controversy in the light of the 'industrial-bureaucratic paradigm', rather than in the light of some 'consumer paradigm'.[21] To have scientists adjudicate disputes, even if they have no apparent conflict of interest, is likely to guarantee at least some bias in favor of a protechnology tradition. It is also likely not to afford any opportunity for creative or innovative resolution of conflict. Scientific decisionmakers, as McGowan points out, are often unable to break out of the professional paradigms dominating their thinking; instead of devising new solutions, he claims that scientists often "solidify the structure that already exists".[22]

Another reason why the two safeguards built into the science-court proposal are not likely to guarantee as much balance and objectivity as might be possible is that the very process of selecting unbiased scientist-judges, in itself, involves numerous value judgments, some of which are clearly political. An unbiased scientist could be one who has taken no public stand on a particular issue, or one who has no close friends or colleagues who have taken a stand on a particular issue, or one who has no direct monetary conflicts of interest, or one who is in favor of a rising GNP but who has no clear opinion on the issue in question, or one who believes that there is no clear right or wrong side on most issues, or one who is likely to compromise in most situations and not to push his own beliefs, and so on. Clearly, the sorts of value judgments one makes in judging which scientists are unbiased is itself likely to provide a bias in the outcome of the adversary procedure.

Apart from the way the political context and the selection criteria for judges are likely to bias even the noblest of scientist-judges, another potential bias arises from the fact that scientific experts are sometimes unable to view problems wholistically or common sensically, even though intelligent laymen often come closer to doing so. It is conceivable, as science becomes bureaucratized, and as the individual scientist deals only with a minute segment of an over-all project, that he could become almost as detached from the implications of his work as Adolf Eichman alleged he was.[23] Harold Laski put this point well in his book, *The Limitations of the Expert:*

> It is one thing to urge the need for expert consultation at every stage in making policy; it is another thing, and a very different thing, to insist that the expert's judgment must be final. For special knowledge and the highly trained mind produce their own limitations which, in the realm of statesmanship, are of decisive importance. Expertise, it may be argued, sacrifices the insight of common sense to intensity of experience. It breeds an inability to accept new views from the very depth of its preoccupation with its own conclusions. It too often fails to see round its subject. It sees results out of perspective by making them the center of relevance to which all other results must be related. Too often, also, it lacks humility, and this breeds in its possessors a failure in proportion which makes them fail to see the obvious which is before their very noses. It has, also, a certain caste-spirit about it, so that experts tend to neglect all evidence which does not come from those who belong to their own ranks. Above all, perhaps, and this most urgently where human problems are concerned, the expert fails to see that every judgment he makes not purely factual in nature brings with it a scheme of values which has no special validity about it. He tends to confuse the importance of his facts with the importance of what he proposes to do about them.[24]

It is also conceivable that scientist-judges might tend to view controversies, not in terms of the merit of the cases of particular scientist-advocates, but in terms of the academic credentials of participants in the debate. Perhaps university professors are more susceptible to this sort of 'appeal to authority' because of the way they often obtain status in their professional communities. Since intelligent laymen are frequently less likely to assess status in terms primarily of academic credentials, it follows that they might be less likely to decide an issue on the merits of the advocates rather than on the quality of the case presented. Admittedly, however, there are reasons for believing that some laymen, too, might be 'cowed' by credentials, even though such credentials are unlikely to figure highly in the family and work communities of which they are members.

If these considerations are at least partially plausible, then there is reason to believe that intelligent citizens, not scientists, might provide the most balanced and objective judgments on technology-related controversies. This is because the public might be more likely to place more importance on public risk and safety, on the consumer's (as opposed to the producer's or the scientist's) point of view, on commonsensical and wholistic considerations as opposed to more narrow concerns, and on the merit of the arguments rather than on the credentials of the spokespersons.

6.2. *The Argument from Democracy*

A related argument in favor of the assumption that scientists alone ought not be responsible for adjudicating technology-related controversies is what I call 'the argument from democracy'. This argument, briefly put, is that democratic ideals require that decisions affecting the people be made by them or their peers and not be alleged experts. A number of considerations support this argument. For one thing, it is premised on the fact that the public is in great danger of being manipulated by technocrats. This premise was perhaps best articulated by Theodore Roszak. Although I do not share his belief in the impossibility of controlling technology, I think that he was correct in his insight, in *The Making of a Counter-Culture* that:

> The key problem we have to deal with is the paternalism of expertise within a socio-economic system which is so organized that it is inextricably beholden to expertise. And, moreover, to an expertise which has learned a thousand ways to manipulate our acquiescence with an imperceptible subtlety.[25]

If indeed Roszak was correct in believing that one of our key contemporary problems is to avoid being manipulated by experts, then there are strong democratic reasons for believing that controversies over technology ought to be adjudicated by intelligent laymen and not just by scientists. Another reason why scientists alone ought not to fill this role is that they were not elected to perform it.[26] Whenever the people of a democracy find that important policy decisions are made by persons other than those whom they have elected to make them, then they ought to question whether they have compromised their notion of democracy.

Of course, one of the main reasons for believing that intelligent citizens ought to adjudicate technology-related controversies is that such controversies almost always involve health, safety, political, or economic risks to the public. Were these controversies purely a matter of esoteric scientific concern, unrelated to human welfare, then there would not be a strong case for 'interference' by a democratically minded public. In other words, the very nature of the technology-related controversies and their consequences to public well-being mean that the public ought to have a say in how they are adjudicated. Without procedures for involving the citizenry in adjudication, then any alleged science court could serve merely as a vehicle for some technocratic bureaucracy. Moreover, if intelligent laymen assumed roles as decisionmakers, then the adversary proceedings of the scientists would likely be more open, more accessible (intellectually), and less secret. This means that having citizens help to adjudicate the adversary proceedings not only would serve the ideals of democracy with respect to final decisionmaking but also would serve them in rendering the public debate, prior to decisionmaking, more understandable to everyone. Because the average citizen has little access at present either to technology-related debate or to final decisionmaking regarding it, his interests are often served only by occasional whistleblowers. The point of the argument from democracy is to insure the public that they will be able to have their interests represented by their peers and that they will not have to depend on occasional whistleblowers within the scientific community.

One reason why the procedure of having scientist-judges might not serve the interests of democracy is that such a process would involve the public at so late a date that its concerns might not be heeded. As the science court was proposed, citizens would likely discover all the details of the adversary proceeding and the resulting decision only after the case was closed. There is no mechanism for incorporating their views and evaluating public opinion during the proceeding. This can only mean that citizens are placed at a disadvantage if they wish to challenge a conclusion which has already been established.

A final reason for the belief that intelligent citizens ought to play a role in adjudicating technology-related controversies is that such a procedure would place the burden of proof on those who did not wish to guarantee democractic process. It could well be that implementing such democratic ideals through a technology tribunal is unrealistic. The point of the argument from democracy, however, is that democratic procedure must be given a chance to work in the case of technology-related controversy. In a democracy, the burden of proof ought not to rest on those who side with democracy. The predisposition ought to be in favor of democratic procedures, at least until it has been established empirically, in the particular case, that such procedures are unworkable.

6.3. *The Argument from Education*

Yet another argument in favor of the assumption that intelligent laymen ought to participate in the adjudication of technology-related controversies after adversary proceedings is that such participation would serve the long-term interests of the people by providing them with a vehicle for public education in matters of the state. The thrust of this argument was eloquently put long ago by Thomas Jefferson:

> I know of no safe depositor of the ultimate powers of the society but the people themselves; and if we think them not enlightened enough to exercise their control with a wholesome discretion, the remedy is not to take it from them, but to inform their discretion.[27]

Instead of assuming that scientists are best equipped to make technology-related decisions, those who accept the argument from education believe that education ought to be given a chance. Only by attempting to educate people in these matters will it become clear whether the technology tribunal could be successful. To act as if the people of the United States are not bright enough to know what they need to know about the modern world is to beg the question of whether the technology tribunal could be successful. What is worse, to act in this manner is to preclude the possibility of their ever being educated so as to be able to make their own decisions. In other words, opponents of the argument from education are likely to generate a self-fulfilling prophecy that education and citizen decisionmaking will not work.

Even if this experiment in education proposed under the name of the 'technology tribunal' did not work, it would at least encourage citizens to evaluate expert opinion and not merely to assume that it is correct. And is that not what real education is all about anyway?

> Today every profession is being challenged by those who believe that trust should rest not on mystique but rather on what the public knows about its exercise of its expertise. Challenging the expert and digging into the facts behind his opinion is the lifeblood of our legal system, whether it is a psychiatrist characterizing a mental disturbance, a physicist testifying on the environmental impact of a nuclear power plant or a Detroit

engineer insisting on the impossibility of meeting legislated automobile exhaust-emission standards by 1975. It is the only way a judge or a jury – or the public – can decide whom to trust.[28]

6.4. *The Argument from Political Realism*

Besides the fact that the public ought to be encouraged to evaluate expert opinion, there is a very practical reason for arguing that citizens ought to adjudicate the adversary proceedings of the technology tribunal. This reason is that allowing scientists alone to decide the issues would cause a radical political split. There would likely by public apprehension, if not opposition, to decisions made only by technical experts. If one of the proposed goals of either the science court or the technology tribunal is to help end needless controversy and achieve some degree of closure on certain technology-related issues, then education and public participation in decisionmaking are necessary to achieve these goals. Even Marx realized that the most desirable revolution in the world would not succeed, over the long term, unless the people themselves were behind it and understood it. The same is surely true of the technology tribunal, and indeed true of any proposal for public policy. In the words of Abraham Lincoln,

> With public sentiment, nothing can fail. Without it nothing can succeed. Consequently, he who moulds public sentiment goes deeper than he who enacts statutes and pronounces decision.[29]

If the public in general tends to be suspicious of decisions made by experts on issues affecting public well-being, then public-interest groups are doubly apprehensive. In the years of discussion following the science court proposal of 1976, consumer groups, environmental societies, and public-interest organizations have expressed almost unequivocal opposition to the science court, and in particular to the proposal that scientists adjudicate the adversary proceedings.[30] To them, the proposal symbolizes "the interest of pro-technology advocates in using the science court to further decisions by a technomanagerial elite with little or no feeling for social values or human needs".[31] Although no particular set of persons ought to be able to dictate procedures for policymaking, the fears of public-interest groups regarding scientist-adjudicators ought not to be completely ignored. Procedures for policymaking ought to be established on rational, well-argued grounds. Political reality dictates, however, that the cooperation of the public is required for the success of the science court or the technology tribunal.

This being so, there are strong practical reasons for advocating democratic, open citizen-adjudication of the adversary proceedings.

6.5. *The Argument from G. E. Moore*

Perhaps the most important epistemological reason for advocating citizen adjudication of technology-related controversies presented in the technology tribunal is that these controversies are ultimately conflicts over values, not scientific facts. Just as proponents of the science court err in believing that facts are separable from values (see section 4 earlier in this chapter), so also they err in committing the 'naturalistic fallacy'. Coined by British ethicist G. E. Moore, this term refers to the error of attempting to replace "ethics by some one of the natural sciences".[32] Proponents of the science court, who favor adjudication by scientists alone, commit the naturalistic fallacy by virtue of assuming that technology-related disputes involve no ethical components and hence can be resolved merely by scientific reasoning.

Arthur Kantrowitz, originator and key advocate of the science court, proposed that it would deal only with matters of facts to be adjudicated by scientists. He claimed, for example, that the question of what can be built is a scientific issue, whereas that of what *should* be built is "not a purely scientific question".[33] In assuming that whether something *can* be built is a purely scientific issue, Kantrowitz commits the naturalistic fallacy by virtue of attempting to reduce the ethical and evaluative components of this question to purely technical considerations.

What are these ethical and evaluative components? Generally they are assumptions based on value-laden inferences. Let us consider what these components might be in a particular case, that of whether a radwaste facility can be built to isolate certain materials from the biosphere for five hundred years. Since no facility has yet been built which has, in fact, completely isolated radwastes from the biosphere for five hundred years, this issue has not been settled empirically, although (given time and appropriate monitoring) it is in principle capable of being settled empirically. Because, in principle it can be settled empirically, however, does not mean that, at present, in practice, it can be decided empirically. Some of the ethical and evaluative components currently involved in resolution of this issue include assumptions about future behavior of materials, based on inferences about their past and present behavior. Other evaluative components include assumptions about future security systems and monitoring, based on evaluative inferences about political organizations, human error, and future scientific discoveries. In other words, the evaluative components might include inductive inferences (which, of course, never 'verify' scientific fact) and ethical inferences about human institutions/behavior capable of interfering with complete isolation of radwaste. More generally, allegedly scientific issues about what *is* the case, or what *can* be done, turn on different assumptions or interpretations of data. These assumptions and interpretations are often not purely scientific, but are evaluative; hence, the issues of which they are a part are not wholly scientific.

In addition to evaluative assumptions and interpretations of data, resolution of allegedly scientific issues is rendered value-laden by virtue of which facts, among the many facts, scientists emphasize or disregard. Any allegedly scientific resolution of the question whether a particular technological goal *can* be achieved will turn on emphasizing some facts over others. It also will likely turn, not merely on whether a particular scientific fact, law, or theory ought to be acceptable, but on whether a certain experiment provides grounds for believing it or for falsifying it. In other words, allegedly scientific controversy is frequently not evaluative by virtue of general disagreement over basic facts or laws, but evaluative by virtue of conflict over the reliability of data, the conditions under which they were generated, and whether an alleged falsification only appears to be such because of improper testing. There are very few 'smoking guns' in science, and none at all which are free from evaluative assumptions, interpretations, inferences, and emphases.

One reason why there are no 'smoking guns', and why many technology-related matters of controversy are interesting and controversial is precisely because they involve the issue of what to do in the face of uncertainty. Take the case of the allegedly scientific question, "what is the per-year, per-reactor probability of a nuclear core melt for US (pressurized water) fission reactors used to generate electricity?". What is controversial about this case is not that some persons have the facts right about the actual probability and that others don't, but that scientists themselves are uncertain about verifying the finite probability of an unlikely occurrence. Given the paucity of empirical data on core melts, due to the short time that fission has been used by utilities, this uncertainty about nuclear reactors can be mitigated with more years of experience and more actuarial data. Nevertheless, the difference between accident frequency and accident probability will always remain, so that even the best data based on actual nuclear performance will never resolve the inherent uncertainty over actual probability. It will always be mathematically and scientifically reasonable to argue that a particular accident frequency is not indicative of the actual probability, for a variety of reasons. In the face of uncertainty caused either by inadequate experience, or incomplete testing, or inability to foresee the future, or inadequate funding, or insufficient work on a particular problem, the most critical technology-related questions involve what to do in cases in which there are in-principle or in-practice obstacles to removing the uncertainty. Does one say, for example, prior to using humans as 'guinea pigs', that evidence of animal carcinogenicity provides scientific evidence of human carcinogenicity? Or, does one say that, since human carcinogenicity has not been proved, there are no legitimate scientific grounds for belief in human carcinogenicity? In other words, is a particular technology innocent of harm to humans until proved guilty, or guilty until proved innocent? To assume that uncertainties like these do not exist, or that all scientific data is not dependent on evaluative inferences, is to commit the naturalistic fallacy by attempting to reduce ethical questions to scientific ones.

To allow scientists to adjudicate disputes brought before the technology tribunal would mean either that one sanctioned the right of scientists alone to make value judgments affecting all citizens, or that one believed that such judgments were purely technical in nature, and therefore that scientists ought to adjudicate them. The demise of the fact-value dichotomy argues for the thesis that no disputes are ever wholly value-free. But if so, then proponents of scientist-adjudication must defend the desirability of scientists alone making decisions which are not purely scientific in nature.

7. ARGUMENTS AGAINST ADJUDICATION BY THE PUBLIC

Why might one defend the claim that scientists alone ought to make decisions about matters which are not purely scientific in nature? One major argument has been made in defense of this position, and it turns on the fact that the public is lacking in basic scientific knowledge necessary for adjudicating technology-related public policy, and that, in the absence of this knowledge, the public is likely to formulate poor policy based on erroneous science. Two of the most famous statements of this argument were given by Columbia University President William McGill and by Canadian philosopher Alex Michalos.[34] Although I too believe that basic scientific knowledge is necessary for adjudicating technology-related disputes, I disagree both with the assumption that most citizens cannot acquire this basic knowledge and with the conclusion that citizens are likely to formulate poor policy based on erroneous science.

7.1. *The Assumption That a Majority of Citizens Cannot Acquire the Requisite Scientific Knowledge*

There are a number of reasons for believing that the public's lack of scientific knowledge is not an insurmountable obstacle to citizen-adjudication of adversary proceedings. For one thing, to support democratic adjudication is not to support the absence of minimum standards for serving as a judge of adversary proceedings. Obviously some criteria would have to be set for insuring that potential judges were capable of understanding and evaluating the relevant issues. In this regard, one might think of the capability of the average member of the League of Women Voters. Surely all of them are not trained in science, and yet many of them might be said to meet some sort of minimum standard for understanding technology-related issues, provided that they had the opportunity to be educated in these matters.

To assume that the public's lack of scientific knowledge is an insurmountable obstacle to citizen-adjudication of adversary proceedings is to underestimate the power of eduction, whether it be through special high school and college courses in technology-related areas, or whether it be through televised coverage of the proceedings of the technology tribunal. As was already argued, the only way to tell if education might work in this regard is to attempt it. Moreover, it has always been a classic objection, to any demand for more democratic procedures, to argue that the people are not capable of exercising their freedom to decide. If one fails to support an adequate testing of the procedure of citizen adjudication through the technology tribunal, then one is simply accepting, without firm evidence, the arguments of fascists throughout history and begging the question as to whether they are right in this particular case. Admittedly, perhaps neither public education nor the technology tribunal would be successful in adjudicating technology-related disputes. They might even make a situation worse. My point, however, is that because of the political and ethical stakes involved, society ought to support adequate testing of this procedure and not assume, *a priori*, that the technology tribunal will not work.

A third point, in considering whether the public might be able to overcome a significant amount of its scientific ignorance, is that many members of the public (who are not trained in science) make technology-related decisions daily. Perhaps the best example in this regard is that of any US Congressional Committee, which typically operates in terms of hearings at which experts testify, but which reserves to itself the right to decide whether or not to bring particular issues to the floor. If Senators and Congresspeople are typical of average, intelligent citizens who are not necessarily educated extensively in science, and if they make informed judgments on technology-related matters, then why could not many similarly educated and intelligent citizens do the same? Both the legislative and administrative branches of federal and state governments employ hearings in which technical experts present testimony on the scientific underpinnings of a public policy issue, and they also often provide recommendations as to the final decisions. Yet, no one really questions whether Congresspeople, for example, have the necessary scientific background for such recommendations.

7.2. *The Assumption That Citizens Will Arrive at Poor Conclusions*

Perhaps the main reason why legislators are able to formulate technology-related policy, even when they have little scientific training, is that they are educated by means of the system of expert witnesses used at Congressional hearings. Likewise, there is reason to believe that citizens could adjudicate technology-related issues because they would have the benefit of adversary proceedings, including cross examination of all advocates. Hence one of the best reasons for rejecting the assumption that citizen adjudication would lead to poor policy and erroneous science is that this adjudication would be preceded by, and would take place in the context of, adversary proceedings enlightened by scientist-experts. This means that the technology tribunal would not be merely a forum for citizen adjudication of technology-related decisions, but also a forum for adjudication by informed citizens of issues which have serious potential consequences for their lives and well-being. For this reason, citizen adjudication ought not to be seen as anti-science, but anti the failure of informed citizens to evaluate science.

Moreover, so long as one admits that scientists working in the area under adjudication ought not to decide particular issues related to it (because of possible conflicts of interest), then it seems that one might as well admit that intelligent laymen ought to take part in adjudication. In other words, under the current proposal for the science court, even the scientist-judges would not be expert in the areas related to a particular controversy. This being so, one can hardly reject the participation of intelligent laymen on the grounds that they lack expertise. If expertise were the only criterion for selecting judges, then one would have to reject all adjudicators except those highly trained in the precise area under consideration. But if expertise is not the only criterion, and if proponents of the science court do not wish it to be the only one, then they should reconsider their arguments (against citizen-adjudication) which turn on lack of expertise.

Perhaps the most basic reason for rejecting the notion, that expertise is the only criterion for selecting adjudicators, turns on the distinction between necessary and sufficient conditions. Because scientific knowledge is *necessary* for adjudication of disputes before the technology tribunal does not mean that it is *sufficient*. This is because ethical and evaluative assumptions and inferences are bound up with resolution of controversies which are allegedly scientific, as was argued earlier. This being so, expertise ought not be taken as the sole criterion for selection of adjudicators.

Another important consideration regarding the possible tendency of citizen judges to formulate bad policy (on the basis of erroneous science) is that poor policy or erroneous scientific conclusions do not necessarily signal the failure of citizen-adjudicated technology tribunals. The real question is whether the tribunals can be amended so as to preclude great numbers of poor decisions, whether people can learn from their mistakes, and whether faulty decisions can easily be reconsidered and, if need be, amended. Consumer activists and environmentalists would likely argue that the present system of technology-related decisionmaking, which provides neither for adversary proceedings, nor for citizen adjudication, chronically arrives at faulty policy based on poor science. One need only look at current energy policy and depletion of fossil fuels for a case in point.

A final reason for believing that democratic adjudication need not produce poor policy recommendations is that good policy is not merely a matter of discovering highly technical *truths*, but also a question of attempting to guarantee *justice*. If one recognizes that the purpose of adversary proceedings is not primarily to establish some empirical point, but instead to attempt to provide conditions under which policy can be established in a fair, orderly, timely, and representative manner, then there seems to be no strong case for claiming that such democratic procedures are likely to yield poor policy.

To the degree that one sees policy formulation as strictly analogous to scientific discovery, then to that extent will one be likely to worry about factual 'errors', if citizens, rather than scientists, adjudicate the controversy. There are a number of reasons, however, why policy formulation is not strictly analogous to scientific discovery. For one thing, the policymaker often must act prior to having complete knowledge of the situation under consideration. The pure scientist, however, in a situation which is uncertain or incompletely known, is never *required*, by practical circumstances, to stipulate or to define something as a 'scientific discovery'. Because of the differences in the way the policymaker and the scientist are limited by uncertainty, by incomplete knowledge, and by the requirements of practical action, their roles are not strictly analogous.

A second difference between policy formulation and scientific discovery arises because the criteria for successful activity in each area are quite different. The scientist is said to be successful if his predictions can be shown to be the case. A policymaker, on the other hand, is not usually said to be successful because he makes correct predictions but because his choices are consistent with current and/or future societal values. Good science is accomplished through empirical methods resulting in (what are thought to be) substantively correct conclusions. Good policy, on the other hand, although it often presupposes certain scientific truths, is not concerned primarily with correct conclusions. It is usually judged in terms of procedural criteria. But if the criteria for distinguishing good science are different from those for assessing good policy, then policy formulation is not strictly analogous to scientific discovery. And if it is not, then there is no strong reason for believing that intelligent and informed citizens will be likely to reach poor conclusions in adjudicating controversies brought before the technology tribunal.

Admittedly there is a danger that, in democratizing policy by means of the technology tribunal, we might submit it to the whims of public opinion. I know of no way to guarantee that this might not occur, at least on occasion. Such dangers are always possible in democratic decisionmaking, and indeed an analogous problem faces any democratic procedure. This being the case, the real argument for the technology tribunal is that the burden of proof ought to be on those who wish to deny democratic procedure and on those who don't wish to give education a chance, to see if it can facilitate greater democracy. In a highly educated, democratic society, *prima facie* plausibility ought to be on the side of the power of education and democracy. To assume otherwise is to fall victim to the erroneous belief that science policy ought to be determined only by scientists, that economic policy ought to be determined only by economists, and that governmental policy ought to be determined only by politicians. But to make this assumption is surely to write a prescription for corruption. Perhaps we need a new Henry George to remind us that, if anybody has the right to control science and technology policy, it is the people themselves.[35]

8. ANOTHER OBJECTION TO THE TECHNOLOGY TRIBUNAL

Some opponents of the technology tribunal might oppose it for reasons other than fear that democratic adjudication of environmental- and technology-related policy might not work. They might claim that the tribunal is not needed, since current methods of democratic decisionmaking involve widespread citizen participation.[36]

To some extent, this is a reasonable objection. Much public policy is subject to democratic control by means of special procedures which are part of administrative and regulatory decisionmaking and by means of administrative and judicial appeal.

Laws and regulations governing *administrative* decisionmaking in most countries stipulate that all affected parties must be notified and consulted in the process of formulating most government policy.[37] Often this notification and consultation takes the form of public hearings on a particular project, e.g., to supply irrigation or hydro-electric power by means of a US Army Corps of Engineers proposal. Democratic procedures built into administrative decisionmaking, however, fall short of the goals of the technology tribunal in two significant ways. First, most public, administrative hearings are not responsive to many of the key interests of citizens. While they address the possible impacts of the decision at hand, they are intended to provide a forum for comments neither on the merits of the specific proposal, per se, nor on the goals of the government in sanctioning the project, nor on the merits of alternative projects.[38] In other words, the allowable scope of remarks considered at administrative hearings is so narrow as to preclude consideration of many of the concerns which are most important to citizens. A second limitation of administrative hearings is that these procedures themselves do not have the power of prohibiting governments from implementing particular actions, no matter how grave the environmental consequences or the public opposition.[39]

More formal than public, administrative hearings, the procedures of regulatory decisionmaking nevertheless suffer from some of the same problems. *Regulatory decisionmaking* includes adjudicatory hearings, presided over by regulatory agencies. These are required for the granting of licenses and construction certificates and for the setting of regulations and standards governing such activities as airport and pipeline construction, power plant siting, and offshore oil drilling.[40] One of the central difficulties of many US regulatory agencies, such as the Atomic Energy Commission and the Food and Drug Administration, has been their historically close relationship to the very interests they were designed to regulate.[41] Inevitably this closeness has sometimes caused the hearings to be conducted in less than a disinterested manner. Since regulatory agencies have considerable discretionary powers to limit the nature of intervenor proceedings, sometimes that power is exercised at the behest of the industry or technology being regulated. The

US Nuclear Regulatory Commission, for example, prohibits citizen intervenors from raising questions as to the adequacy of safety standards for US reactors. It limits them, instead, to questioning the ability of a particular nuclear utility to meet those standards.[42] A second limit on the democratic effectiveness of regulatory decisionmaking is that, although industries proposing a particular technological project may hire expert witnesses and attorneys for the rulemaking hearings, comparable citizen involvement is almost impossible because of the high cost of participation. Several US regulatory agencies, e.g., the US Consumer Product Safety Commission, have sought to lower this barrier to public participation by developing compensatory programs of financial assistance to citizen groups who take part in their hearings. Unfortunately, however, legislative proposals to authorize all US regulatory agencies to reimburse citizens for their participation in hearings have been defeated in the US Congress.[43] At least part of this reluctance to compensate citizens stems from industrial unwillingness to accept more democratic decisionmaking. As the Vice-President of one oil company put it: "public hearings and public participation in general act as brakes, and if you apply these brakes too long, you'll discourage industrial investments".[44]

Citizen recourse to the legal system for resolving technology- and environment-related policy disputes likewise provides for democratic decisionmaking by means of *administrative* and *judicial appeal*. Citizen groups have gone to court, for example, to block the construction of energy facilities and to contest governmental enforcement of environmental regulations. As an alternative to the technology tribunal, however, the legal process is far less accessible to the average citizen, primarily because of the costs of litigation.[45] This is especially true when risks are high that, if citizens are unsuccessful in their appeal, they will have to bear all legal costs. (In the *Alyeska* decision, the Supreme Court denied the federal courts general power to award litigation costs to successful litigants representing unorganized public interests.)[46]

Opponents of the technology tribunal, who believe that it is unnecessary, might also claim that decisionmaking is already democratic, as is evidenced by the existence of a number of governmental scientific advisory bodies. These are largely composed of citizens, and they are responsible for the formulation and enforcement of scientific research guidelines.[47] City councils in Cambridge, Massachusetts, San Diego, California, and Ann Arbor, Michigan, for example, have taken a number of initiatives in forming such citizen boards. In Cambridge, the City Council authorized its City Manager to appoint a citizen review board to evaluate the safety procedures required by the US National Institute of Health (NIH) for recombinant DNA research. Both the City Council and the City Commissioner of Health and Hospitals unanimously approved the recommendations of the citizen review board.[48] As a tool for democratic policymaking, however, such citizen boards are somewhat limited. For one thing, they ordinarily have no operating budget. This means that they are unable both to reimburse expert witnesses and to develop a long-range plan for developing a cumulative base of information. *Second*, in the absence of a required system of adversary proceedings, complete with cross examination of expert witnesses from opposing sides, members of citizen review panels have been required to educate themselves with respect to the issues under consideration.[49] This means that, at least in theory, they are probably less well equipped to deal with technical issues than is the technology tribunal. *Third*, like recourse to the legal system, use of citizen panels in major technology- or environment-related controversies is purely optional. This means that, beneficial as they might be, such panels are not now required and hence cannot be used as evidence for the claim that citizen participation is already a part of routine decisionmaking regarding technology.

From this brief examination of democratic procedures built into administrative and regulatory decisionmaking, administrative and judicial appeal through the legal system, and citizen review panels, it is clear that the technology tribunal, as previously described, would provide a number of benefits of democratic procedure not currently available through any other means. *First*, the tribunal would require that funding be given to all sides involved in a particular dispute about technology or environmental policy. Thus it would avoid the difficulty faced by administrative, regulatory, and legal procedures, all of which are severely limited by the prohibitively high cost of citizen participation. *Second*, the technology tribunal would make consideration of alternative positions, through adversary proceedings, a requirement of democratic decisionmaking rather than a luxury accessible only in a small number of communities or only to those financially able to participate in administrative hearings or legal appeals. *Third*, unlike administrative and regulatory hearings, the procedures of the technology tribunal would be decisive. If the jury of the tribunal recommended it, projects under consideration could even be prohibited. (Note that nothing has been said regarding the size of the jury of the technology tribunal; conceivably such a jury could be of traditional size or as large as several hundred or several thousand members. Jury size is a practical question to be determined, once citizens have agreed to try the technology tribunal experiment.) *Fourth*, unlike regulatory and administrative decisionmaking, the adversary procedures of the technology tribunal would be less likely to be co-opted by representatives of the industry or technology whose interests are at stake in the project under consideration. This is because the adversary proceedings would not be controlled by a regulatory agency capable of exercising discretionary powers, but by a tribunal whose members were chosen precisely because they had no conflict of interest with respect to the issue under consideration.

9. CONCLUSION

Admittedly all these considerations do not demonstrate that democratic adjudication of disputes brought before the technology tribunal is desirable At least, they do establish the plausibility of a technology tribunal experiment, complete with citizen adjudication. I, for one, am a believer in the importance of this experiment, although I am an agnostic regarding its ultimate success. Even if this democratic tribunal does not provide any final answers, I am convinced that we as a society should at least examine the political, educational, and scientific consequences of adversary proceedings carried out in democratic, rather than elitist, fashion. The very debate itself, with funding for various advocates, could provide sufficient enlightenment to justify the experiment. And, if we were fortunate, we might find ourselves delivered, at least in part, from "two-armed scientists". Senator Muskie was said to have yearned for a "one-armed scientist", an expert who did not follow each assertion with the statement, "on the other hand. . .".[50]

Of the basic premises which were part of the 1976 Task Force recommendation for establishing a science-court experiment, I have proposed that we amend two. The original report was built on five main theses:

1. that science-related controversies are best resolved by adversary methods;
2. that the science component is separable from the political/ethical/evaluative component in these conflicts;
3. that distinguished scientists should adjudicate the scientific facts at issue in these controversies;
4. that the roles of the advocates should be separated from those of adjudicators in the proceedings; and
5. that the court should be conducted in full public purview.

Agreeing in full with premises 1, 4, and 5, I have argued that this procedure be renamed the 'technology tribunal', and that, instead of premises 2 and 3, the following theses be accepted:

2′. that the science component is not separable from the politicial/ethical/evaluative component of technology-related controversies and, as a consequence,

3′. that intelligent and educated citizens, informed by expert opinion, adjudicate these controversies.

As mentioned earlier, I have not argued in general for the plausibility of the three remaining premises, although many persons are likely to find them controversial. I urge that these persons read the classic essays by Michalos and Kantrowitz on the science-court experiment.[51] Had I the time to do so, my strategy for defending these three premises would be that of John Stuart Mill. He set forth the ultimate rationale of the experiment I propose:

> The only way in which a human being can make some approach to knowing the whole of a subject is by hearing what can be said about it by persons of every variety of opinion, and studying all modes in which it can be looked at by every character of mind. No wise man ever acquired his wisdom in any mode than this; nor is it in the nature of human intellect to become wise in any other manner.[52]

NOTES

[1] See, for example, W. Häfele, 'Energy', in *Science, Technology, and the Human Prospect* (edited by C. Starr and P. Ritterbush), Pergamon, New York, 1979, p. 139; hereafter cited as: Häfele, Energy, in *Science*. See also M. Maxey, 'Managing Low-Level Radioactive Wastes', in *Low-Level Radioactive Waste Management* (ed. by J. E. Watson), Health Physics Society, Williamsburg, Virginia, 1979, pp. 410–417; hereafter cited as: Maxey, Wastes.

[2] See, for example, R. S. Banks, 'The Science Court Proposal in Retrospect: A Literature Review and Case Study', *Critical Reviews in Environmental Control* 10, (2), (August 1980), 111; hereafter cited as: Banks, SCP.

[3] Alex C. Michalos, 'A Reconsideration of the Idea of a Science Court', *Research in Philosophy and Technology* (ed. by P. T. Durbin), vol. 3, JAI Press, Greenwich, Connecticut, 1980, pp. 14, 26; hereafter cited as: Michalos, Reconsideration.

[4] Michalos, Reconsideration, p. 19; see also J. A. Martin 'The Proposed Science Court', *Michigan Law Review* 75 (5–6), (April–May 1977), 1059; hereafter cited as: SC.

[5] See J. A. Martin, SC, pp. 1059–1060 and B. M. Casper, 'Technology Policy and Democracy', *Science* 194 (4260), (1 October–1976), 34; hereafter cited as: Policy.

[6] Ian Barbour, *Technology, Environment, and Human Values*, Praeger, New York, 1980, p. 129.

[7] *Science* 193 (4254), (20 August 1976), 653–656; hereafter cited as: Task Force, Report.
[8] For further details on the science court and its history, see A. Kantrowitz, 'Controlling Technology Democratically', *American Scientist* 63 (5), (September–October 1975), 506–507; hereafter cited as: Controlling. See also Casper, Policy, p. 29 and B. M. Casper and P. C. Wellstone, 'The Science Court on Trial in Minnesota', *Hastings Center Report* 8 (4), (August 1978), 5–7; hereafter cited as: Trial. Finally, see R. S. Bank, SCP, and K. G. Nichols and the OECD for Science, Technology, and Industry, *Technology on Trial*, Organization for Economic Co-operation and Development, Paris, 1979, pp. 97–101; hereafter cited as: Nichols, *TOT*.
[9] See Michalos, Reconsideration, pp. 12–27, and A. Kantrowitz, 'The Science Court Experiment: Criticisms and Responses', *Bulletin of the Atomic Scientists* 133 (4), (April 1977), 44–47; hereafter cited as: Response.
[10] Michalos, Reconsideration, p. 13; see also E. Callen, 'The Science Court', *Science* 193 (4257), (September 1976), 948–951, and L. Lipson, 'Technical Issues and the Adversary Process', *Science* 194 (4268), November 1976), 890.
[11] Michalos, Reconsideration, p. 14.
[12] Michalos, Reconsideration, pp. 15, 20, 26.
[13] Michalos, Reconsideration, pp. 14–15.
[14] Michalos, Reconsideration, pp. 20–21.
[15] Michalos, Reconsideration, p. 26.
[16] At the recent conference on the science court, Nancy Abrams of the US Office of Technology Assessment made exactly this suggestion for use of television to educate the public. See 'General Discussion', in Commerce Technical Advisory Board, *Proceedings of the Colloquium on the Science Court, Held at Leesburg, Virginia, on September 19–21, 1976* (ed. by Commerce Technical Advisory Board), American Association for the Advancement of Science, Washington, D.C., January 1977, PB-261 305, NTIS, p. 105; hereafter cited as: Commerce Technical, *Proceedings*.
[17] This is one of Casper's and Wellstone's main points. See Casper and Wellstone, Trial, p. 7. H. Wheeler, 'Bringing Science Under Law', in *The Establishment and All That* (ed. by R. M. Hutchins), Center for the Study of Democratic Institutions, Santa Barbara, 1970, p. 149; hereafter cited as: Wheeler, Science, and Hutchins, *Establishment*.
[18] Kantrowitz, Response, p. 653.
[19] See note 7 in this chapter.
[20] R, M. Hutchins, 'The Center in the Sixties and Seventies', in Hutchins, *Establishment*, p. 10.
[21] J. N. Martin, 'Procedures for Decisionmaking under Conditions of Scientific Uncertainty: The Science Court Proposal., *Harvard Journal on Legislation* 16 (2), (1979), 464; hereafter cited as: Decisionmaking.
[22] Alan McGowan, 'The Science Court', in Commerce Technical, *Proceedings*, p. 91.
[23] Wheeler, Science, p. 140, makes a similar point.
[24] Fabian Society, London, 1931, Fabian Tract no. 235. See also Kantrowitz, Controlling, p. 506.
[25] Quoted in Kantrowitz, Controlling, p. 505.
[26] Martin, SC, p. 1068, makes this same point.
[27] Cited in D. Bazelon, 'Risk and Responsibility', *Science* 205 (4403), (1979), 277–280.

[28] D. L. Bazelon, 'Psychiatrists and the Adversary Process', *Scientific American* 230 (6), (1974), 18.
[29] Quoted in Casper, Policy, p. 33.
[30] See Banks, SCP, p. 111.
[31] Banks, SCP, 111.
[32] Moore, *Principia Ethica*, University Press, Cambridge, 1929, p. 40.
[33] Controlling, p. 506.
[34] See Banks, SCP, p. 112, and Michalos, Reconsideration, p. 26.
[35] This same point is made by Wheeler, Science, p. 147.
[36] Dr. Stan Carpenter and Dr. Paul Durbin raised exactly this point when an earlier version of this proposal was presented in September, 1983 at the New York Colloquium on Philosophy and Technology.
[37] Nichols, *TOT*, p. 82.
[38] Nichols, *TOT*, p. 83.
[39] Nichols, *TOT*, p. 85.
[40] Nichols, *TOT*, p. 86.
[41] See K. S. Shrader-Frechette, *Nuclear Power and Public Policy*, Reidel, Boston, 1980, p. 12; hereafter cited as: *NPPP.*
[42] See Shrader-Frechette, *NPPP*, pp. 91–93; see also Nichols, *TOT*, p. 87.
[43] Nichols, *TOT*, p. 90.
[44] Nichols, *TOT*, p. 91.
[45] Nichols, *TOT*, pp. 94–95.
[46] See K. S. Shrader-Frechette, *Environmental Ethics*, Boxwood, Pacific Grove, 1981, pp. 144–145.
[47] Nichols, *TOT*, p. 99.
[48] Nichols, *TOT*, p. 100.
[49] Nichols, *TOT*, p. 100.
[50] Cited in Martin, SC, p. 1090.
[51] See note 9.
[52] Cited by Casper, Policy, p. 32.

INDEX OF NAMES

(All numbers refer to pages, unless otherwise specified)

INDEX OF SUBJECTS

(All numbers refer to pages, unless otherwise specified)

WITHDRAWN